Protection Strategy against Spruce Budworm

Protection Strategy against Spruce Budworm

Special Issue Editor

David A. MacLean

MDPI • Basel • Beijing • Wuhan • Barcelona • Belgrade

MDPI

Special Issue Editor
David A. MacLean
University of New Brunswick
Canada

Editorial Office
MDPI
St. Alban-Anlage 66
4052 Basel, Switzerland

This is a reprint of articles from the Special Issue published online in the open access journal *Forests* (ISSN 1999-4907) from 2018 to 2019 (available at: https://www.mdpi.com/journal/forests/special_issues/Spruce_Budworm)

For citation purposes, cite each article independently as indicated on the article page online and as indicated below:

LastName, A.A.; LastName, B.B.; LastName, C.C. Article Title. *Journal Name* **Year**, *Article Number*, Page Range.

ISBN 978-3-03928-096-4 (Pbk)
ISBN 978-3-03928-097-1 (PDF)

Cover image courtesy of David A. MacLean.

Contents

About the Special Issue Editor

David A. MacLean (Dr.) is Emeritus Professor of Forest Ecology at the University of New Brunswick, where he served as Dean from 1999–2009. Prior to that, he spent 21 years as a Research Scientist with the Canadian Forest Service, researching spruce budworm impacts, modeling, and decision support systems. Through the 1990s, Dr. MacLean coordinated Canada-wide research networks to (1) develop GIS-based decision support systems for four of Canada's major insect pests, and (2) determine silvicultural approaches to integrated insect management. In 2008, Dr. MacLean was awarded the Canadian Forestry Scientific Achievement Award by the Canadian Institute of Forestry. Dr. MacLean retired from teaching at the University of New Brunswick in 2017, but is still active in leading two large spruce budworm and forest management research projects. He is the Lead Scientist for the $75 million project Early Intervention Strategies to Suppress a Spruce Budworm Outbreak, funded from 2018–2022 by Natural Resources Canada and by the forest industry and the provincial governments of the four Canadian Atlantic provinces. Other current positions include President of the Fundy Model Forest, Chair of the J.D. Irving, Limited Forest Research Advisory Committee, and Chair of the Research Oversight Committee of the Genome Canada bioSAFE research network. Dave began his career in the midst of the 1970s–1980s budworm outbreak and is excited to experience a second outbreak. Dr. MacLean has published over 150 refereed journal papers and 75 technology transfer publications.

forests

Editorial

Protection Strategy against Spruce Budworm

David A. MacLean

Faculty of Forestry and Environmental Management, University of New Brunswick, POB 4400, Fredericton, NB E3B 5A3, Canada; macleand@unb.ca

Received: 9 December 2019; Accepted: 10 December 2019; Published: 12 December 2019

Abstract: Spruce budworm is one of the most significant forest insects worldwide, in terms of outbreak extent, severity, and economic impacts. As a defoliator, spruce budworm larvae are susceptible to insecticide protection, and improvements in efficacy and reductions in non-target environmental effects have made such protection attractive. In this Special Issue, 12 papers describe the advances in spruce budworm protection, most notably an 'early intervention strategy' approach that after six years of trials in New Brunswick, Canada, shows considerable success to date in reducing budworm outbreak occurrence and severity.

Keywords: early intervention strategy; foliage protection; defoliation; monitoring; insecticide application

1. Introduction

Spruce budworm (*Choristoneura fumiferana* (Clem.)) outbreaks are a dominant natural disturbance in forests of Canada and northeastern USA. The last major spruce budworm outbreak in eastern Canada in the 1970s–1980s peaked at 52 million hectares in 1975 [1,2]. Widespread, severe defoliation by this native insect results in large-scale mortality and growth reductions of spruce (*Picea* sp.) and balsam fir (*Abies balsamea* (L.) Mill.) forests, and largely determines future age-class structure and productivity. Repeated annual defoliation typically lasts about 10 years during outbreaks, resulting in growth reductions up to 90% [3], mortality averaging 85% in mature balsam fir stands [4], and changes in regeneration and succession [5,6]. Spruce budworm outbreaks also cause substantial losses in timber and economic production [7,8] and increase the risk of forest fire [9,10]. Several papers have discussed spruce budworm population dynamics during outbreaks [11–13], tree mortality [4], and effects on stand development and ecosystem functioning [14,15].

The province of Nova Scotia, Canada decided not to protect forests with insecticide treatments during the severe 1970s–1980s spruce budworm outbreak, and suffered an average of 87% mortality in mature balsam fir stands [16]. Mortality on Cape Breton Island, Nova Scotia covered 629,900 ha, reduced the growing stock of spruce and fir by 70% or 21.5 million m^3 [17], and increased the hardwood covertype from 16% to 36% [18]. In total, spruce budworm defoliation during eastern Canada's last major outbreak caused timber losses estimated at 107 million m^3 $year^{-1}$ from 1977–1981 and 81 million m^3 $year^{-1}$ from 1982–1987 [19,20]. To put these amounts in perspective, they were equivalent to 50%–70% of the total 156 million m^3 timber harvested in Canada in 2016 [21].

Management to deal with spruce budworm outbreaks has emphasized forest protection by spraying registered insecticides to prevent defoliation and keep trees alive [7]. Other tactics can include salvage harvesting, altering harvest schedules to remove the most susceptible stands, or reducing future susceptibility by planting or thinning [7]. Chemical insecticides are no longer used, and protection strategies use the biological insecticides *Bacillus thuringiensis* (B.t.) or tebufenozide, an insect-specific growth regulator. To prevent extensive tree mortality caused by spruce budworm defoliation, from 1970 to 1983, the eastern Canadian province of New Brunswick treated an average of two million hectares of forest per year with insecticide, at an average cost of $7.7 million per year [22]. In comparison, it is estimated that a similar forest insecticide protection program covering two million hectares today would cost between $90 and $160 million per year, due to increased insecticide active ingredient and application costs [22]. Without insecticide protection, timber harvest reductions are estimated at 18%–25% [7], equivalent to a reduction in timber supply of 2.4–3.3 million m^3 year^{-1} in the Atlantic Canada region [23]. The direct and indirect economic losses resulting from an Atlantic Canada region outbreak would be $10.8–$15.3 billion CAD, depending on outbreak severity [8,23]. Regional job losses over 30 years could total 46,000–56,000 person-years, or approximately 1500–1900 jobs per year [24]. This analysis underestimates job losses during periods of temporary mill closures or in communities where mills could permanently close due to a lack of timber supply.

A large-scale spruce budworm outbreak would also have massive carbon sequestration and greenhouse gas implications [25]. The total potential wood supply loss from a future spruce budworm outbreak in Atlantic Canada projected over 30 years is estimated at 96 million m^3, which would generate approximately 66 Mt CO_2 emissions [26]. On an annual basis, the emissions from dead and dying trees would be on average 2.21 MT CO_2e, equal to the emissions of an additional 466,000 passenger vehicles [26].

In addition to the compelling economic case for forest protection intervention against spruce budworm outbreaks, there is also considerable public support, as documented in a 2007 public survey [27], which found that 94% of New Brunswick respondents supported funding research and development on pest control, and 82% supported controlling future spruce budworm outbreaks.

Over the last five years, a $30 million research project has tested another possible management tactic, termed an early intervention strategy, aimed at area-wide management of spruce budworm populations [28]. This includes intensive monitoring to detect 'hotspots' of rising budworm populations before defoliation occurs, targeted insecticide treatment to prevent spread, and detailed research into effects on target and non-target insects [28,29].

2. Description of Papers in This Special Issue

The objective of this *Protection Strategy against Spruce Budworm* Special Issue of *Forests* was to compile recent research on protection strategies and related topics about detection, monitoring, impacts, population dynamics, and integrated pest management of spruce budworm. The issue includes 12 papers that describe the results and prospects for the use of an early intervention strategy in spruce budworm and other insect management, as well as related topics. A brief description of the content and main findings of the 12 papers in this *Protection Strategy against Spruce Budworm* Special Issue is as follows.

The first six papers are all directly related to the application and testing of an early intervention strategy:

1   Johns et al. [28] described a conceptual framework for an early intervention strategy against spruce budworm, including all of the core components needed for such a program to be viable. Early intervention and foliage protection strategies against spruce budworm are not necessarily mutually exclusive and core elements are relevant to population control for other insect pests that show hotspot outbreak dynamics [28]. Components required for a spruce budworm protection program to be successful include hotspot monitoring, population control, cost–benefit analyses, and proactive communications with stakeholders [28].

2   MacLean et al. [29] reported positive results after five years of early intervention strategy trials conducted by a consortium of government, forest industry, researchers, and other partners. Following over 420,000 ha of treatments of low but increasing spruce budworm populations, second instar larvae (L2) levels across northern New Brunswick, Canada were considerably lower than populations in adjacent Québec [29]. Blocks treated with *Bacillus thuringiensis* or tebufenozide insecticide consistently had reduced budworm levels, generally did not require treatment in the subsequent year, and areas with moderate or higher L2 populations declined by over 90% reductions in 2018, while they continued to increase in Québec.

3   Liu et al. [30] investigated the potential economic impacts of future spruce budworm outbreaks on 2.8 million ha of Crown land in New Brunswick and compared early intervention and foliage protection approaches. They found that timber harvest supply from 2017 to 2067 was projected to be reduced by 29 to 43 million m^3 by uncontrolled moderate or severe budworm outbreaks, which would reduce total economic output by $25 billion (CAD) to $35 billion [30]. Depending upon outbreak severity, the early intervention strategy was projected to have benefit/cost ratios of 3.8 to 6.4 and net present values of $186 million to $353 million, both higher than foliage protection strategies [30].

4   Régnière et al. [31] reported on detailed observations of the dynamics of low but rising spruce budworm populations, the target for early intervention. Results showed strong density-dependent survival between early larval stages and adult emergence, explained by natural enemy impacts and overcrowding, and inverse density-dependence of apparent fecundity, with a net immigration into lower-density populations and net emigration from higher populations at a threshold of about 25% defoliation [31]. This supported the conclusion that immigration, to elevate budworm above a threshold density of about four L4 larvae branch^{-1} was required for a population to increase to outbreak density [31], which helps set a target treatment density.

5   Régnière and Nealis [32] found strong evidence of density-dependent emigration in both eastern and western spruce budworms, and concluded that migration was not random, but was density-dependent.

6   Zhang et al. [33] tested the influence of a gradient of balsam fir-hardwood species composition on the defoliation of fir during the first five years of a spruce budworm outbreak. Fir defoliation was significantly lower as hardwood content increased, but the relationship varied with overall defoliation severity each year [33]. Results helped to set a fir-hardwood threshold below which insecticide protection is not used.

Four papers were related to specific aspects of spruce budworm management:

7   Li et al. [34] used spatial autocorrelation analyses to determine patterns of spruce budworm defoliation of trees (clustered, dispersed, or random) and plots. About one-quarter to one-half of plots had significantly clustered defoliation, and data on plot-level defoliation and tree basal area were sufficient for modeling individual tree defoliation [34].

8   Rahimzadeh-Bajgiran et al. [35] assessed the use of Landsat-5 and Landsat-MSS data to detect and map spruce budworm defoliation. A combination of three vegetation indices derived from Landsat data were able to detect and classify defoliation in three classes with an accuracy of 52%–77%.

9,10 Régnière et al. [36] described the effects of temperature constraints in an individual-based model of spruce budworm moth migration that was parameterized with observations from moths captured in traps or observed migrating under field conditions. A related paper [37] incorporated crepuscular (evening) circadian rhythms of moth flight activity as influenced by evening temperatures into the model. Given the importance of density-dependent emigration [32] and the requirement for moth immigration to elevate budworm above a threshold for outbreak initiation [31], methods to model and map moth flights are important for budworm monitoring for early intervention.

The final two papers dealt with elements of integrated management of spruce budworm:

11 Régnière et al. [38] reported results of trials of aerial applications of a registered formulation of synthetic spruce budworm female sex pheromone to disrupt mating in populations. Although the pheromone application reduced the capture of male budworm moths in pheromone-baited traps by 90% and reduced mating success of virgin females held in individual cages at mid-crown, results showed that populations of eggs or overwintering larvae in the following generation were not reduced, possibly because of the immigration of mated females [38].

12 Quiring et al. [39] tested the influence of a foliar endophyte and budburst phenology on budworm survival. Survival of budworm larvae to pupation and to adult emergence was 13%–17% lower on endophyte positive trees, suggesting that endopytes inoculated into spruce seedlings could limit the spruce budworm population as part of an early intervention strategy [39].

3. Conclusions

Collectively, the 12 papers comprising the *Protection Strategy Against Spruce Budworm* Special Issue of *Forests* describe a promising new method to reduce the occurrence or severity of defoliation in outbreaks. Early intervention strategy research continues in New Brunswick, and the most recent (autumn 2019) budworm L2 monitoring data show that populations remain at low levels (www.healthyforestpartnership.ca), while budworm populations in the adjacent province of Québec continued to increase in 2019 [40]. So far, after six years of trials, the early intervention strategy appears to be working.

Funding: This research was funded by the Atlantic Canada Opportunities Agency, Natural Resources Canada, Government of New Brunswick, and forest industry in New Brunswick.

Acknowledgments: The Early Intervention Against Spruce Budworm research project was overseen by the Healthy Forest Partnership, a consortium of researchers, landowners, forestry companies, governments, and forest protection experts. Many scientists and staff of industry and government agencies have made important contributions, without which the project could not have proceeded.

Conflicts of Interest: The author declares no conflict of interest.

References

1. Kettela, E.G. *A Cartographic History of Spruce Budworm Defoliation 1967 to 1981 in Eastern North America*; Inf. Rep. DPC-X-14; Canadian Forestry Service; Environment Canada: Ottawa, ON, Canada, 1983; p. 9.
2. Canadian Council of Forest Ministers. *National Forestry Database: Forest Insects and Forest Fires Statistics*; Canadian Council of Forest Ministers; Natural Resources Canada; Canadian Forest Service: Ottawa, ON, Canada, 2019. Available online: http://nfdp.ccfm.org/ (accessed on 8 December 2019).
3. Ostaff, D.P.; MacLean, D.A. Patterns of balsam fir foliar production and growth in relation to defoliation by spruce budworm. *Can. J. For. Res.* **1995**, *25*, 1128–1136. [CrossRef]
4. MacLean, D.A. Vulnerability of fir-spruce stands during uncontrolled spruce budworm outbreaks: A review and discussion. *For. Chron.* **1980**, *56*, 213–221. [CrossRef]
5. Virgin, G.V.; MacLean, D.A. Five decades of balsam fir stand development after spruce budworm-related mortality. *For. Ecol. Manag.* **2017**, *400*, 129–138. [CrossRef]
6. Baskerville, G.L. Spruce budworm: Super silviculturist. *For. Chron.* **1975**, *51*, 138–140. [CrossRef]
7. Hennigar, C.; Erdle, T.; Gullison, J.; MacLean, D. Re-examining wood supply in light of future spruce budworm outbreaks: A case study in New Brunswick. *For. Chron.* **2013**, *89*, 42–53. [CrossRef]
8. Chang, W.-Y.; Lantz, V.A.; Hennigar, C.R.; MacLean, D.A. Economic impacts of spruce budworm (*Choristoneura fumiferana* Clem.) outbreaks and control in New Brunswick, Canada. *Can. J. For. Res.* **2012**, *42*, 490–505. [CrossRef]
9. Stocks, B.J. Fire potential in the spruce budworm-damaged forests of Ontario. *For. Chron.* **1987**, *63*, 8–14. [CrossRef]
10. James, P.M.A.; Robert, L.-E.; Wotton, B.M.; Martell, D.L.; Fleming, R.A.; Robert, L. Lagged cumulative spruce budworm defoliation affects the risk of fire ignition in Ontario, Canada. *Ecol. Appl.* **2017**, *27*, 532–544. [CrossRef]
11. Régnière, J.; Nealis, V.G. Ecological mechanisms of population change during outbreaks of the spruce budworm. *Ecol. Entomol.* **2007**, *32*, 461–477. [CrossRef]
12. Johns, R.C.; Flaherty, L.; Carleton, D.; Edwards, S.; Morrison, A.; Owens, E. Population studies of tree-defoliating insects in Canada: A century in review. *Can. Entomol.* **2016**, *148*, S58–S81. [CrossRef]
13. Royama, T.; Eveleigh, E.S.; Morin, J.R.B.; Pollock, S.J.; McCarthy, P.C.; McDougall, G.A.; Lucarotti, C.J. Mechanisms underlying spruce budworm outbreak processes as elucidated by a 14-year study in New Brunswick, Canada. *Ecol. Monogr.* **2017**, *87*, 600–631. [CrossRef]

14. Kneeshaw, D.; Sturtevant, B.R.; Cooke, B.; Work, T.; Pureswaran, D.; DeGrandpre, L.; MacLean, D.A. Insect disturbances in forest ecosystems. Chapter 7. In *Routledge Handbook of Forest Ecology*; Peh, K.S.-H., Corlett, R.T., Bergeron, Y., Eds.; Routledge: Oxon, UK, 2015; pp. 93–113.

15. MacLean, D.A. Impacts of insect outbreaks on tree mortality, productivity, and stand development. *Can. Entomol.* **2016**, *148*, S138–S159. [CrossRef]

16. MacLean, D.A.; Ostaff, D.P. Patterns of balsam fir mortality caused by an uncontrolled spruce budworm outbreak. *Can. J. For. Res.* **1989**, *19*, 1087–1095. [CrossRef]

17. Nova Scotia Department of Lands and Forests. *The Current Status of the Softwood Resource on Cape Breton Island*; Nova Scotia Department of Lands and Forests: Truro, NS, Canada, 1982; p. 2.

18. Nova Scotia Department of Natural Resources. *Impact of the 1974-81 Spruce Budworm Infestation on the Forests of Cape Breton Island*; Nova Scotia Department of Natural Resources: Halifax, NS, Canada, 1994; p. 8.

19. Sterner, T.E.; Davidson, A.G. *Forest Insect and Disease Conditions in Canada, 1981*; Canadian Forest Service: Ottawa, ON, Canada, 1982.

20. Power, J.M. National data on forest pest damage. In *Canada's Timber Resources*; Inf. Rep., PI-X-101; Brand, D.G., Ed.; Canadian Forest Service: Ottawa, ON, Canada, 1991; pp. 119–129.

21. Natural Resources Canada. Statistical Data on Canada's Forest Resources. 2019. Available online: https://www.nrcan.gc.ca/forests/resources/13507 (accessed on 8 December 2019).

22. Forest Protection Limited. *Spruce Budworm Aerial Treatment Program Areas and Costs, 1970–1993*; Forest Protection Limited: Fredericton, NB, Canada, 1993.

23. MacLean, D.A. *Potential Economic Losses from the Next Spruce Budworm Outbreak*; Report prepared for Forest Protection Limited; Forest Protection Limited: Fredericton, NB, Canada, 2013; p. 15, unpublished work.

24. MacLean, D.A. *Potential Regional Employment Losses from an Uncontrolled Spruce Budworm Outbreak*; Report prepared for New Brunswick Spruce Budworm Technical Committee; New Brunswick Spruce Budworm Technical Committee: Fredericton, NB, Canada, 2013; p. 1, unpublished work.

25. Dymond, C.; Neilson, E.; Stinson, G.; Porter, K.; MacLean, D.A.; Gray, D.; Campagna, M.; Kurz, W. Future spruce budworm outbreak may create a carbon source in eastern Canadian forests. *Ecosystems* **2010**, *13*, 917–931. [CrossRef]

26. MacLean, D.A. *Effects of the Next Spruce Budworm Outbreak on Greenhouse Gases and Climate Change*; Report prepared for the Healthy Forest Partnership; The Healthy Forest Partnership: Fredericton, NB, Canada, 2017; p. 6, unpublished work.

27. Chang, W.-Y.; Lantz, V.A.; MacLean, D.A. Public attitudes about forest pest outbreaks and control options: Case studies in two Canadian provinces. *For. Ecol. Manag.* **2009**, *257*, 1333–1343. [CrossRef]

28. Johns, R.C.; Bowden, J.J.; Carleton, D.R.; Cooke, B.J.; Edwards, S.; Emilson, E.J.S.; James, P.M.A.; Kneeshaw, D.; MacLean, D.A.; Martel, V.; et al. A conceptual framework for the spruce budworm early intervention strategy: Can outbreaks be stopped? *Forests* **2019**, *10*, 910. [CrossRef]

29. MacLean, D.A.; Amirault, P.; Amos-Binks, L.; Carleton, D.; Hennigar, C.; Johns, R.; Régnière, J. Positive results of an early intervention strategy to suppress a spruce budworm outbreak after five years of trials. *Forests* **2019**, *10*, 448. [CrossRef]

30. Liu, E.Y.; Lantz, V.A.; MacLean, D.A.; Hennigar, C. Economics of early intervention to suppress a potential spruce budworm outbreak on Crown land in New Brunswick, Canada. *Forests* **2019**, *10*, 481. [CrossRef]

31. Régnière, J.; Cooke, B.J.; Béchard, A.; Dupont, A.; Therrien, P. Dynamics and management of rising outbreak spruce budworm populations. *Forests* **2019**, *10*, 748. [CrossRef]

32. Régnière, J.; Nealis, V.G. Density dependence of egg recruitment and moth dispersal in spruce budworms. *Forests* **2019**, *10*, 706. [CrossRef]

33. Zhang, B.; MacLean, D.A.; Johns, R.C.; Eveleigh, E.S. Effects of hardwood content on balsam fir defoliation during the building phase of a spruce budworm outbreak. *Forests* **2018**, *9*, 530. [CrossRef]

34. Li, M.; MacLean, D.A.; Hennigar, C.R.; Ogilvie, J. Spatial-temporal patterns of spruce budworm defoliation within plots in Québec. *Forests* **2019**, *10*, 232. [CrossRef]

35. Rahimzadeh-Bajgiran, P.; Weiskittel, A.R.; Kneeshaw, D.; MacLean, D.A. Detection of annual spruce budworm defoliation and severity classification using Landsat imagery. *Forests* **2018**, *9*, 357. [CrossRef]

36. Régnière, J.; Delisle, J.; Sturtevant, B.R.; Garcia, M.; Saint-Amant, R. Modeling migratory flight in the spruce budworm: Temperature constraints. *Forests* **2019**, *10*, 802. [CrossRef]

37. Régnière, J.; Garcia, M.; Saint-Amant, R. Modeling migratory flight in the spruce budworm: Circadian rhythm. *Forests* **2019**, *10*, 877. [CrossRef]
38. Régnière, J.; Delisle, J.; Dupont, A.; Trudel, R. The impact of moth migration on apparent fecundity overwhelms mating disruption as a method to manage spruce budworm populations. *Forests* **2019**, *10*, 775. [CrossRef]
39. Quiring, D.; Adams, G.; Flaherty, L.; McCartney, A.; Miller, J.D.; Edwards, S. Influence of a foliar endophyte and budburst phenology on survival of wild and laboratory-reared eastern spruce budworm, *Choristoneura fumiferana* on white spruce (*Picea glauca*). *Forests* **2019**, *10*, 503. [CrossRef]
40. Québec Ministère des Forêts, de la Faune et des Parcs. *Aires Infestées par la Tordeuse des Bourgeons de L'épinette au Québec en 2019*; Gouvernement du Québec, Direction de la Protection des Forêts: Quebec City, QC, Canada, 2019; p. 32.

forests

MDPI

Article

A Conceptual Framework for the Spruce Budworm Early Intervention Strategy: Can Outbreaks be Stopped?

Robert C. Johns [1,*], Joseph J. Bowden [2], Drew R. Carleton [3], Barry J. Cooke [4], Sara Edwards [5], Erik J. S. Emilson [4], Patrick M. A. James [6], Dan Kneeshaw [7], David A. MacLean [8], Véronique Martel [9], Eric R. D. Moise [2], Gordon D. Mott [10], Chris J. Norfolk [3], Emily Owens [1], Deepa S. Pureswaran [9], Dan T. Quiring [11], Jacques Régnière [9], Brigitte Richard [12] and Michael Stastny [1]

[1] Natural Resources Canada, Canadian Forest Service, Atlantic Forestry Centre, Fredericton, NB E3B 5P7, Canada; emily.owens@canada.ca (E.O.); michael.stastny@canada.ca (M.S.)
[2] Natural Resources Canada, Canadian Forest Service, Atlantic Forestry Centre, Corner Brook, NL A2H 5G4, Canada; joseph.bowden@canada.ca (J.J.B.); eric.moise@canada.ca (E.R.D.M.)
[3] New Brunswick Department of Energy and Resource Development, 1350 Regent Street, Fredericton, NB E3C 2G6, Canada; drew.carleton@gnb.ca (D.R.C.); chris.norfolk@gnb.ca (C.J.N.)
[4] Natural Resources Canada, Canadian Forest Service, Great Lakes Forestry Centre, Sault Ste. Marie, ON P6A 2E5, Canada; barry.cooke@canada.ca (B.J.C.); erik.emilson@canada.ca (E.J.S.E.)
[5] Forest Protection Ltd., 2502 Route 102 Highway, Lincoln, NB E3B 7E6, Canada; sara.edwards@unb.ca
[6] Faculty of Forestry, University of Toronto, Toronto, ON M5S 3E8, Canada; patrick.james@utoronto.ca
[7] Université de Montréal, Département des Sciences Biologiques, Pavillon Marie-Victorin, C.P. 6128, Succursale Centre-Ville Montréal, QC H3C 3J7, Canada; kneeshaw.daniel@uqam.ca
[8] Faculty of Forestry and Environmental Management, University of New Brunswick, Fredericton, NB E3B 5A3, Canada; macleand@unb.ca
[9] Natural Resources Canada, Canadian Forest Service, Laurentian Forestry Centre, Québec City, QC G1V 4C7, Canada; veronique.martel@canada.ca (V.M.); deepa.pureswaran@canada.ca (D.S.P.); jacques.regniere@canada.ca (J.R.)
[10] U.S. Forest Service (retired), 42 Damon Pasture Lane, Lakeville, ME 04487, USA; forester@AlmanacMtn.US
[11] Population Ecology Group, Faculty of Forestry and Environmental Management, University of New Brunswick, Fredericton, NB E3B 6C2, Canada; dquiring@mac.com
[12] Natural Resources Canada, Communications and Portfolio Sector, Atlantic Forestry Centre, Fredericton, NB E3B 5P7, Canada; Brigitte.richard2@canada.ca
*   Correspondence: rob.johns@canada.ca; Tel.: +1-506-260-5457

Received: 13 August 2019; Accepted: 10 October 2019; Published: 16 October 2019

Abstract: The spruce budworm, *Choristoneura fumiferana*, Clem., is the most significant defoliating pest of boreal balsam fir (*Abies balsamea* (L.) Mill.) and spruce (*Picea* sp.) in North America. Historically, spruce budworm outbreaks have been managed via a reactive, foliage protection approach focused on keeping trees alive rather than stopping the outbreak. However, recent theoretical and technical advances have renewed interest in proactive population control to reduce outbreak spread and magnitude, i.e., the Early Intervention Strategy (EIS). In essence, EIS is an area-wide management program premised on detecting and controlling rising spruce budworm populations (hotspots) along the leading edge of an outbreak. In this article, we lay out the conceptual framework for EIS, including all of the core components needed for such a program to be viable. We outline the competing hypotheses of spruce budworm population dynamics and discuss their implications for how we manage outbreaks. We also discuss the practical needs for such a program to be successful (e.g., hotspot monitoring, population control, and cost–benefit analyses), as well as the importance of proactive communications with stakeholders.

Keywords: foliage protection; population control; monitoring; area-wide management; science communication; economic and ecological cost: benefit analyses

1. Introduction

Ecological disturbances such as forest fires and insect outbreaks play a crucial role in shaping productivity, structure, and successional dynamics of forest ecosystems [1]. Despite these essential functions, disturbances sometimes reach levels that harm local ecosystems or socioeconomic interests, thus justifying human intervention [2]. Where outright prevention is impossible or impractical, intervention efforts tend to track one of two strategic pathways. One strategy is to manage the disturbance proactively, deploying large-scale suppression efforts to stop the disturbance before it spreads. The alternative reactive strategy is to let the disturbance run its natural course while only protecting the most valuable resources in its path. For managing insect pests, we often refer to these proactive and reactive strategies, respectively, as 'population control' and 'plant protection' [3]. Both strategies can be useful in pest management but require very different conceptual frameworks, action criteria and thresholds, as well as cost–benefit trade-offs that ultimately determine their relative suitability, feasibility, and efficacy.

Historically, pest management programs for forest insects have favored reactive plant protection over proactive population control. This is certainly the case for the spruce budworm, *Choristoneura fumiferana* Clem., the foremost defoliating pest of balsam fir (*Abies balsamea* (L.) Mill.) and spruce (*Picea* sp.) throughout the North American boreal and eastern mixedwood forest [2]. The plant protection strategy (aka, Foliage Protection strategy) for managing spruce budworm first came to prominence nearly 70 years ago and arose in part as a response to failed attempts at population control. During the early 1950s, researchers leading the first large-scale efforts to manage spruce budworm were optimistic that aggressive use of DDT (dichlorodiphenyltrichloroethane), a powerful broad-spectrum insecticide, could reduce populations to pre-outbreak levels [4,5]. This conviction was tested in 1952 with the aerial application of DDT to over 186,000 ha of budworm-infested forest in northern New Brunswick, Canada at an application rate of ~0.45 kg/ha [4,5]. Despite inflicting substantial larval mortality, as high as 99% in some stands, the outbreak continued to expand and populations quickly rebounded in treated areas [4,6]. Almost immediately, researchers abandoned large-scale population control efforts and recalibrated operations for more localized, fine-scale Foliage Protection [5]. In making this shift, they scaled back DDT application rates to merely limit defoliation and thereby prevent tree death. Over time other broad-spectrum insecticides were adopted (e.g., fenitrothion, matacil) to address the significant environmental concerns around the use of DDT and to reduce the probability of selection for insecticide resistance [7]. These broad-spectrum, topical insecticides were eventually banned entirely and replaced with new types of ingestible insecticides that specifically targeted larval Lepidopterans (*Bacillus thuringiensis* var. *kurstaki* (Btk) and tebufenozide) [8–10]. Protection efforts were limited mainly to high-value spruce-fir stands with more than two years of moderate defoliation and high budworm densities [5,11]. In eastern Canada, Foliage Protection has remained the dominant management strategy for spruce budworm for the better part of three outbreaks since the 1950s, including for the current outbreak in eastern Québec [12].

In recent decades, substantial advances in population theory and experimentation, insecticide specificity, as well as surveillance and treatment technologies have provided researchers with renewed opportunity to test proactive population control strategies [3,13]. In this article, we explore the conceptual basis for developing a proactive population control strategy for spruce budworm, the so-called Early Intervention Strategy (EIS). The EIS aims to stop the expansion of spruce budworm outbreaks by controlling emerging 'hotspots' as soon as they arise. This approach is informed by the success of area-wide management programs for invasive pest species, such as the gypsy moth [14], and draws on ecological theories behind vertebrate population management (e.g., [15,16]). Development and testing of EIS has been ongoing in the eastern Canadian province of New Brunswick since 2014 and so far appears to be effective for containing outbreak spread [13]. To our knowledge, this program constitutes the first ever attempt to develop an outbreak containment program for an endemic forest insect pest.

Here, we lay out the conceptual framework for our novel EIS approach and its essential components. First, we describe the underlying population dynamics that might make spruce budworm amenable to population control. We then outline the practical components needed for such a program to succeed (i.e., hotspot monitoring, population control, cost–benefit analyses, and public engagement). This EIS framework provides the basis for guiding effective management, including methods to evaluate EIS efficacy, determining under what conditions it might work best, and identifying knowledge and technical gaps for future research.

2. Conceptual Framework

2.1. Population Dynamics

Effective population management requires an understanding of how outbreaks start (Figure 1). For spruce budworm, the initiation of outbreaks has been a topic of debate largely centered on two competing theories, the 'oscillatory hypothesis' and the 'double-equilibrium hypothesis' ([3], Table S1). Several recent reviews have synthesized the historical and theoretical details of these hypotheses [2,3], so here we will highlight some of the core arguments and their implications for managing spruce budworm outbreaks.

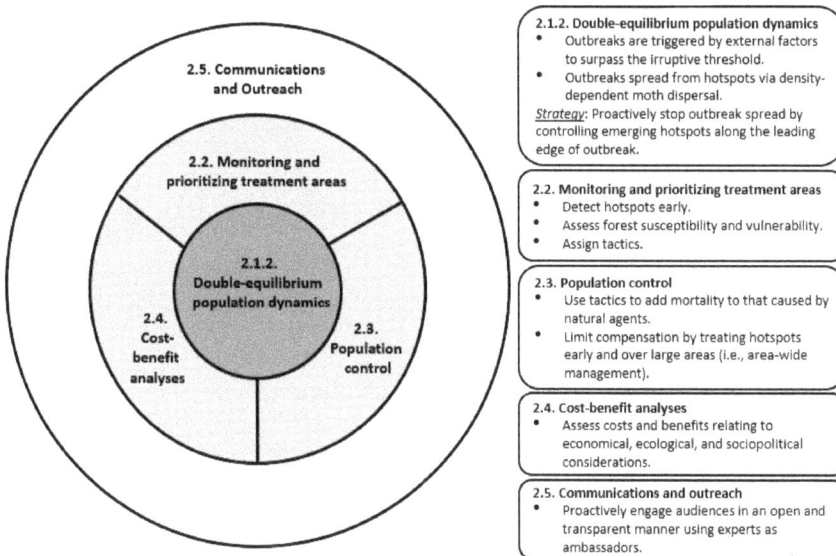

Figure 1. A conceptual framework for Spruce budworm Early Intervention Strategy (EIS) program illustrating the relationships between its different components. Double-equilibrium population dynamics provides the core ecological justification for EIS. In turn, the aims of EIS dictate monitoring and treatment prioritization protocol, population control practices and tactics, and the criteria used in cost–benefit analyses. These particular components are highly dependent upon one another, in that challenges or innovations in one component will likely influence the efficacy or feasibility of the others. Proactive communications and outreach are essential for disseminating information and garnering social license to allow all other aspects of the program to operate. Numbers denote the section of each topic within the body of the article.

2.1.1. Oscillatory Dynamics

Certainty around the inevitability of outbreaks was bolstered in the 1980s by the development of the so-called oscillatory hypothesis (Table S1). In brief, Royama [17,18] argued that budworm outbreaks are part of a slow, cyclical oscillation of density-dependent mortality in late-instar larval and pupal stages shaped by predators, parasitoids, and disease. Outbreaks arise only after spruce budworm become so scarce that natural enemy populations collapse. As mortality rates decline, spruce budworm populations grow rapidly, and in some instances reach densities that strip trees of all new foliage. But, as spruce budworm populations grow so does natural enemy abundance, albeit with a lag of a few years. The resurgence of natural enemies and diseases drive budworm populations back to low density, and the cycle begins anew.

A core tenet of the oscillatory hypothesis is that the cycles occur over relatively large areas, with disparate pockets rising more or less synchronously across the region [19–21]. Moth dispersal, although frequent and often conspicuous during outbreaks [22–24], does not drive outbreak spread per se but rather acts in tandem with weather to draw regional outbreak trends into closer synchrony [20,25]. According to the oscillatory hypothesis, moth dispersal does not start outbreaks, but can hasten or slow population growth and create "noise" in an otherwise smooth predator–prey cycle.

These arguments provide the justification for Foliage Protection. If this oscillatory dynamic is ubiquitous throughout spruce budworm's range, range-wide outbreaks are essentially inevitable. Population control efforts (including EIS) against rising populations would be futile as inherently high growth rates and strong regional synchronization of outbreaks would perpetually swamp control efforts and quickly return treated sites to outbreak levels. Instead, a reactive Foliage Protection strategy becomes the obvious solution to reduce the multi-year cumulative defoliation that leads to tree mortality.

2.1.2. Double-Equilibrium Dynamics and EIS

It is worth emphasizing that the oscillatory explanation for how and why outbreaks start did not arise from field data. When it was conceptualized in the early 1980s, detailed life table data for spruce budworm were only available for peak and declining outbreak stages [3]. Recent studies have sought to fill this knowledge gap, which has led to skepticism around the 'inevitability' of regional spruce budworm outbreaks. Our current understanding of outbreak dynamics more strongly supports an updated version of an older idea, the so-called double-equilibrium hypothesis [26–28] (see also Table S1).

In essence, the double-equilibrium hypothesis argues that spruce budworm outbreak dynamics are driven by irruptive population shifts between lower and upper equilibrium states. At the lower (endemic) equilibrium, populations remain low due to heavy mortality caused by natural enemies [29], poor larval dispersal success, inclement weather events, and reduced mating success among females (i.e., "Allee effects" or "depensation" [30]). Outbreaks occur when mortality rates decline or recruitment increases, thus triggering rapid population growth that may eventually approach the local carrying capacity. At this stage, populations enter the upper (epidemic) equilibrium and persist there until natural enemies and diseases, acting alone or synergistically with declining foliage availability, drive populations back to low density [31]. Thus, outbreaks are not the product of simple predator–prey cycles, but instead follow more of an irruption–collapse dynamic with populations fluctuating between two density extremes.

A key implication of the double-equilibrium hypothesis is that moth dispersal does not merely synchronize population growth—it is also an active driver of outbreak spread. Outbreaks may start in small, localized patches (aka hotspots), perhaps "spontaneously" from enhanced survival in response to local factors (e.g., [28]), but these rapidly rise and begin to emit moths into neighboring forests in a density-dependent fashion [24]. This outflow of moths triggers rapid population rise such that growth quickly outstrips the capacity of local mortality agents to keep populations low. These new hotspots become sources of moths leading to explosive propagation of the epidemic, which continues until resources collapse and/or natural agents can regain control. This dynamic implies that in the absence of significant moth dispersal, outbreaks would remain relatively geographically isolated.

This interpretation of spruce budworm dynamics has several important implications for population management (Figure 2). Ideally, if we could predict or detect an outbreak early, we could use an EIS program to proactively control hotspots at their origin, thus preventing a regional outbreak. Alternatively, if an outbreak has already surpassed levels that might be manageable via population control—as is currently the case in Québec—we might use EIS to control hotspots along the leading outbreak edge as a way to contain further spread or reduce the magnitude of outbreak. This is akin to "slow-the-spread" strategies used with exotic invasive pests [14]. A key underpinning of EIS is the existence of an irruption threshold below which populations tend to be constrained at the lower equilibrium state by natural enemies [29] and a paucity of mates [30] (i.e., depensatory pressures). Depending on how robust and persistent these depensatory pressures are, populations driven below this threshold may remain low or go locally extinct even without further intervention [32]. Recent work by Régnière et al. [28] suggests that this threshold for spruce budworm in relatively pure balsam fir stands lies around four fourth-instar feeding larvae per 45 cm branch tip, which roughly translates to seven second-instar larvae (L2) per branch. This is likely conservative, as stands with higher hardwood content would have a higher density threshold owing to population dilution and more diverse and robust natural enemy communities [33–36]. Thus, the aim (and challenge) of EIS is to detect local populations as they surpass the irruption threshold (i.e., become hotspots) and cause sufficient mortality to return them to the lower equilibrium state (Figure 2).

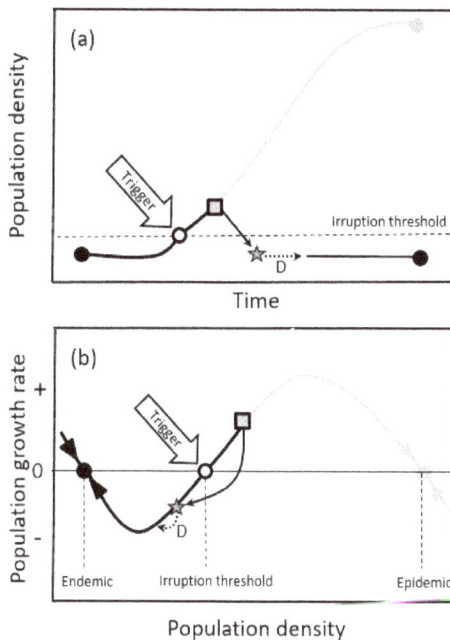

Figure 2. Hypothetical population trend (**a**) and recruitment curve (**b**) for rising spruce budworm populations under the 'double-equilibrium' hypothesis [26,27]. White circles represent the unstable irruption threshold, whereas the filled circles are stable. Gray boxes represent the starting point of a control treatment (e.g., with an insecticide), with its potential impacts on population density or growth rate indicated by stars, followed by expected population responses. Under this dynamic, sufficient treatment mortality (i.e., that is additive to natural mortality) may return populations to the lower equilibrium (dotted lines, grey stars), where depensatory pressure ('D') may keep populations low.

2.2. Monitoring and Prioritizing Treatment Areas

EIS requires efficient methods for detecting hotspots and identifying priority areas for spruce budworm control [13]. Many of the same proxies for population density used in Foliage Protection programs are also used in EIS [37]. However, the sampling intensity and action thresholds used to define treatment areas in EIS differ substantially. The basic protocol for defining treatment priority areas in EIS involves: (1) detecting hotspots; (2) assessing forest susceptibility; and (3) selecting and assigning control tactics.

2.2.1. Detecting Hotspots

EIS detects and monitors hotspots through annual sampling of overwintering L2. These data provide the first consideration in prioritizing areas for treatments. L2 monitoring is useful for several reasons. First, L2 density acts as a fair proxy of population density relative to irruption threshold where populations seemingly shift into their outbreak phase [28]. Second, L2 sampling occurs in the fall and winter, which provides sufficient time for budget and treatment planning for the upcoming season. Finally, L2 monitoring has been the convention for monitoring annual budworm densities for Foliage Protection since the early 1980s. As such, the techniques and infrastructure for collecting and extracting larvae from branches are well-established and relatively low cost [37].

Monitoring via L2 densities is much more intensive in EIS than in Foliage Protection as the program requires a much higher spatial resolution to ensure that potential hotspots are not missed [13]. EIS also uses secondary proxies of population density to identify areas of interest for follow-up L2 branch collections, including pheromone trapping of male moths [38] and defoliation assessment via aerial surveys and branch sampling [39]. Instances of abnormally high moth or defoliation levels trigger a second round of L2 sampling in the affected area, which further help to define treatment area boundaries. All areas with known spruce budworm L2 densities above the irruption threshold are set to high priority to ensure their treatment.

L2 monitoring efficiency could be further enhanced in the future through integration with modelling tools to predict moth dispersal patterns via weather forecasting [40] or radar [23], or through defoliation assessment using remote monitoring approaches [41,42].

2.2.2. Assessing Forest Susceptibility

Although L2 densities are the basis for interpolating treatment areas, forest composition provides a secondary criterion to optimize insecticide applications. This procedure involves incorporating stand composition into treatment-priority area calculations through differential weighting of stands based on species composition [13].

The relative abundance of mature balsam fir is likely to substantially impact outbreak severity and population growth trends (Figure 3). Stands dominated by balsam fir are particularly susceptible and vulnerable to spruce budworm. During outbreaks balsam fir suffer an average of ~85% mortality in mature stands and ~40% mortality in immature stands [43]. In comparison, the other host species—white spruce (*Picea glauca* (Moench) Voss), red spruce (*Picea rubens* Sarg.), and black spruce (*Picea mariana* (Mills.) B.S.P.)—sustain, respectively, 72%, 41%, and 28% as much defoliation as balsam fir [44]. Especially during the early rising phase of the outbreak targeted by the EIS, balsam fir sustains the highest intensity of defoliation even when other hosts are available [45].

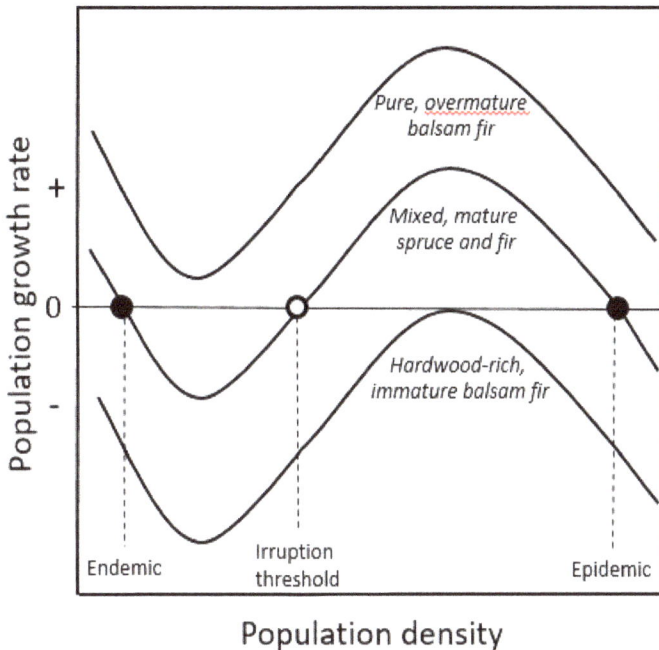

Figure 3. Hypothetical recruitment curves for spruce budworm showing how population density and forest type might interact to influence forest susceptibility and action priorities (adapted from [19]).

Several studies also suggest that during spruce budworm outbreaks, stand-level and landscape-level defoliation and growth loss in balsam fir may decline with increasing proportions of non-host, hardwood species [46–48]. Reduced population growth rates and associated outbreak severity in mixed or hardwood-dominated areas (Figure 3) may reflect increased parasitoid community complexity and pressure [33–36] or increased dispersal mortality during early instars. Zhang et al. [48] also showed, however, that hardwood effects are most evident during the rising phase of an outbreak and become muted as overall defoliation severity increases.

In the current EIS program, spruce-fir and hardwood contents are the only forest composition factors used in refining the treatment priority layer [13]. Future iterations of this process could also incorporate thresholds associated with less susceptible species (e.g., black or red spruce), though these have not been tested to date. Forest landscape structure, including forest configuration and fragmentation, could also be incorporated as they have small but significant effects on outbreak severity [49,50]. Additionally, these factors play a role in determining the geographic locations of outbreak development (e.g., hotspots). Recently, Bouchard and Auger [51] demonstrated that outbreaks tended to start in low elevation, high-density host stands. Refined models of forest susceptibility and vulnerability to spruce budworm attack will be essential to fully integrate the influence of forest spatial heterogeneity into the EIS.

2.2.3. Selecting and Assigning Control Tactics

After establishing the treatment priority area, the final step is to determine which insecticide to use and where. Two registered insecticides are currently available for spruce budworm in Atlantic Canada: *Btk* [9] and tebufenozide [10]. *Btk* is a bacterial agent that upon ingestion by larvae causes the breakdown of the insect's gut wall and ultimately death [52]. Tebufenozide is a synthetic chemical that mimics the juvenile molting hormone. When ingested, tebufenozide triggers premature molting or other developmental and reproductive complications during later stages that end in death or sterility [53]. A third registered product, a synthesized spruce budworm pheromone, could disrupt mating but has not yet proven efficacious for population control [54,55]. Because *Btk* and tebufenozide are both applied aerially using the same application technologies, and are similarly efficacious [28], neither product has particular operational advantages over the other. However, tebufenozide has additional label restrictions prohibiting application in designated protected watersheds and residential areas, where *Btk* is therefore assigned automatically. In all other scenarios, product choice focuses on the logistics required to store both products at multiple bases and flight costs to treatment areas. To date, EIS has tended to use both insecticides in roughly equal measure [13], in part to reduce the likelihood of spruce budworm becoming resistant to either product.

2.3. Population Control

EIS does not aim to eradicate spruce budworm from the forest. Nor is population control achieved simply by inflicting maximal mortality in high-density populations, as was assumed during experiments with DDT in the early 1950s [4]. Populations respond in a variety of ways following insecticide treatments and knowing how and why they respond in certain ways is the key to effective population control (Figure 4). In the ideal scenario, EIS is able to impose sufficient additive mortality to reduce populations below the irruption threshold, where depensatory pressures may continue to ensnare the population (e.g., [29,30]; Figure 2). However, there are also management scenarios where treatment mortality is insufficiently additive or where populations compensate, resulting in rebound (Figure 4). Which scenario plays out depends largely on what types of tactics are used and how these tactics are used. Population control programs for invasive organisms offer insights on how to more efficiently enhance additive mortality and minimize compensation [14–16]. These provide the basis for the tactics and practices used in EIS [13].

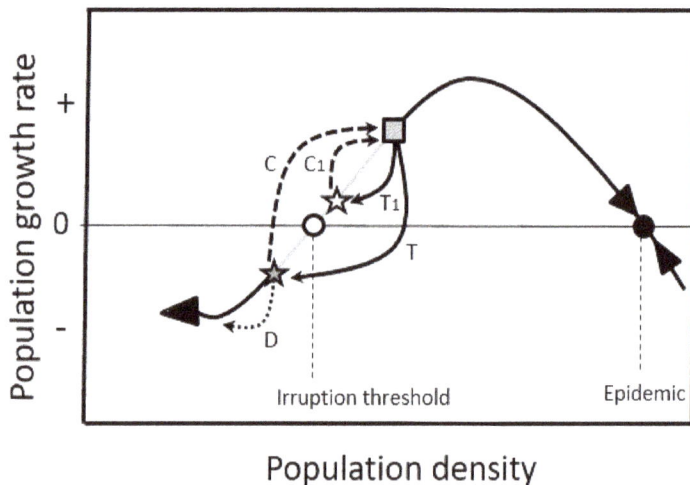

Figure 4. Hypothetical recruitment curve for rising spruce budworm populations with different intensities of treatment mortality and possible population responses. The gray box represents the starting point of a treatment (e.g., with an insecticide) with its potential impacts on population density and growth rate indicated by the stars. Relatively high (and additive) treatment mortality (T) may return populations to the lower equilibrium (gray star) where depensatory pressure (D) should keep populations low in the absence of a strong compensatory response (C). On the other hand, inadequate treatment mortality (T_1) and even modest compensation (C_1) could allow populations to rebound (thick dashed lines) and continue the outbreak.

2.3.1. Enhance Additive Mortality

To reduce populations below the irruption threshold, tactics used in EIS must enhance mortality above that occurring naturally, and this additive mortality must be high enough to reduce populations to below the irruption threshold (Figure 4). In other words, treatment mortality cannot completely overlap or interfere with mortality from natural agents. Because of their particular biological modes of action and the timing of their application, neither *Btk* nor tebufenozide should directly affect the more than 78 reported parasitoids and entomopathogens, or its myriad vertebrate and invertebrate predators [34,56–58]. Some concerns have been raised that by killing spruce budworm larvae or other species of caterpillars harboring parasitoid larvae there could be indirect effects on parasitoid populations that could reduce natural control in subsequent generations. However, for *Btk* at least this risk is probably low, as parasitism tends to suppress feeding rates making parasitized larvae less likely to ingest a lethal insecticide dose [59,60]. Indeed, it has been shown that ingestible insecticides sprayed later in the season (i.e., when the feeding rates of parasitized budworm are dropping) can help to minimize redundant mortality [61], thus potentially boosting additive mortality from *Btk* as well as natural enemy impacts across generations. Whether similar interactions occur for parasitized larvae that have ingested tebufenozide remains unknown.

Development or integration of other tactics that can add mortality while avoiding interference with natural enemies (e.g., tree resistance, biological control, semiochemical control) could further increase EIS efficiency and efficacy [62,63].

2.3.2. Limit Compensation

Compensation may occur after treatments if mortality is insufficient to reduce populations below the irruptive threshold (Figure 4). Because of the inherently high reproductive rate of spruce budworm, even a modest proportion of survivors can rapidly replenish a treated area. Adding to this, inflicting mortality on populations that are too far above the outbreak threshold is likely to alleviate density-dependent mortality resulting from competition. Such was the case in the DDT trials of the early 1950s, where severe early-instar larval mortality was offset by enhanced survival and fecundity in later life stages, thus allowing populations to bounce back within just a year or two [6]. This rebound occurred despite high parasitism rates, which in comparison to untreated sites were either unaffected or higher following DDT treatments [6,58]. To counter this aspect of outbreak dynamics in spruce budworm, EIS focuses on controlling hotspots as soon as they arise. Although it is not yet clear what population level is "too high" for EIS, there does appear to be a threshold at around 28% defoliation (or ~10 fourth-instar larvae per 45 cm branch tip) above which hotspots begin to emit emigrant moths to surrounding areas [28]. In general, catching hotspots early should reduce the probability of compensation occurring.

Pest reinvasion via dispersal is another compensatory response that can allow populations to rebound even when treatments are highly efficacious (Figure 4). Spruce budworm are prolific dispersers and certain weather conditions can produce long-distance mass-exodus flights that disperse millions of egg-laying or mating moths throughout the region [23]. Although the impact of these mass dispersal events on regional demographic trends remain uncertain, there is clear evidence that moth dispersal from high-density populations (e.g., hotspots) results in successful reproduction [64] and can promote local outbreak spread and reinvasion of treated sites [6,28,65]. Compensation through moth reinvasion is a classic problem in pest management and has been one of the major drivers behind the historical use of area-wide management programs [66]. Such regional-scale management programs involve coordinated monitoring and control over large areas as a means of reducing the ability of pests to spread or reinvade treated areas. Commitment and sufficient funding by the participating jurisdictions are necessary for ongoing monitoring and treatments in all or nearly all areas where spruce budworm populations exceed the threshold. Ultimately, whether EIS is successful in Atlantic Canada will depend on whether immigration intensity and frequency from Québec exceeds the capacity to detect and control hotspots. After 5 years of trials in northern New Brunswick, however, moth dispersal has so far not thwarted EIS efforts [13].

2.4. Costs and Benefits

2.4.1. Economic

The economic feasibility of EIS will ultimately depend on how management costs compare with potential losses from an uncontrolled spruce budworm outbreak. Spruce budworm outbreaks cause massive timber supply and economic losses in part because they cover large areas (52 million ha in Canada in 1975; Canadian Council of Forest Ministers 2014) and cause high tree mortality (e.g., > 85% of balsam fir; [43,67]. Hennigar et al. [68] estimated that an uncontrolled spruce budworm outbreak could reduce future spruce-fir harvests in New Brunswick, from 2013 to 2052, by as much as 18%–25%, depending on outbreak severity. In a related study, Chang et al. [69,70] estimated that the associated economic declines (from 2012 to 2041) could be as much as $3.3–4.7 Billion (CND) depending on the severity of outbreak.

Insect outbreaks can also influence local economies through their impact on non-timber forest products (e.g., Christmas trees, mushrooms, berries, flowers, shrubs), though the economic value of such losses can be harder to quantify (e.g., [71]). Large areas of defoliation and tree mortality can also have impacts on ecotourism and recreation activities such as hiking and camping in provincial and national parks. While all these impacts are dwarfed by the costs of the outbreak to the commercial forestry sector, their mitigation is one of the potential benefits of stopping outbreak spread.

A strong economic case has been made for EIS in New Brunswick and the rest of Atlantic Canada. Using a 50-year timber supply model provided by the Province of New Brunswick, Liu et al. [72] estimated that an uncontrolled outbreak could cause harvest reductions of 29–43 million m^3, with associated direct and indirect reductions in economic output of $25–35 billion. Scenarios of Foliage Protection covering 20% of susceptible forest area resulted in losses of 6–17 million m^3 and $0.5–4.1 Billion CND. Depending upon outbreak severity, the economic benefits of EIS were 3.8–6.4 times higher than the total potential costs, including the costs of running the program under each scenario [72]. Based on extrapolations of impacts and regional harvest levels, a similar economic case has been made for the rest of Atlantic Canada [13] but whether it applies to other regions within the outbreak range of spruce budworm will require further analyses.

A key difference between EIS and Foliage Protections is the likelihood and magnitude of long-term secondary impacts on forests. Because of budgetary and logistical constraints, Foliage Protection often cannot protect all affected areas and many go untreated, especially during large outbreaks. Thus, even under a perfectly implemented Foliage Protection program, some level of impacts on spruce and fir (i.e., growth loss, mortality) is inevitable. Moreover, severe spruce budworm defoliation on unprotected forests can often invite secondary impacts from bark beetles [73,74], disease [75,76], and forest fire [77–79], which may spill over into adjacent protected stands. For example, out of the 8.2 million ha defoliated to date in Québec [80], budgets have so far only permitted about 250,000–400,000 ha to be protected each year (i.e., < 5% of the outbreak area). Ideally, EIS could limit the scale of an outbreak and prevent cumulative defoliation, thereby diminishing risks of secondary mortality factors over large scales.

2.4.2. Ecological

Concerns over the ecological costs of insecticide applications that arose in the aftermath of DDT have shaped public perception of pest management, and specifically large-scale insecticide-based spruce budworm management [81,82]. Once an outbreak extends throughout a region, Foliage Protection often necessitated extensive application: at their peak in 1976, insecticide applications in New Brunswick covered over 3.8 million ha [83]. In contrast, through limiting the range of outbreak spread, EIS is expected to substantially limit the scale and duration of annual treatments. In addition to keeping defoliation to a minimum, the success of EIS will thus depend on verification of minimal ecological costs—direct and indirect. Efforts to minimize non-target effects have led to the development, regulation and strategic application of narrow-spectrum ingestible insecticides [84], which have been further supported by rigorous research on their toxicology and persistence. Both tebufenozide and *Btk* have minimal toxicological effects on a wide range of organisms [9,10], apart from larval Lepidopterans (moths and butterflies) that might ingest treated foliage. Even at levels exceeding current application rates, few detectable direct effects of tebufenozide on aquatic taxa have been found [85,86], although potential indirect effects are more nuanced for some groups. For instance, declines in molting and reproduction of midges were not detected in laboratory assays but have been observed under field mesocosm conditions [87]. Reducing these risks involves controlled insecticide application that minimizes drift, buffering around aquatic habitats, and monitoring of insecticide residue and persistence in nearby water bodies. Ongoing research on non-target effects through food webs involving spruce budworm natural enemies is explicitly included in the ongoing EIS project [88].

While the ecological risks associated with EIS treatments are well studied, the flipside—the indirect ecological cost of not intervening against defoliation, and conversely, the ecosystem benefits of outbreak prevention through EIS versus strategies like Foliage Protection—have received little attention. These ecological costs should be highest in stands suffering severe, multi-year defoliation and associated tree mortality, and where salvage logging occurs. EIS aims to avoid this scenario by suppressing pest densities before sustained high defoliation can negatively affect ecosystems. As large-scale disturbances, insect outbreaks can significantly disrupt timber supply but also indirectly affect ecological goods and services provided by forests, including the provision of water resources and flood control, nutrient cycling, and habitat for both terrestrial and aquatic species [89]. Some of these indirect ecological costs translate to direct economic costs, through increased water treatment requirements [90]. Tree mortality and salvage logging following severe defoliation can affect hydrologic regimes and alter soil nutrient cycling, resulting in increased flow and nutrient runoff, and erosion and sedimentation [90–92]. Coupled with these associated hydrologic changes, the loss of streamside canopy cover through defoliation can increase stream temperature, and collectively alter aquatic food webs and in turn, the availability and quality of critical habitat for cold-water fish, including Atlantic Salmon and Brook Trout [91]. Reduced canopy cover and shifts in vegetation may impair habitat for birds, including threatened species (e.g., Canada Warbler) that prey on spruce budworm [93]. Although spruce budworm is a native defoliator with historical periodic impacts on these ecosystems, other ongoing changes and stressors (e.g., commercial logging, climate change, and altered biota) may affect their resilience and functioning in the face of this major disturbance. Because Foliage Protection focuses mainly on economically important forest stands, critical habitats such as those described above typically remain unprotected. In contrast, through limiting the extent of outbreaks, and therefore avoiding significant defoliation, EIS could provide a regional strategy to help maintain the ecological integrity of critical habitats.

2.4.3. Sociopolitical

Forest management practices, including EIS, must accommodate differing value systems around the critical roles forests play in cultural and spiritual practices, as well as in human health [94–97]. These sociopolitical dynamics can be difficult to quantify, but are nonetheless a key determinant of whether an EIS program can work. In Foliage Protection these issues tend to be less prominent as the majority of treatments take place on Crown Lands [11]. In contrast, an area-wide management program such as EIS aims to control all areas with hotspots and thus must account for the values of a diverse range of affected stakeholders. In many jurisdictions there are limits to the governing authority to apply treatments onto forests not under their direct control. In New Brunswick, for example, the Crown Lands and Forests Act does not provide the authority to treat federal lands, and allows private landowners to opt-out of the EIS program. Additionally, parks, protected areas, and other conservation zones often have the explicit objective to allow natural processes (including disturbance regimes such as outbreaks and forests fires) to unfold without human intervention. Canada has committed to the permanent conservation of 17% of its terrestrial area by 2020 and the risk that those areas may present a barrier to treatment must be considered when assessing the feasibility of EIS. Although it has not been an issue in the program yet, Provincial and National Park managers may eventually have to weigh potential risks of an outbreak to ecotourism or sensitive habitats against potential impacts on rare species or the natural processes that these areas were set aside to preserve (which may be difficult to know with certainty). Similarly, Indigenous peoples in Canada (e.g., the Mi'gmaw, Wolasoqiyik, Peskotomuhkati in Atlantic Canada, and Innu of Labrador) must be formally consulted if any EIS activities potentially affect Aboriginal or Treaty rights. Thus far, there has been no attempt to implement approaches such as *Etuaptmumk* ("Two-Eyed Seeing"; [98]) that merge lessons from traditional knowledge with modern science when making management decisions. It remains unclear how much spruce-fir forest might be set aside for exclusions in the future or what effects that might have on the viability of EIS, although it so far has not been a detriment to ongoing experiments [13].

2.5. Communication and Outreach

Even if all aspects of EIS work as intended and are highly cost effective, without public support the program could not be sustainable at levels needed for successful regional outbreak control. Controversy has long surrounded insecticide usage for spruce budworm [81,82,99] and concerns linger to this day, despite the replacement of broad-spectrum insecticides with more ecologically benign alternatives. A variety of audiences must be engaged and consulted before treatments can occur, including governmental decision-makers, industry, environmental groups, Indigenous Peoples, landowner organizations, provincial and federal parks, municipalities, provincial and local media, and residents. A core component of EIS is a proactive (and bilingual) communications approach that improves knowledgeability about EIS science and management, spruce budworm ecology, and the tactics used in the past and present. This proactive approach contrasts the more reactive approach that has underpinned most past Foliage Protection operations [99]. In Atlantic Canada, a consortium of scientists, government, industry, landowner organizations, and communications experts from throughout the region (i.e., the Health Forest Partnership [100]) meets regularly to discuss and oversee the EIS communications strategy.

2.5.1. Communications Strategy

A few core principles guide the approach. First, scientists and other experts are the communications ambassadors for EIS management and science. This approach gives audiences opportunities to engage with experts directly and reduces the likelihood of miscommunications around the underlying science or ongoing management efforts. Second, timely updates are communicated on all aspects of EIS, including on spruce budworm population trends, location and timing of treatments, and the ongoing progress in the management and scientific research. Finally, scientists and other experts address all public inquiries and concerns directly and openly, and where possible provide reference material from available scientific literature.

2.5.2. Key Messages

Several key messages have become central to EIS communications, in part because they reflect the most frequently asked questions.

Public presentations by scientists and other experts emphasize the important ecological role spruce budworm plays in forest ecology and succession and highlight the historical record of outbreaks as a natural disturbance in our forests since at least the Holocene [101]. This context helps to convey the message that the aim of EIS is not spruce budworm eradication, but rather control of populations to limit the scale and magnitude of the regional impacts. As we often state in public forums, we are in effect playing a game of hotspot "whack-a-mole" in northern New Brunswick with the expectation that it saves us from having to control spruce budworm to the south.

Concerns around the possible environmental and health impacts of insecticides are the most common and persistent issue raised. A significant portion of EIS outreach therefore focuses on explaining past and present insecticide usage, including details on biological modes of action, environmental persistence, and potential non-target impacts of the two narrow-spectrum insecticides employed in EIS. The messages emphasize that the insecticides need only add a small amount of mortality to that occurring naturally to be effective. By aggressively targeting hotspots where they arise near the ongoing outbreak, EIS may prevent outbreaks in the rest of Atlantic Canada.

For groups or managers considering whether to authorize EIS on their lands (e.g., First Nations Lands, Federal and Provincial Parks, Protected Natural Areas), communications efforts focus on explaining the details of the program and establishing collaborations for spruce budworm monitoring. The goal is not to convince such groups to participate in EIS, but rather to provide the tools, data, and information needed to foster productive conversations and support informed decisions once hotspots appear.

2.5.3. Outreach Tactics

EIS communications uses a variety of tools and approaches for dissemination of information and outreach. These include direct engagement through scientific talks, lectures, meetings, exhibits, and roundtables. Information is also made available through an active website [100] with written and video Blogs, informative videos, infographics, and live updates, as well as through proactive media engagement. Scientists typically address questions during talks, panel discussions, and forums, but are also accessible through direct e-mails, phone calls, or by submitting questions through the website. One particularly effective outreach tool has been a Budworm Tracker community science program, which serves the dual purposes of providing data on regional moth density and dispersal, as well as community engagement [102].

In terms of the efficacy of this approach, since 2015 there have been >165 individual stories published >300 times about EIS in various local and national media outlets. Coverage to date has been universally positive. Moreover, of the ~3175 private landowners that have been given the option to opt out of EIS treatments in New Brunswick, <4% have done so.

3. Conclusions

Our EIS framework illustrates the fundamental components of managing spruce budworm through a proactive population control strategy (Figure 1). This framework offers a roadmap to jurisdictions considering proactive EIS vs. reactive Foliage Protection for spruce budworm. These strategies are not necessarily mutually exclusive and there may be areas or conditions where one strategy is more viable than the other. Although the details of this framework are built around stopping spruce budworm outbreaks, its core elements are relevant to population control for insect pests that show a hotspot-style outbreak dynamic. This might include other species of outbreak-prone budworm species (e.g., *Choristoneura* sp. and *Acleris* sp.), though the extent to which these systems fit within our EIS framework remains uncertain.

While this framework has to date proven highly effective for controlling outbreak spread [13], there remain areas of uncertainty and opportunity that warrant further investigation. There is already evidence for climate change induced shifts in the spruce budworm outbreak range and dynamics [103–105]. While these are not likely relevant to the current ongoing outbreak, future range shifts could have significant implications for where EIS might be most efficacious. Moreover, any innovation that improves our fundamental understanding or the efficacy of tools and practices in the EIS framework are likely to increase its viability. This could include the development of more efficient hotspot monitoring and detection protocols, new or additional tactics for population control, or new economic markets or disturbances that further enhance the cost–benefit case for using EIS. Another area of uncertainty is that the proposed EIS framework assumes an absence of genetic barriers to spruce budworm dispersal or establishment. Also unknown is the role of spatial adaptive genetic variation in spruce budworm, including variation in voltinism, genomic legacies of post-glacial expansion, or trade-offs between different host trees. Although these are interesting and promising areas of active research, given the regional (vs. continental) scale at which EIS is likely to be applied we would not expect such variation to be a major hindrance to an EIS program.

Aside from its pest management implications, the EIS is also a rare test of applied population theory. Population dynamics sits at the core of EIS and decades of research (and debate) have revealed an eruptive-spread dynamic that should be conducive to EIS. Attempting to manage outbreaks through an EIS also constitutes a test of predictions emanating from the double-equilibrium versus oscillatory hypotheses at very large scales. In particular, EIS tests predictions underlying cyclic vs. irruptive-collapse dynamics and the extent to which outbreaks reflect range-wide processes vs. local irruptions with contagious spread. It is worth acknowledging that these "experiments" lack true replication or experimental controls, i.e., there are not multiple New Brunswicks available that can be randomly assigned to treatments and controls. On the other hand, large-scale studies such as EIS may offer more realistic results than might arise from small but well-replicated mesocosm experiments [106]. EIS offers a unique opportunity to improve our fundamental understanding of spruce budworm spatiotemporal dynamics and to what extent intervention can alter those dynamics for better management outcomes.

Supplementary Materials: The following are available online at http://www.mdpi.com/1999-4907/10/10/910/s1, Table S1: Patterns and causes of periodic outbreaks, according to two contrasting theories, and how to optimally manage populations under each set of assumptions.

Author Contributions: Conceptualization, R.C.J., V.M., D.A.M., D.S.P. and J.R.; Writing—original draft preparation, R.C.J. and D.T.Q. with significant contributions for individual sections by R.C.J., J.J.B., D.R.C., B.J.C., S.E., E.J.S.E., P.M.A.J., D.K., D.A.M., V.M., E.R.D.M., G.D.M., C.J.N., E.O., D.S.P., D.T.Q., J.R., B.R. and M.S.; Writing—review and editing, all authors; Visualization, R.C.J., S.E., J.R. and B.J.C.; Funding acquisition, R.C.J., J.J.B., B.J.C., V.M., D.A.M., E.J.S.E., D.S.P., J.R. and M.S.

Funding: This research was funded by Atlantic Canada Opportunities Agency, Natural Resources Canada, Government of New Brunswick, and forest industry in New Brunswick.

Acknowledgments: We are also grateful to S. Butterson and B. Pike and two anonymous reviewers for constructive comments on earlier versions of this manuscript. The EIS research was overseen by the Healthy Forest Partnership, a consortium of researchers, landowners, forestry companies, governments, and forest protection experts. Many scientists and staff of industry and government agencies have made important contributions without which the project could not have proceeded.

Conflicts of Interest: The authors declare no conflict of interest.

References

1. Risser, P.G. Landscape ecology: State of the art. In *Landscape Heterogeneity and Disturbance*; Turner, M.G., Ed.; Springer: New York, NY, USA, 1987; pp. 3–14.
2. Sturtevant, B.R.; Cooke, B.J.; Kneeshaw, D.D.; MacLean, D.A. Modelling insect disturbance across forested landscapes: Insights from spruce budworm. In *Simulation Modelling of Forest Landscape Disturbances*; Perera, A.H., Sturtevant, B.R., Buse, L.J., Eds.; Springer International Publishing: Geneva, Switzerland, 2015; pp. 93–134.
3. Pureswaran, D.; Johns, R.C.; Heard, S.B.; Quiring, D. Paradigms in eastern spruce budworm (Lepidoptera: Tortricidae) population ecology: A century of debate. *Environ. Entomol.* **2016**, *45*, 1333–1342. [CrossRef] [PubMed]
4. Webb, F.E. *Studies of Aerial Spraying Against the Spruce Budworm in New Brunswick: A Preliminary Report on the 1952 Upsalquitch Project*; Interim Report; Forest Biology Laboratory: Fredericton, NB, Canada, 1952; pp. 1–9.
5. Kettela, E.G. Aerial spraying for protection of forests infested by spruce budworm. *For. Chron.* **1975**, *51*, 141–142. [CrossRef]
6. MacDonald, D.R.; Webb, F.E. Insecticides and the spruce budworm. *Mem. Entomol. Soc. Can.* **1963**, *95*, 288–310. [CrossRef]
7. Randall, A.P. Evidence of DDT resistance in populations of spruce budworm, *Choristoneura fumiferana* (Clem.), from DDT-sprayed areas of New Brunswick. *Can. Entomol.* **1965**, *97*, 1281–1293. [CrossRef]

8. Van Frankenhuyzen, K. Development and current status of *Bacillus thuringiensis* for control of defoliating forest insects. In *Forest Insect Pests in Canada*; Armstrong, J.A., Ives, W.G.H., Eds.; Canadian Forest Service: Ottawa, ON, Canada, 1995; pp. 315–325.

9. Durkin, P. *Control/Eradication Agents for the Gypsy Moth: Human Health and Ecological Risk Assessment for Bacillus thuringiensis var. Kurstaki (B.T.K.)*; Syracuse Environment Research Associates, Inc.: Fayetteville, NY, USA, 2004; p. 152.

10. Durkin, P.; Klotzbach, J. *Control/Eradication Agents for the Gypsy Moth - Human Health and Ecological Risk Assessment for Tebufenozide (Mimic)*; Syracuse Environment Research Associates, Inc.: Fayetteville, NY, USA, 2004; p. 161.

11. Fuentealba, A.; Dupont, A.; Hébert, C.; Berthiaume, R.; Quezada-García, R.; Bauce, É. Comparing the efficacy of various aerial spraying scenarios using *Bacillus thuringiensis* to protect trees from spruce budworm defoliation. *For. Ecol. Manag.* **2019**, *432*, 1013–1021. [CrossRef]

12. Ministère des Forêts, de la Faune et des Parcs. *Aires Infestées par la Tordeuse des Bourgeons de L'épinette au Québec en 2015—Version 1.0*; Gouvernement du Québec, Direction de la Protection des Forêts: Quebec, QC, Canada, 2015.

13. MacLean, D.A.; Amirault, P.; Amos-Binks, L.; Carleton, D.; Hennigar, C.; Johns, R.C.; Régnière, J. Positive results of an Early Intervention Strategy to suppress a spruce budworm outbreak after five years. *Forests* **2019**, *10*, 448. [CrossRef]

14. Sharov, A.A.; Leonard, D.; Liebhold, A.M.; Roberts, E.A.; Dickerson, W. "Slow the spread": A national program to contain the gypsy moth. *J. For.* **2002**, *100*, 30–36.

15. Bomford, M.; O'Brien, P. Eradication or control for vertebrate pests? *Wildl. Soc. Bull.* **1995**, *23*, 249–255.

16. Sandercock, B.K.; Nilsen, E.B.; Brøseth, H.; Pedersen, H.C. Is hunting mortality additive or compensatory to natural mortality? Effects of experimental harvest on the survival and cause-specific mortality of willow ptarmigan. *J. Anim. Ecol.* **2011**, *80*, 244–258. [CrossRef]

17. Royama, T. Population dynamics of the spruce budworm *Choristoneura fumiferana*. *Ecol. Monogr.* **1984**, *54*, 429–462. [CrossRef]

18. Royama, T.; Eveleigh, E.S.; Morin, J.R.B.; Pollock, S.J.; McCarthy, P.C.; McDougall, G.A.; Lucarotti, C.J. Mechanisms underlying spruce budworm outbreak processes as elucidated by a 14-year study in New Brunswick, Canada. *Ecol. Monogr.* **2017**, *87*, 600–631. [CrossRef]

19. Régnière, J.; Lysyk, T.J. Population dynamics of the spruce budworm, *Choristoneura fumiferana*. In *Forest Insects Pests in Canada*; Armstrong, J.A., Ives, W.G.H., Eds.; Natural Resources Canada, Canadian Forest Service Publication: Ottawa, ON, Canada, 1995; pp. 95–105.

20. Williams, D.W.; Liebhold, A.M. Spatial synchrony of spruce budworm outbreaks in eastern North America. *Ecology* **2000**, *81*, 2753–2766. [CrossRef]

21. Royama, T.; MacKinnon, W.E.; Kettela, E.G.; Carter, N.E.; Hartling, L.K. Analysis of spruce budworm outbreak cycles in New Brunswick, Canada, since 1952. *Ecology* **2005**, *86*, 1212–1224. [CrossRef]

22. Greenbank, D.O. The role of climate and dispersal in the initiation of outbreaks of the spruce budworm in New Brunswick. I. The role of climate. *Can. J. Zool.* **1956**, *34*, 453–476. [CrossRef]

23. Boulanger, Y.; Fabry, F.; Kilambi, A.; Pureswaran, D.S.; Sturtevant, B.R.; Saint-Amant, R. The use of weather surveillance radar and high-resolution three dimensional weather data to monitor a spruce budworm mass exodus flight. *Agric. Forest Meteorol.* **2017**, *234*, 127–135. [CrossRef]

24. Régnière, J.; Nealis, V.G. Density dependence of egg recruitment and moth dispersal in spruce budworms. *Forests* **2019**, *10*, 706. [CrossRef]

25. Cooke, B.J.; Nealis, V.G.; Régnière, J. Insect defoliators as periodic disturbances in northern forest ecosystems. In *Plant Disturbance Ecology: The Process and the Response*; Johnson, E.A., Miyanishi, K., Eds.; Elsevier Academic Press: Burlington, MA, USA, 2007; pp. 487–526.

26. Morris, R.F. The dynamics of epidemic spruce budworm populations. *Mem. Entomol. Soc. Can.* **1963**, *95*, 7–12. [CrossRef]

27. Ludwig, D.; Jones, D.D.; Holling, C.S. Qualitative analysis of insect outbreak systems: The spruce budworm and forest. *J. Anim. Ecol.* **1978**, *47*, 315–332. [CrossRef]

28. Régnière, J.; Cooke, B.J.; Béchard, A.; Dupont, A.; Therrien, P. Dynamics and management of rising outbreak spruce budworm populations. *Forests* **2019**, *10*, 748. [CrossRef]

29. Bouchard, M.; Régnière, J.; Therrien, P. Bottom-up factors contribute to large-scale synchrony in spruce budworm populations. *Can. J. For. Res.* **2018**, *48*, 277–284. [CrossRef]

30. Régnière, J.; Delisle, J.; Pureswaran, D.; Trudel, R. Mate-finding Allee effect in spruce budworm population dynamics. *Entomol. Exp. Appl.* **2013**, *146*, 112–122. [CrossRef]

31. Régnière, J.; Nealis, V.G. Ecological mechanisms of population change during outbreaks of the spruce budworm. *Ecol. Entomol.* **2007**, *32*, 461–477. [CrossRef]

32. Liebhold, A.; Bascompte, J. The Allee effect, stochastic dynamics and the eradication of alien species. *Ecol. Lett.* **2003**, *6*, 133–140. [CrossRef]

33. Cappuccino, N.; Lavertu, D.; Bergeron, Y.; Régnière, J. Spruce budworm impact, abundance and parasitism rate in a patchy landscape. *Oecologia* **1998**, *114*, 236–242. [CrossRef] [PubMed]

34. Eveleigh, E.S.; McCann, K.S.; McCarthy, P.C.; Pollock, S.J.; Lucarotti, C.J.; Morin, B.; McDougall, G.A.; Strongman, D.B.; Huber, J.T.; Umbanhowar, J.; et al. Fluctuations in density of an outbreak species drive diversity cascades in food webs. *Proc. Natl. Acad. Sci. USA* **2007**, *104*, 16976–16981. [CrossRef]

35. Marrec, R.; Pontbriand-Paré, O.; Legault, S.; James, P.M.A. Spatiotemporal variation in drivers of parasitoid metacommunity structure in continuous forest landscapes. *Ecosphere* **2018**, *9*, e02075. [CrossRef]

36. Legault, S.; James, P.M.A. Parasitism rates of spruce budworm larvae: Testing the enemy hypothesis along a gradient of forest diversity measured at different spatial scales. *Environ. Entomol.* **2018**, *47*, 1083–1095. [CrossRef]

37. Miller, C.A.; Kettela, E.G.; McDougall, G.A. *A Sampling Technique for Overwintering Spruce Budworm and Its Applicability to Population Surveys*; Canadian Forestry Service: Fredericton, NB, Canada, 1971; p. 11.

38. Sanders, C.J. Monitoring spruce budworm population density with sex pheromone traps. *Can. Entomol.* **1988**, *120*, 175–183. [CrossRef]

39. MacLean, D.A.; MacKinnon, W.E. Accuracy of aerial sketch-mapping of spruce budworm defoliation in New Brunswick. *Can. J. For. Res.* **1996**, *26*, 2099–2108. [CrossRef]

40. Régnière, J.; Delisle, J.; Sturtevant, B.; Garcia, M.; St-Amant, R. Modeling migratory flight in the spruce budworm: Temperature Constraints. *Forests* **2019**, *10*, 802. [CrossRef]

41. Rahimzadeh-Bajgiran, P.; Weiskittel, A.; Kneeshaw, D.; MacLean, D.A. Detection of annual spruce budworm defoliation and severity classification using Landsat imagery. *Forests* **2018**, *9*, 357. [CrossRef]

42. Goodbody, T.R.H.; Coops, N.C.; Hermosilla, T.; Tompalski, P.; McCartney, G.; MacLean, D.A. Digital aerial photogrammetry for assessing cumulative spruce budworm defoliation and enhancing forest inventories at a landscape-level. *ISPRS J. Photogramm. Remote Sens.* **2018**, *142*, 1–11. [CrossRef]

43. MacLean, D.A. Vulnerability of fir-spruce stands during uncontrolled spruce budworm outbreaks: A review and discussion. *For. Chron.* **1980**, *56*, 213–221. [CrossRef]

44. Hennigar, C.R.; MacLean, D.A.; Quiring, D.T.; Kershaw, J.A. Differences in spruce budworm defoliation among balsam fir and white, red, and black spruce. *For. Sci.* **2008**, *54*, 158–166.

45. Bognounou, F.; De Grandpré, L.; Pureswaran, D.S.; Kneeshaw, D. Temporal variation in plant neighborhood effects on the defoliation of primary and secondary hosts by an insect pest. *Ecosphere* **2017**, *8*, e01759. [CrossRef]

46. Su, Q.; Needham, T.D.; MacLean, D.A. The influence of hardwood content on balsam fir defoliation by spruce budworm. *Can. J. For. Res.* **1996**, *26*, 1620–1628. [CrossRef]

47. Campbell, E.M.; MacLean, D.A.; Bergeron, Y. The severity of budworm-caused growth reductions in balsam fir/spruce stands varies with the hardwood content of surrounding forest landscapes. *For. Sci.* **2008**, *54*, 195–205.

48. Zhang, B.; MacLean, D.A.; Johns, R.C.; Eveleigh, E.S. Effects of hardwood content on balsam fir defoliation during the building phase of a spruce budworm outbreak. *Forests* **2018**, *9*, 530. [CrossRef]

49. Robert, L.E.; Kneeshaw, D.; Sturtevant, B.R. Effects of forest management legacies on spruce budworm (*Choristoneura fumiferana*) outbreaks. *Can. J. For. Res.* **2012**, *42*, 463–475. [CrossRef]

50. Robert, L.E.; Sturtevant, B.R.; Cooke, B.J.; James, P.M.; Fortin, M.J.; Townsend, P.A.; Wolter, P.T.; Kneeshaw, D. Landscape host abundance and configuration regulate periodic outbreak behavior in spruce budworm *Choristoneura fumiferana*. *Ecography* **2018**, *41*, 1556–1571. [CrossRef]

51. Bouchard, M.; Auger, I. Influence of environmental factors and spatio-temporal covariates during the initial development of a spruce budworm outbreak. *Landsc. Ecol.* **2014**, *29*, 111–126. [CrossRef]

52. Bauce, É.; Carisey, N.; Dupont, A.; van Frankenhuyzen, K. *Bacillus thuringiensis* subsp. *kurstaki* aerial Spray prescriptions for balsam fir stand protection against spruce budworm (Lepidoptera: Tortricidae). *J. Econ. Entomol.* **2004**, *97*, 97–1624. [CrossRef] [PubMed]

53. van Frankenhuyzen, K.; Régnière, J. Multiple effects of tebufenozide on the survival and performance of the spruce budworm (Lepidoptera: Tortricidae). *Can. Entomol.* **2017**, *149*, 227–240. [CrossRef]

54. Régnière, J.; Delisle, J.; Dupont, A.; Trudel, R. The impact of moth migration on apparent fecundity overwhelms mating disruption as a method to manage spruce budworm populations. *Forests* **2019**, *10*, 775. [CrossRef]

55. Rhainds, M.; Kettela, E.G.; Silk, P.J. Thirty-five years of pheromone-based mating disruption studies with *Choristoneura fumiferana* (Clemens) (Lepidoptera: Tortricidae). *Can. Entomol.* **2012**, *144*, 379–395. [CrossRef]

56. Crawford, H.S.; Titterington, R.W.; Jennings, D.T. Bird predation and spruce budworm populations. *J. For.* **1983**, *81*, 433–478.

57. Jennings, D.T.; Houseweart, M.W. Predation by eumenid wasps (Hymenoptera: Eumenidae) on spruce budworm (Lepidoptera: Tortricidae) and other lepidopterous larvae in spruce-fir forests of Maine. *Ann. Entomol. Soc. Am.* **1984**, *77*, 39–45. [CrossRef]

58. Waage, J.K.; Hassell, M.P.; Godfray, H.C.J. The dynamics of pest-parasitoid-insecticide interactions. *J. Appl. Ecol.* **1985**, *22*, 825–838. [CrossRef]

59. Nealis, V.; van Frankenhuyzen, K. Interactions between *Bacillus thuringiensis* Berliner and *Apanteles fumiferanae* Vier. (Hymenoptera: Braconidae), a parasitoid of the spruce budworm, *Choristoneura fumiferana* (Clem.) (Lepidoptera: Tortricidae). *Can. Entomol.* **1990**, *122*, 585–594. [CrossRef]

60. Nealis, V.; van Frankenhuyzen, K.; Cadogan, B.L. Conservation of spruce budworm parasitoids following application of *Bacillus thuringiensis* var. *kurstaki* Berliner. *Can. Entomol.* **1992**, *124*, 1085–1092. [CrossRef]

61. Cooke, B.J.; Régnière, J. An objectoriented, process-based stochastic simulation model of *Bacillus thuringiensis* efficacy against spruce budworm, *Choristoneura fumiferana* (Lepidoptera: Tortricidae). *Int. J. Pest Manag.* **1996**, *42*, 291–306. [CrossRef]

62. Quiring, D.; Flaherty, L.; Adams, G.; McCartney, A.; Miller, J.D.; Edwards, S. An endophytic fungus interacts with crown level and larval density to reduce the survival of eastern spruce budworm, *Choristoneura fumiferana* (Lepidoptera: Tortricidae), on white spruce (*Picea glauca*). *Can. J. For. Res.* **2019**, *49*, 221–227. [CrossRef]

63. Williams, M.; Eveleigh, E.; Forbes, G.; Lamb, R.; Roscoe, L.; Silk, P. Evidence of a direct chemical plant defense role for maltol against spruce budworm. *Entomol. Exp. Appl.* **2019**, *167*, 755–762. [CrossRef]

64. Larroque, J.; Legault, S.; Johns, R.; Lumley, L.; Cusson, M.; Renaut, S.; Levesque, R.C.; James, P.M.A. Temporal variation in spatial genetic structure during population outbreaks: Distinguishing among different potential drivers of spatial synchrony. *Evol. Appl.* **2019**. [CrossRef]

65. Dobesberger, E.J.; Lim, K.P.; Raske, A.G. Spruce budworm (Lepidoptera: Tortricidae) moth flight from New Brunswick to Newfoundland. *Can. Entomol.* **1983**, *115*, 1641–1645. [CrossRef]

66. Elliott, N.C.; Onstad, D.W.; Brewer, M.J. History and ecological basis for areawide pest management. In *Area-wide Pest Management: Theory and Implementation*; Koul, O., Cuperus, G.W., Elliott, N., Eds.; CABI International: Oxford, UK, 2008; pp. 15–33.

67. MacLean, D.A.; Ostaff, D.P. Patterns of balsam fir mortality caused by an uncontrolled spruce budworm outbreak. *Can. J. Forest Res.* **1989**, *19*, 1087–1095. [CrossRef]

68. Hennigar, C.R.; Erdle, T.A.; Gullison, J.J.; MacLean, D.A. Re-examining wood supply in light of future spruce budworm outbreaks: A case study in New Brunswick. *For. Chron.* **2013**, *89*, 42–53. [CrossRef]

69. Chang, W.Y.; Lantz, V.A.; Hennigar, C.R.; MacLean, D.A. Benefit-cost analysis of spruce budworm (*Choristoneura fumiferana* Clem.) control: Incorporating market and non-market values. *J. Environ. Manag.* **2012**, *93*, 104–112. [CrossRef]

70. Chang, W.Y.; Lantz, V.A.; Hennigar, C.R.; MacLean, D.A. Economic impacts of spruce budworm (*Choristoneura fumiferana* Clem.) outbreaks and control in New Brunswick, Canada. *Can. J. For. Res.* **2012**, *42*, 490–505. [CrossRef]

71. Cathro, J.; Mulkey, S.; Bradley, T. A bird's eye view of small tenure holdings in British Columbia. *J. Ecosyst. Manag.* **2007**, *8*, 58–66.

72. Liu, E.Y.; Lantz, V.; MacLean, D.A.; Hennigar, C. Economics of early intervention to suppress a potential spruce budworm outbreak on Crown land in New Brunswick, Canada. *Forests* **2019**, *10*, 481. [CrossRef]

73. Belyea, R.M. Death and deterioration of balsam fir weakened by spruce budworm defoliation in Ontario. *Can. Entomol.* **1952**, *84*, 325–335. [CrossRef]

74. Ostaff, D.P.; MacLean, D.A. Spruce budworm populations, defoliation, and changes in stand condition during an uncontrolled spruce budworm outbreak on Cape Breton Island, Nova Scotia. *Can. J. For. Res.* **1989**, *19*, 1077–1086. [CrossRef]

75. Stillwell, M.A. Pathological aspects of severe spruce budworm attack. *For. Sci.* **1956**, *2*, 174–180.

76. Ostaff, D. *D. A Wood Quality Study of Dead and Dying Balsam Fir—The Incidence of Armillaria Root Rot*; Technical Note 82; Environment Canada, Canadian Forestry Service, Maritimes Forest Research Centre: Fredericton, NB, Canada, 1983.

77. Stocks, B.J. Fire potential in the spruce budworm-damaged forests of Ontario. *For. Chron.* **1987**, *63*, 8–14. [CrossRef]

78. James, P.M.A.; Fortin, M.J.; Sturtevant, B.R.; Fall, A.; Kneeshaw, D. Modelling spatial interactions among fire, spruce budworm, and logging in the boreal forest. *Ecosystems* **2011**, *14*, 60–75. [CrossRef]

79. James, P.M.A.; Robert, L.E.; Wotton, B.M.; Martell, D.L.; Fleming, R.A. Lagged cumulative spruce budworm defoliation affects the risk of fire ignition in Ontario, Canada. *Ecol. Appl.* **2017**, *27*, 532–544. [CrossRef]

80. QMFFP: Québec Ministère des Forêts, de la Faune et des Parcs. *Aires Infestées par la Tordeuse des Bourgeons de L'épinette au Québec en 2018*; Gouvernement du Québec, Direction de la Protection des Forêts: Québec City, QC, Canada, 2018; p. 20.

81. Carson, R. *Silent Spring*; Houghton Mifflin: New York, NY, USA, 1962.

82. May, E. *Budworm Battles*; Four East Publications Ltd.: Halifax, NS, Canada, 1982.

83. Armstrong, J.A.; Cook, C.A. *Aerial Spray Applications on Canadian Forests: 1945–1990*; Information Report; Forestry Canada: Ottawa, ON, Canada, 1993.

84. Holmes, S.B.; MacQuarrie, C.J.K. Chemical control in forest pest management. *Can. Entomol.* **2016**, *148*, S270–S295. [CrossRef]

85. Soin, T.; Smagghe, G. Endocrine disruption in aquatic insects: A review. *Ecotoxicology* **2007**, *16*, 83–93. [CrossRef]

86. Kreutzweiser, D.P.; Gunn, J.M.; Thompson, D.G.; Pollard, H.G.; Faber, M.J. Zooplankton community responses to a novel forest insecticide, tebufenozide (RH-5992), in littoral lake enclosures. *Can. J. Fish. Aquat. Sci.* **1998**, *55*, 639–648. [CrossRef]

87. Tassou, K.T.; Schulz, R. Low filed-relevant tebufenozide concentrations affect reproductions in *Chironomus riparius* (Diptera: Chironomidae) in a long-term toxicity test. *Environ. Sci. Pollut. Res.* **2013**, *20*, 3735–3742. [CrossRef] [PubMed]

88. Martel, V.; Johns, R.C.; Eveleigh, E.; McCann, K.; Pureswaran, D.; Sylvain, Z.; Morrison, A.; Morin, B.; Owens, E.; Hébert, C. Landscape level impacts of EIS on SBW, other herbivores and associated natural enemies (ACOA RD100 2.2.2). In Proceedings of the SERG-I Workshop, Fredericton, NB, Canada, 7–8 February 2017; pp. 112–118.

89. Schowalter, T.D. Insect herbivore effects on forest ecosystem services. *J. Sustain. For.* **2012**, *31*, 518–536. [CrossRef]

90. Postel, S.L.; Thompson, B.H., Jr. Watershed protection: Capturing the benefits of nature's water supply services. *Nat. Resour. Forum* **2005**, *29*, 98–108. [CrossRef]

91. Dhar, A.; Parrott, L.; Heckbert, S. Consequences of mountain pine beetle outbreak on forest ecosystem services in western Canada. *Can. J. For. Res.* **2016**, *46*, 987–999. [CrossRef]

92. Redding, T.; Winkler, R.; Teti, P.; Spittlehouse, D.; Boon, S.; Rex, J.; Dubé, S.; Moore, R.D.; Wei, A.; Carver, M.; et al. Mountain pine beetle and watershed hydrology. *BCJ. Ecosys. Manag.* **2008**, *9*, 33–50.

93. Venier, L.A.; Holmes, S.B. A review of the interaction between forest birds and eastern spruce budworm. *Environ. Rev.* **2010**, *18*, 191–207. [CrossRef]

94. Crewe, J.; Johnstone, J.F. *Plant Use in Vuntut Gwitchin Territory*; Vuntut Gwitchin First Nation: Old Crow, YT, Canada, 2008; pp. 1–49.

95. Kanowski, P.J.; Williams, K.J.H. The reality of imagination: Integrating the material and cultural values of old forests. *For. Ecol. Manag.* **2009**, *258*, 341–346. [CrossRef]

96. Nowak, D.J.; Hirabayashi, S.; Bodine, A.; Greenfield, E. Tree and forest effects on air quality and human health in the United States. *Environ. Pollut.* **2014**, *193*, 119–129. [CrossRef]

97. Chang, W.Y.; Lantz, V.A.; MacLean, D.A. Public attitudes about forest pest outbreaks and control: Case studies in two Canadian provinces. *For. Ecol. Manag.* **2009**, *257*, 1333–1343. [CrossRef]
98. Two-Eyed Seeing. Available online: http://www.integrativescience.ca/Principles/TwoEyedSeeing/ (accessed on 12 August 2019).
99. Miller, A. Conventional problem solving. In *Environmental Problem Solving: Psychosocial Barriers to Adaptive Change*, 1st ed.; Alexander, D.E., Ed.; Springer: New York, NY, USA, 1999; pp. 82–123.
100. Healthy Forest Partnership. Available online: http://www.healthyforestpartnership.ca (accessed on 12 August 2019).
101. Simard, I.; Morin, H.; Lavoie, C. A millennial-scale reconstruction of spruce budworm abundance in Saguenay, Quebec, Canada. *Holocene* **2006**, *16*, 31–37. [CrossRef]
102. Budworm Tracker Program. Available online: http://healthyforestpartnership.ca/budworm-tracker/about-the-program/ (accessed on 23 September 2019).
103. Régnière, J.; St-Amant, R.; Duval, P. Predicting insect distributions under climate change from physiological responses: Spruce budworm as an example. *Biol. Invasions* **2012**, *14*, 1571–1586. [CrossRef]
104. Pureswaran, D.S.; De Grandpré, L.; Paré, D.; Taylor, A.; Barrette, M.; Morin, H.; Régnière, J.; Kneeshaw, D. Climate-induced changes in host tree–insect phenology may drive ecological state-shift in boreal forests. *Ecology* **2015**, *96*, 1480–1491. [CrossRef]
105. Pureswaran, D.S.; Neau, M.; Marchand, M.; De Grandpré, L.; Kneeshaw, D. Phenological synchrony between eastern spruce budworm and its host trees increases with warmer temperatures in the boreal forest. *Ecol. Evol.* **2018**, *9*, 576–586.
106. Schindler, D.W. Replication versus realism: The need for ecosystem-scale experiments. *Ecosystems* **1998**, *1*, 323–334. [CrossRef]

![forests logo] *forests*

MDPI

Article

Positive Results of an Early Intervention Strategy to Suppress a Spruce Budworm Outbreak after Five Years of Trials

David A. MacLean [1,*], Peter Amirault [2], Luke Amos-Binks [3], Drew Carleton [3], Chris Hennigar [1,3], Rob Johns [4] and Jacques Régnière [5]

[1] Faculty of Forestry and Environmental Management, University of New Brunswick, Fredericton, NB E3B 5A3, Canada; Chris.Hennigar@gnb.ca
[2] Forest Protection Limited, Lincoln, NB E3B 7E6, Canada; PAmirault@ForestProtectionLimited.com
[3] New Brunswick Department of Energy and Resource Development, Fredericton, NB E3B 5H1, Canada; Luke.Amos-Binks@gnb.ca (L.A.-B.); Drew.Carleton@gnb.ca (D.C.)
[4] Canadian Forest Service, Atlantic Forest Centre, Fredericton, NB E3B 5P7, Canada; rob.johns@canada.ca
[5] Canadian Forest Service, Laurentian Forest Centre, Ste. Foy, QC G1V 4C7, Canada; jacques.regniere@canada.ca
*   Correspondence: macleand@unb.ca; Tel.: +1-506-458-7552

Received: 24 April 2019; Accepted: 21 May 2019; Published: 23 May 2019

Abstract: Spruce budworm (*Choristoneura fumiferana* Clem.; SBW) outbreaks are one of the dominant natural disturbances in North America, having killed balsam fir (*Abies balsamea* (L.) Mill.) and spruce (*Picea* sp.) trees over tens of millions of hectares. Responses to past SBW outbreaks have included the aerial application of insecticides to limit defoliation and keep trees alive, salvage harvesting of dead and dying trees, or doing nothing and accepting the resulting timber losses. We tested a new 'early intervention strategy' (EIS) focused on suppressing rising SBW populations before major defoliation occurs, from 2014 to 2018 in New Brunswick, Canada. The EIS approach included: (1) intensive monitoring of overwintering SBW to detect 'hot spots' of low but rising populations; (2) targeted insecticide treatment to prevent spread; and (3) proactive public communications and engagement on project activities and results. This is the first attempt of area-wide (all areas within the jurisdiction of the province of New Brunswick) management of a native forest insect population. The project was conducted by a consortium of government, forest industry, researchers, and other partners. We developed a treatment priority and blocking model to optimize planning and efficacy of EIS SBW insecticide treatment programs. Following 5 years of over 420,000 ha of EIS treatments of low but increasing SBW populations, second instar larvae (L2) SBW levels across northern New Brunswick were found to be considerably lower than populations in adjacent Québec. Treatments increased from 4500 ha in 2014, to 56,600 ha in 2016, and to 199,000 ha in 2018. SBW populations in blocks treated with *Bacillus thuringiensis* or tebufenozide insecticide were consistently reduced, and generally did not require treatment in the subsequent year. Areas requiring treatment increased up to 2018, but SBW L2 populations showed over 90% reductions in that year. Although this may be a temporary annual decline in SBW population increases, it is counter to continued increases in Québec. Following 5 years of tests, the EIS appears to be effective in reducing the SBW outbreak.

Keywords: insect population management; spruce budworm; early intervention; defoliation; economic losses; decision support system; optimized treatment design

---

1. Introduction

The spruce budworm (*Choristoneura fumiferana* Clem.; SBW) outbreak in eastern Canada and Maine from 1967 to 1993 was the dominant natural disturbance in the region, peaking at over 50 million hectares of defoliation [1]. Outbreaks (repeated annual defoliation typically lasting up to 10 years) results in growth reduction of up to 90% [2], tree mortality in balsam fir (*Abies balsamea* (L.) Mill.)-spruce (*Picea* sp.) forests often exceeding 85% [3,4], and changes in regeneration patterns [5]. SBW outbreaks also affect forest landscape structure (i.e., stand species composition and spatial configuration) with consequences for forest succession [6], timber production [7], and the risk of future disturbances such as fire [8]. Several papers have reviewed SBW and other insect outbreak effects on tree mortality [3], stand development and ecosystem responses [9,10], and ecological mechanisms of SBW population changes during outbreaks [11–13]. Defoliation associated with larval feeding caused timber volume losses estimated at up to 44 million m^3 per year, or 30% of the total Canadian timber harvest in 2012. To limit timber supply shortfalls and the economic impact of SBW, at the peak of the last outbreak, 6.9 million hectares of forest was treated with insecticide in Canada in 1976, primarily in the provinces of Québec and New Brunswick [14]. In Québec, mortality losses during the 1967 to 1992 SBW outbreak were estimated at 238 million m^3 of spruce and balsam fir, with an estimated similar additional amount of reduced growth [15]. The total losses from the SBW outbreak in Québec had an estimated commercial value of $12.5 billion [16].

Forest species composition affects SBW defoliation in several ways, and understanding these effects is important in setting criteria and prioritizing areas for SBW control treatments. SBW defoliation differs among host species [17], with balsam fir the most defoliated and white spruce (*Picea glauca* [Moench] Voss), red spruce (*Picea rubens* Sarg.), and black spruce (*Picea mariana* [Mill.] B.S.P.) having approximately 72%, 41%, and 28% as much defoliation as balsam fir, respectively [17]. In addition, several studies have reported lower SBW defoliation of balsam fir, and lower resulting growth reduction and mortality, in stands or landscapes with higher proportions of broadleaved, hardwood species [3,18,19]. In 25 plots in northern New Brunswick over a 5-year period in the last stages of the 1970s–1990s SBW outbreak, defoliation of balsam fir was <15% with >80% hardwood content, compared to 58%–71% when hardwood content was <40% [19]. Tree-ring analysis also showed that SBW-caused growth reductions were twice as high (40%) in stands with <50% hardwood content, compared to 20% in stands with >50% hardwood content [20]. Fir-hardwood stands (~30% hardwood content) also sustained 14%–30% less SBW-caused fir mortality than in fir-dominated stands [3,18].

Forests in New Brunswick are composed of 85% species susceptible to SBW [21] and have undergone defoliation of up to 3.6 million ha in 1975 [1]. As a result, there has been a strong commitment to insecticide treatment in this jurisdiction, with an average of 2.0 million hectares per year treated from 1970–1983, at an average cost of $4 per hectare or $7.7 million per year. Today, owing to inflation and increased pest control product and application costs (currently $40 or $80 per hectare, depending on whether one or two applications per year), a similar protection strategy on 2 million hectares would cost between $80–$160 million per year. In a 2007 survey of the New Brunswick public [22], 94% of respondents supported funding research and development on pest control, and 82% supported controlling future SBW outbreaks.

Two detailed studies have quantified the potential timber supply and economic impacts of SBW outbreak scenarios in New Brunswick, which provided much of the rationale for continued pest control research on the topic. Hennigar et al. [21] determined that timber harvest reductions, relative to a no defoliation case, for the 3.0 million ha of Crown land in New Brunswick were projected to reach 18% and 25% by 2052, under moderate and severe outbreak defoliation scenarios from 2012–2032. Up to 30% to 50% of these reductions were projected to be avoided through insecticide treatments, depending on the outbreak scenario. Peak wood supply reduction of 25% was projected during the period of defoliation, but impacts also greatly reduced the large increases in wood supply projected from 2042–2062 that would otherwise result from long-term silviculture. Chang et al. [7,23] estimated the costs of SBW outbreak scenarios and benefits of treatments based on the Hennigar et al. [21] timber supply projections. Under uncontrolled moderate and severe SBW outbreak scenarios, total output in the New Brunswick economy over the 2012–2041 period was projected to decline in present value terms by $3.3 billion (CAD) and $4.7 billion, respectively. SBW control was projected to reduce the negative impacts on economic output by up to 66% when protecting 40% of susceptible forests. Combining SBW control with re-scheduling harvests and a salvage strategy under moderate and severe outbreaks was projected to reduce the negative impacts on output by a further 1%–18%, depending on the level of control implemented [7].

Eastern North America is now undergoing another SBW outbreak that began in northern Québec in about 2005 [24]. In the past, SBW outbreaks have been managed through a reactive "foliage protection" approach focused on keeping trees alive, whereby areas are treated following some defoliation but before tree mortality occurs [25–27]. This approach usually has required that at least 2 years of moderate-severe current-year defoliation occur before allowing insecticide treatment, because it typically takes 4–5 years of defoliation to kill trees [3]. The main goal of SBW control programs in eastern Canada is to protect the trees' current-year foliage (target 50% current foliage retained in Québec, 60% in New Brunswick) in order to ensure tree survival and limit wood losses during outbreaks [28]. Insecticide applications every 2 years in balsam fir and white spruce stands and every 3 years in black spruce-dominated stands provides an adequate level of protection to reduce growth losses (maintain the residual photosynthetic capacity above 39%), while reducing the number of required annual insecticide applications [28].

While a foliage protection strategy will reduce SBW-caused tree mortality, it cannot suppress the overall rise or spread of outbreaks. As an alternative to this long-standing approach, we are testing an Early Intervention Strategy (EIS) to suppress SBW populations in New Brunswick, which involves: (i) intensive monitoring and early detection of low-level increases of SBW populations, before substantial defoliation occurs; and (ii) small area, target-specific application of insecticides to locations with rising SBW populations. Recent advances in our understanding of SBW population dynamics [29] have prompted efforts to develop this new EIS approach to managing SBW. It is the first attempt of area-wide management (within the currently funded trial area of Atlantic Canada) of a native defoliating insect. EIS focuses on controlling relatively low-density populations along the leading edge of outbreaks as a way of containing outbreak spread. Important science considerations addressed include what SBW density to initiate an EIS; what insecticide products are effective; what are the consequences of treatments on natural enemy populations attacking SBW in subsequent years; what new decision-support tools and technology need to be developed to optimize treatments; and the assessment of costs and benefits.

The EIS program shares many characteristics with area-wide containment programs used to contain invasive species, such as the "Slow the Spread" program for gypsy moth (*Lymantria dispar* Linnaeus) in the United States [30,31]. Many practical and theoretical considerations underlie the development of a pest containment program, which in essence is a population control program. These include how to monitor and decide when and where to treat hot spots; what pest control products should be used; whether pest control treatments result in additive mortality (i.e., mortality in addition to what would otherwise occur naturally) and thereby drive population decline; and whether mass moth dispersal beyond the leading edge of the outbreak might offset treatment efficacy.

In addition, given that natural enemies (parasitoids in particular) are a major source of natural SBW control [32], evaluation of whether treatments adversely affect natural enemy populations and thereby reduce natural mortality rates of SBW is required. Most of the key parasitoids thought to control SBW are generalists that attack other herbivores when SBW densities are low and these may be adversely affected if a low-density population is treated [32,33]. Unwanted impacts of treatments on the general parasitoid community could promote SBW in years following treatment, if parasitoids are reduced by insecticide-induced SBW mortality or through alternative hosts mortality.

From 2014–2018, EIS research trials were conducted by a consortium termed the Healthy Forest Partnership, encompassing the Governments of Canada and New Brunswick, Natural Resources Canada, universities, and forest industry (www.healthyforestpartnership.ca). In addition to SBW monitoring and EIS control measures designed to suppress populations, the research included longer-term understanding of effects of natural enemies, factors affecting outbreak initiation, inoculation of seedlings with endophytic fungi to increase host resistance [34,35] and improving decision support capabilities to facilitate planning. The project engaged participation of the region's leading forestry companies, universities, and federal and provincial research agencies. We put considerable effort into clear, timely communication of the details of the infestation, treatments, impacts to human health and ecosystems, and research results to the public and stakeholders.

We developed a new treatment priority and blocking model to optimize planning of annual EIS SBW insecticide treatment programs, by directing treatments to the highest priority areas to maximize reductions of SBW L2 populations. Herein we use the term 'block' to designate a contiguous area, typically rectangular in shape, designed for treatment with insecticide using a series of aircraft flight swaths. The model aims to minimize the cost and effort required to achieve a given pest control objective, optimizing use of *Bacillus thuringiensis* var. Kurstaki (*Btk*) or tebufenozide pest control products and application technologies. It was based upon elements of the SBW Decision Support System (DSS), which includes stand and forest-level models and a GIS that projects effects of SBW defoliation and management/treatment strategies on stand growth, timber supply, and economic indicators [29,36–38]. The DSS permits users to integrate forest harvest planning, protection using pesticides, and salvage, within a spatial optimization framework, to reduce losses to SBW [37]. The most recent version of the SBW DSS is termed the *Accuair Forest Protection Optimization System* (*ForPRO*) [39], which optimizes treatment schedules to reduce losses, prioritizes areas to be treated, determines impacts on harvest levels, and integrates salvage activities with protection.

In this paper, we will present and discuss 5-year interim results of EIS SBW monitoring and treatment trials conducted in New Brunswick from 2014–2018. Objectives are: (1) to develop and test an EIS SBW insecticide treatment priority and blocking model (where 'blocks' refer to typically rectangular areas defined for aircraft delivery of insecticide) to optimize planning and direct treatments to the highest priority areas and maximize SBW population reductions; and (2) evaluate the effectiveness of EIS control treatments conducted from 2014 to 2018 and their impacts on SBW population trends. Our underlying hypothesis is that intensive monitoring and treatment of rising SBW 'hot spots' with insecticide, before defoliation occurs, can delay, prevent, or reduce the severity of a native insect outbreak. If successful, an EIS approach has the potential to reduce or eliminate the high levels of defoliation that can only be reduced by a foliage protection approach.

2. Methods

2.1. Monitoring and Detection of SBW 'Hot Spots'

Monitoring of SBW populations for EIS treatments was conducted using a combination of SBW pheromone trapping, intensive second-instar larval (L2) population surveys based on branch sampling, and aerial defoliation surveys. L2 population surveys were the primary data source to indicate rising SBW populations, because they directly measure the overwintering larvae that cause defoliation in the following summer. Pheromone traps were located in susceptible forests and were helpful for identifying areas where additional L2 monitoring plots might be needed. A large number of points (1136–1964 per year; Table 1) were sampled, with emphasis on northern New Brunswick due to its proximity to the Québec SBW outbreak (Figure 1). The L2 branch sampling was done in the fall/winter and thus years in Table 1 and Figure 1 relate to that period, but determine what was treated in the following summer. Each sample point consisted of sampling one mid-crown branch from each of three balsam fir or spruce trees. A key feature of the EIS sampling was that forest industry crews assisted New Brunswick Department of Energy and Resource Development (NB ERD) staff in collecting branches for L2 sampling, as part of their contribution to resources for the project. SBW overwinters as L2 in a hibernaculum spun under bark scales and lichen on the host tree, and the hibernaculum can be destroyed and larva washed from the foliage with a sodium hydroxide solution [40]. Sampled branches were bagged and transported to the NB ERD lab, where they underwent a sodium hydroxide wash, filtering, and counting under a microscope [40] to determine the number of overwintering L2 SBW per branch, as an estimate of populations in the subsequent season. The annual SBW L2 data are publicly available at www.healthyforestpartnership.ca. A threshold of 7 L2/branch (rounded; an actual mean of three branches >6.5 L2/branch) was proposed to plan treatments because above this threshold, populations were expected to increase. The threshold was estimated based on SBW population data collected in the Lower St-Lawrence region of Québec between 2012 and 2015, during the rise of the current outbreak in that area. It was calculated from the average L4 density that led to an annual population growth rate just under 1, given the observed density dependence of generation survival, an average apparent fecundity of 60 eggs per surviving adult, and average mortality from egg to L4 in the next generation. The threshold was then modified to translate from L4 to overwintering L2.

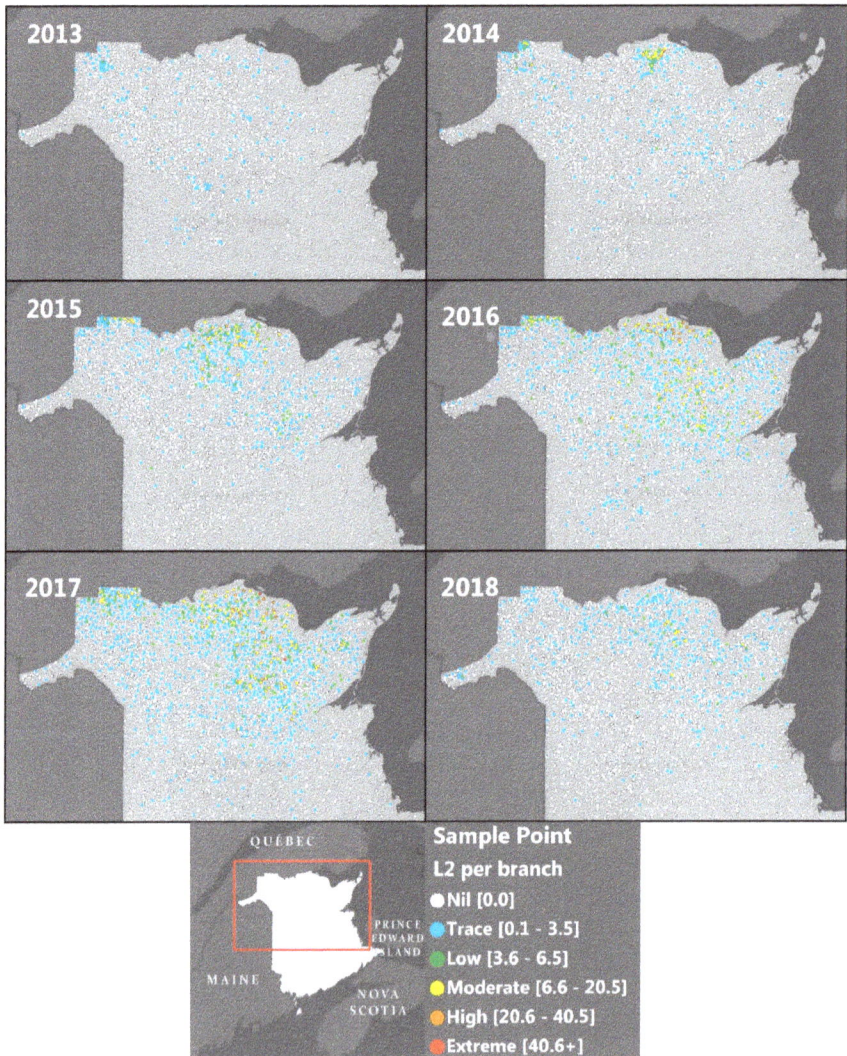

Figure 1. SBW L2 sample points in New Brunswick from 2013 to 2018. These L2 point data were spatially interpolated and used to plan treatment areas (generally ≥ 7 L2/branch), using an optimized blocking algorithm described in Section 2.3.

Table 1. Percentage of SBW L2 samples in New Brunswick in six second instar larvae (L2) classes, each year from 2013–2018. Three mid-crown branches were sampled at each sample point each year. Locations of sample points are shown in Figure 1.

Year	% of L2 Samples by L2/Branch Class (in Parentheses)						No. Sample Points
	Nil (0)	Trace (0.1–3.5)	Low (3.6–6.5)	Moderate (6.6–20.5)	High (20.6–40.5)	Extreme (>40.5)	
2013	83	17	0.0	0.0	0.0	0.0	1136
2014	82	17	0.5	0.5	0.0	0.0	1503
2015	68	26	3.4	2.2	0.1	0.0	1561
2016	48	40	6.4	4.5	0.6	0.1	1649
2017	43	44	7.1	4.7	0.7	0.1	1964
2018	74	25	0.9	0.5	0.0	0.0	1851

2.2. Incorporation of Effects of Forest Species Composition on SBW Dynamics

The results of two recent studies have helped focus use of tree species in our treatment priority algorithm. Zhang et al. [41] tested effects of hardwood composition on defoliation during the initiation phase (first 5 years) of a SBW outbreak in a gradient of 27 fir-hardwood plots selected to represent three percent hardwood basal area classes (0%–25%, 40%–65%, and 75%–95%). Fir defoliation was significantly lower ($p < 0.001$) as hardwood content increased, but the relationship varied with overall defoliation severity each year. Annual plot defoliation in fir-hardwood plots, estimated using Random Forests prediction incorporating 11 predictor variables, yielded a correlation of 0.92 compared to measured defoliation. Average defoliation severity in softwood plots and % hardwood content were the most influential variables. Bognounou et al. [42] compared stands dominated by highly vulnerable balsam fir, stands dominated by low vulnerability black spruce, and mixed composition stands (fir and black spruce). They found resource concentration effects on the primary host (balsam fir) during the increasing outbreak phase in fir-dominated and mixed stands. Balsam fir, the most susceptible species, depended more on immediate neighboring trees and thus associational effect, whereas black spruce, the less preferred host, showed a greater resource dilution effect from neighboring trees [42]. A stand spruce-fir content threshold of above 20% was selected for use in our treatment priority algorithm and fir-spruce differences could potentially be incorporated.

2.3. Development of the Optimum Pest Control Treatment Priority Model

The biggest difference between our new treatment priority and blocking model and the current *ForPRO* [39] was that *ForPRO* and past SBW DSS iterations have been based on estimated timber supply or harvest level impacts (m^3 losses) of defoliation, whereas EIS planning is based on SBW population levels. Our model used spatial heuristic algorithms to estimate effects of alternative EIS control strategies, specifically determining the most cost-effective application of insecticide to minimize SBW L2 levels. The spray treatment priority raster combines an interpolated SBW L2 sample with a % spruce-fir forest composition layer, as inputs into the blocking tool (Figure 2), which analyzes cells (originally 1 ha 100 × 100 m, but changed to 80 × 80 m in 2018, to coincide with aircraft application swath width), along with tests of alternative desired treatment flight directions, to produce an optimal treatment area.

The blocking tool uses information about aircraft speed, turn times, insecticide hopper and fuel capacities, and treatment swath width to constrain treatment area and determine whether the aircraft should turn or continue to fly to the next high priority area when passing over excluded or low priority treatment areas. The objective is to maximize cumulative treatment priority score over the entire program, with penalties assigned for simulated product deposition losses and high ratios of aircraft flight-time to boom-on-time. Limits on total area blocked are determined externally by the program budget (area to be treated or funding and treatment cost per hectare). Spatial treatment priority score can be determined based on minimizing the expected volume losses, or in the case of EIS, an interpolated and scaled SBW L2 raster layer (described below). Simulated deposition losses result when: (1) boom-on-off events occur (a penalty at the beginning or end of the treatment line, because of application lag); and (2) a flight line is not flanked by adjacent lines (a penalty for increasing likelihood of incomplete line deposition from product drift). The combination of these penalties acts to spatially aggregate areas targeted for treatment to make blocks operationally realistic; e.g., isolated high L2 areas will be less likely to be included compared to aggregated high L2 areas.

The Process

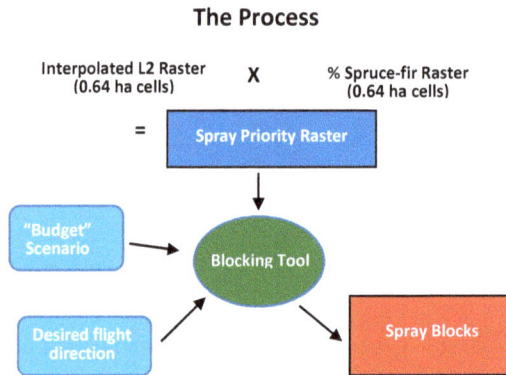

Figure 2. The EIS SBW treatment priority and blocking tool process.

Several spatial interpolation methods were evaluated to produce a continuous population model from the L2 point data, and we selected averaging the output of four interpolation methods: empirical Bayesian kriging, inverse distance weighting, radial basis function, and kernel smoothing (Figure 3A). Raster cell size was set equal to the most common aircraft swath width (80 m) used so each cell becomes a 'yes' or 'no' treatment decision in the blocking tool. The interpolated L2 layer was transformed and scaled (1–100) to put more emphasis on moderate-high L2 populations and then added to the % spruce and fir (scaled from 1–10). Transformation of interpolated L2 was performed using the TfLarge object method available as part of the Spatial Analyst extension in the ArcPy Python site package; transformed $L2 = 1/(1 + (L2/midpoint)^{-spread})$, where midpoint = 7 and spread = 3. Alternative transformation methods were tested to identify what objective function would result in a treatment block solution similar to that expected if created manually by NB ERD experts. The objective was to target treatment of high L2 areas, but not waste insecticide on high hardwood content stands where L2 survival on non-host species is low. The resulting algorithm was based on consensus from expert panel reviews by researchers and NB ERD staff of alternate spray priority weighting rules. The % spruce-fir represents the proportion of merchantable volume in mature stands and relative abundance based on density, stocking and canopy closure for the immature forest. Areas with higher spruce-fir content were expected to yield more L2 per unit area compared to low spruce-fir content areas (Figure 3C). All areas with moderate or higher SBW L2 populations (≥7 L2/branch) were set to high priority to ensure that they were treated. Areas with <20% spruce-fir had spray priority value set to zero, based on results of Zhang et al. [41]. Habitation and other operational setbacks were excluded from treatment in all scenarios (Figure 3B). Together, these methods define the treatment priority model used by the blocking algorithm (Figure 3D).

The optimum treatment priority blocking tool was used by Forest Protection Limited (FPL) and NB ERD staff in designing the EIS protection trials from 2016 to 2018. The blocking algorithm was run on the treatment priority model with flight lines oriented north-south and east-west. Treatment priority input layers were then rotated 45 degrees and the composite priority layer was rebuilt to allow the blocking algorithm to build blocks for northeast-southwest and northwest-southeast flight directions. In total, four different spatial blocking solutions from the different flight orientations were produced, and these four layers were combined to yield areas eligible for treatment. Areas selected for treatment four times were more likely to be good treatment candidates (four 'votes') than areas selected less often. This composite treatment area (Figure 4A,B) was reviewed by NB ERD staff and sent to FPL for development of the final treatment blocks (Figure 4C). FPL staff converted the identified treatment solutions into digitized flight lines in a process that selected the solution from the composite treatment area that best matched the desired flight direction, which is largely determined by proximity to residential areas, infrastructure, and topography. Final treatment blocks were edited to respect exclusion areas in accordance with environmental permits, and to remove areas that were not operationally feasible due to anticipated flight line orientation. The final operational blocks and flight lines closely resembled the eligible treatment area (Figure 4B versus Figure 4C).

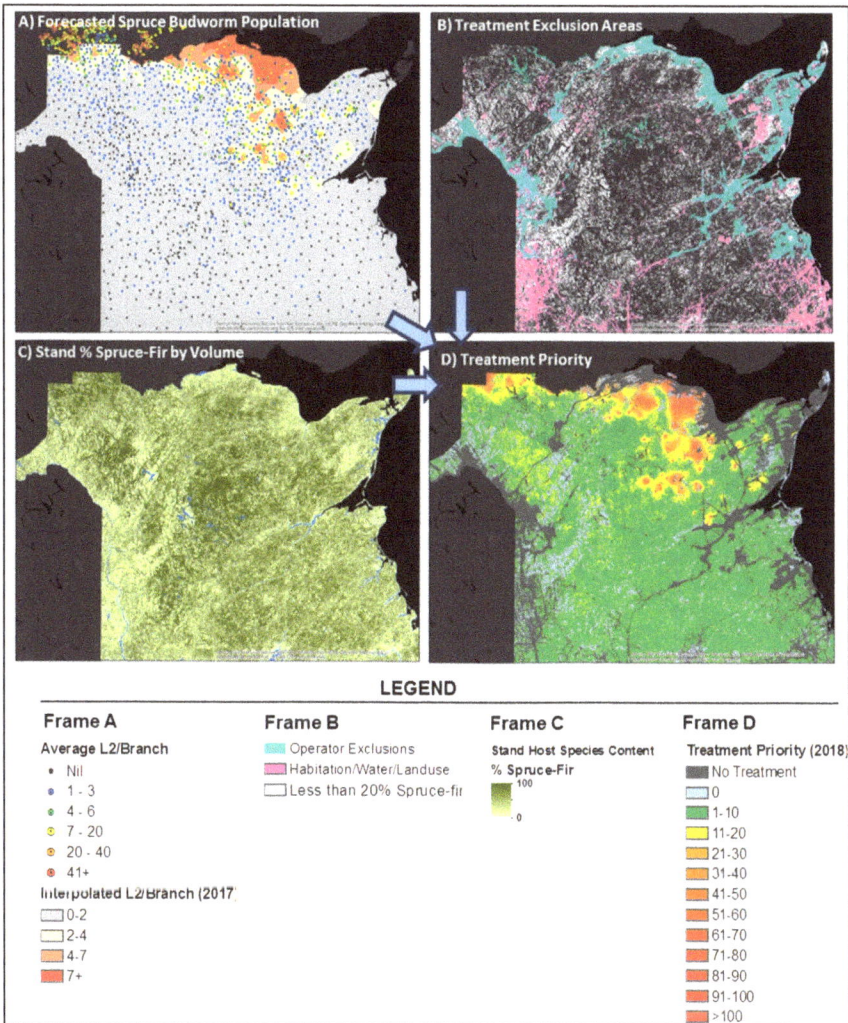

Figure 3. Inputs required for developing the 2018 treatment priority raster map, including: (**A**) interpolated L2 population, (**B**) treatment exclusion areas, (**C**) stand percent spruce-fir, and (**D**) the final treatment priority map as a function of scaled L2 plus percent spruce-fir content with treatment exclusion areas removed.

2.4. Insecticide Treatments for EIS against SBW

Given that EIS is area-wide SBW population management, we attempted to treat all SBW area with L2 ≥ 7, regardless of land ownership. The New Brunswick Crown Lands and Forests Act contains provisions allowing landowners to opt out of planned provincial pest management programs, if desired, and all potentially affected landowners (several hundred per year in 2017 and 2018) were notified and given an opportunity to opt out. Fewer than 5% of landowners opted out of treatments, largely owing to extensive communication efforts to inform media, politicians, landowners, and other stakeholders of objectives and results of the EIS research.

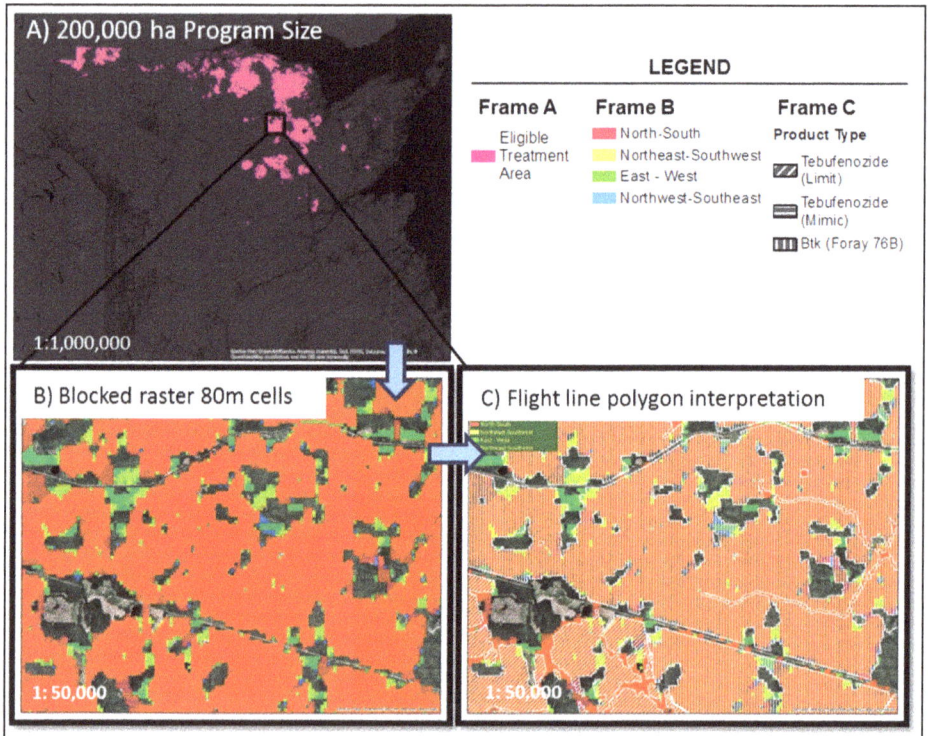

Figure 4. Optimum blocked raster solution (**A**,**B**) generated by *ForPRO II* for the 2018 EIS SBW treatment priority layer (Figure 3D) and (**C**) adjusted final polygon blocks digitized by FPL analysts.

In total, three bases of operation, namely Charlo, Miramichi, and Boston Brook, were used as staging areas for application flights. Treatments targeted later larval instars (3.5 or later), initially based on locally-calibrated degree-day models of SBW larval development, but confirmed for individual treatment blocks by SBW larval sampling to monitor insect development (or bud development as a proxy for insect stage). Insecticide applications therefore were optimally timed with insect development, commencing once the appropriate development stage was reached, during periods of favorable meteorological conditions (low wind, no rain, etc.). Costs of the *Btk* and tebufenozide treatments have averaged about $40 per hectare per application, or over $17 million of direct treatment costs for the 425,000 ha treated from 2014 to 2018 (Table 2). Costs will vary by jurisdiction and are lower than usual in New Brunswick because Forest Protection Limited owns aircraft, rather than having to contract them.

Population surveys carried out in treated versus untreated areas [43] provided estimates of spring-feeding larval density estimates (i.e., L4), which allowed us to assess how efficient L2 estimates from the previous year were for determining the treatment layer. In brief, we selected sites within treated and untreated areas in 2017 (53 sites) and 2018 (96 sites) and collected 15 branches per site during the L4 stage (~mid June).

Treatment efficacy was assessed using the annual L2 survey data used to identify hotspots (described above in Section 2.1). These L2 density data were separated by 'time' (i.e., pre-spray and post-spray) and by 'treatment' where we compared density in treated areas with those in untreated areas within 3 km of blocks or within 3–6 km from blocks. All statistical analyses on these data were conducted in R version 3.4.0 [44]. To determine the effect of treatments on population growth, for each year, we carried out a linear model assessing the effects on L2 density of 'time' (i.e., pre vs. post treatment) and treatment (treated areas vs. untreated areas < 3 km vs. untreated 3–6 km), as well as their interaction. Prior to analysis, measures of L2 density were transformed based on Tukey's ladder of power using the *transformTukey* function from the *rcompanion* package [45]. Because of significant time x treatment interactions, we did not try to interpret the main effects from these models and instead have focussed on how they interacted with one another. For each year we reported the difference in L2 density from the start to end of year for each treatment to indicate the direction of change (+ or −) and conducted a post-hoc interaction contrast using the *testInteractions* function from the *phia* package [46], to determine if these differences were significant.

3. Results

3.1. EIS Insecticide Treatments from 2014 to 2018

The area treated with insecticides increased steadily over the 5 years tested, from less than 5000 ha in 2014 to nearly 200,000 ha in 2018 (Figure 5, Table 2). Treatments in 2018 included about 23,000 ha of trial double or triple applications of *Btk* in the highest population areas (L2 > 20/branch; Table 2). Initial results after one year indicated that one application of *Btk* was as effective as two applications and that three applications were unnecessary, but this needs to be tested further. The 2018 treatment size was decided based on simulations of a range of potential treatment program size and budget scenarios ranging from 150,000 ha ($6 million cost) to 300,000 ha ($12 million) and resulting projected L2 reduction efficiencies. In total, treatments were 55% *Btk*, 45% tebufenozide, and 0.6% trials of SBW pheromone (Table 2). The pheromone trials were experimental, and have not yet achieved sufficiently satisfactory results for use. The geographic extent of treatment areas was primarily in northwest and north-central New Brunswick, and expanded southward over the years (Figure 5). In general, areas were not treated in successive years (Figure 6A). Areas treated that overlapped areas treated in the preceding year were 0%, 6.7%, 14.6%, and 26.3% from 2015 to 2018, respectively. The main areas that required repeated treatments were close to the Québec border, where the major SBW outbreak expanded from 4.3 million ha in 2014 to 8.2 million ha in 2018 [24]. Of the total of 112 L2 points that had ≥ 7 L2/branch in autumn 2017, only 12% of those fell within areas that had been treated with insecticide in summer 2017 (Figure 6B), and none of the 10 points in 2018 that will require treatment in 2019 occurred in areas treated in 2018.

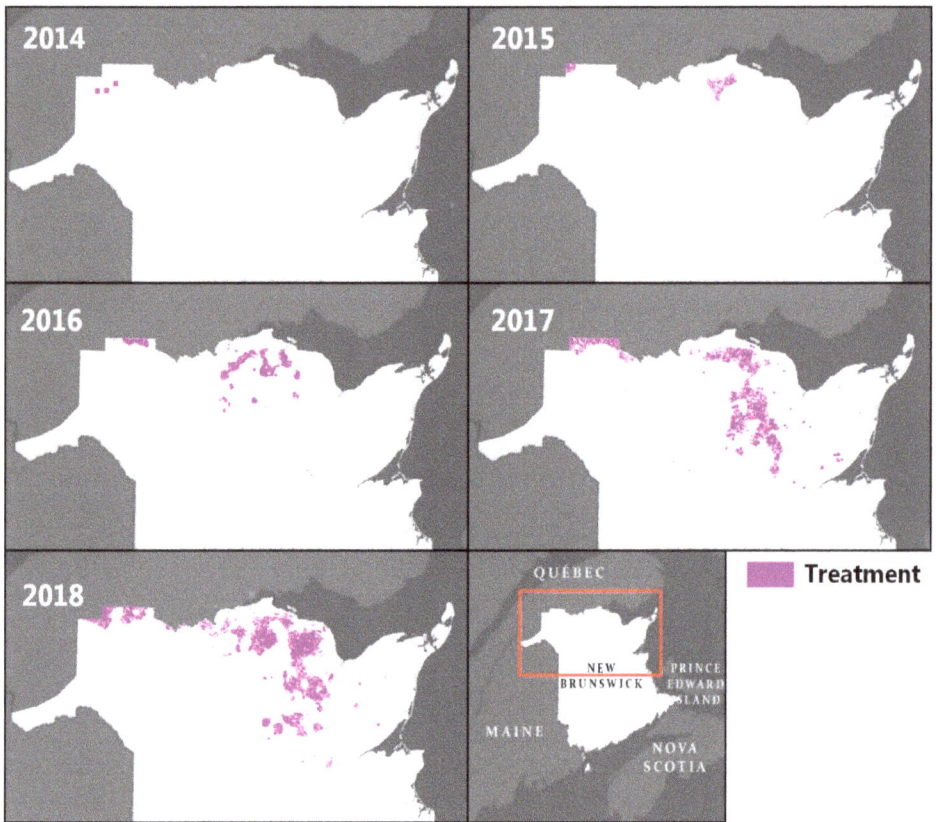

Figure 5. Areas treated with insecticides (*Btk* or tebufenozide), or pheromone in EIS SBW trials each year from 2014 to 2018 in New Brunswick.

Table 2. Area treated by active ingredient during the EIS-SBW Project (2014–2018).

Year	Area Treated by Active Ingredient (ha)			
	Bacillus thuringiensis K.	Tebufenozide	Pheromone	Total
2014	169	4472	490	5131
2015 [1]	12,093	3263	271	15,627
2016 [2]	36,889	19,719	1000	57,608
2017	79,088	68,142	0	147,230
2018 [3]	104,660	94,403	633	199,696
Total	232,899	189,999	2394 [4]	425,292

[1] Consisted of 12,093 ha of double application of *Btk*. [2] Included 5000 ha of double application of *Btk*. [3] Included 22,220 ha of double application and 734 ha of triple application of *Btk*. [4] Pheromone trials were experimental and have not yet had sufficient efficacy for more widespread use.

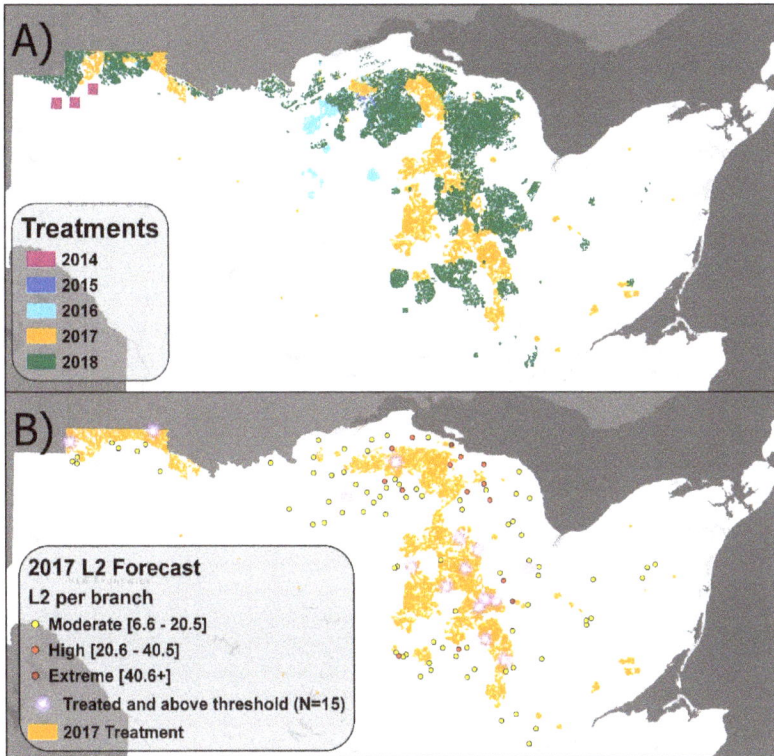

Figure 6. (**A**) Successive areas treated with *Btk* or tebufenozide insecticides each year from 2014 to 2018. (**B**) Comparison of 2017 treatment areas and autumn 2017 L2 samples with ≥ 7 L2/branch.

3.2. Efficacy of L2 Monitoring and Blocking Approach

Results from SBW population surveys carried out on 15 branches per site during the L4 stage within treated and untreated areas in 2017 (53 sites) and 2018 (96 sites) [43] were used to assess how efficient L2 estimates from the previous year were for determining the treatment layer. Treatment areas for 2017 and 2018 included 75% and 76% of the sites sampled with L4 densities above the >6.5 L2/branch treatment threshold. In almost all instances, the other 24%–25% of sites that were above the treatment threshold but not included in the spray area were intentionally excluded due to buffer restrictions for waterways or residences.

3.3. Efficacy of Treatments for Suppressing Population Growth

For each year, there was a significant time x treatment interaction (Table 3), which was attributed to differences in both the magnitude and direction of changes between pre- and post-spray L2 densities among treatments. In general, mean L2 densities from pre-treatment to post-treatment periods, comparing L2 points within treated blocks, untreated points within 3 km of blocks, and untreated points 3–6 km from blocks showed that mean L2 values within treated blocks declined by 38%–39% in 2015–2016, by 60% in 2017, and by 96% in 2018 (Table 3). In contrast, mean L2 of samples outside but within 3 km of treatment blocks increased by 75%–105% and those 3–6 km from blocks increased by 146%–300% from 2015–2017. The year 2018 differed, however, in that L2 samples outside treated blocks declined by 43%–63%, whereas treated samples declined by 96% (Table 3). SBW survival in 2018 was clearly low in untreated as well as treated samples. Overall, these results indicate that treatments were effective in reducing populations in all years and that moth immigration did not appear to offset treatment mortality (i.e., treatments resulted in additive mortality). Ongoing life table studies in treated and untreated areas of the EIS program have offered further support for this contention [43]. These results also indicate that successful containment may be possible even when some areas with densities above the action threshold are necessarily excluded from spray areas (e.g., see Section 3.2 above).

Table 3. Mean and standard error (SEM) number of SBW second instar larvae (L2) per branch from samples taken before and after budworm protection treatment, each year from 2015–2018, comparing treated samples with untreated samples within 3 km and 3–6 km adjacent to treated blocks. Treated values are shown in bold and % difference from pre- to post-treatment is shown.

Year	Treatment [1]	Pre-Treatment [2] L2/Branch			Post-Treatment [2] L2/Branch			Difference (Pre-Post Treatment) [4]	F [5]
		N [3]	Mean	SEM	N [3]	Mean	SEM		
2015	**Treated (0 km)**	**201**	**2.9**	**0.31**	**65**	**1.8**	**0.32**	**−1.1**	**3.4 ***
	Untreated < 3 km	116	2.0	0.30	46	3.5	0.74	+1.4	8.9 ***
	Untreated 3–6 km	33	0.47	0.14	30	2.0	0.44	+1.6	6.6 *
2016	**Treated (0 km)**	**77**	**6.2**	**0.63**	**84**	**3.8**	**0.70**	**−2.4**	**15 *****
	Untreated < 3 km	156	1.8	0.23	121	3.7	0.63	+1.9	10 ***
	Untreated 3–6 km	95	1.1	0.15	73	3.8	0.91	+2.7	12 ***
2017	**Treated (0 km)**	**149**	**7.5**	**0.68**	**158**	**3.0**	**0.48**	**−4.5**	**65 *****
	Untreated < 3 km	171	2.4	0.35	195	4.6	0.48	+2.2	25 ***
	Untreated 3–6 km	149	1.3	0.17	192	3.2	0.34	+1.8	20 ***
2018	**Treated (0 km)**	**209**	**7.3**	**0.53**	**209**	**0.34**	**0.05**	**−7.0**	**619 *****
	Untreated < 3 km	262	2.4	0.24	191	0.88	0.14	−1.5	75 ***
	Untreated 3–6 km	185	1.4	0.12	142	0.80	0.15	−0.62	48 ***

[1] Treated = *Btk* or tebufenozide treatment; Untreated = surrounding area within 3 km or 3–6 km. [2] Pre-treatment L2 were sampled in the previous autumn, and post-treatment L2 were sampled in the autumn of the year after treatment. [3] Total N = 3309 points sampled over 4 years, each consisting of 3 branch samples per point. [4] Pre-treatment mean L2/branch minus post-treatment mean, so decreases are negative and increases are positive. [5] * $p < 0.05$, ** $p < 0.01$, *** $p < 0.001$

3.4. Comparison with SBW Populations and Defoliation in Adjacent Québec

Following 5 years of EIS treatments of low but increasing SBW, L2 levels across northern New Brunswick were considerably lower than adjacent SBW populations across the provincial border in Québec (Figure 7A). The SBW outbreak (defined in terms of area of aerially-detected defoliation) began in about 2004 in Québec, when 33,700 ha of defoliation were detected. This increased to 133,600 ha by 2008, 1,643,000 ha by 2011, 4,275,000 ha by 2014, and 8,181,000 ha in 2018 [24]. SBW control treatments in Québec over this time period covered a maximum of several hundred thousand hectares per year, because of cost. The huge scale of the outbreak in Québec renders an EIS approach impossible. Québec is much larger than New Brunswick (1,668,000 km² versus 72,908 km²), but the regions of Québec that are closest to New Brunswick (south of the St. Lawrence River in Figure 7; Bas-Saint-Laurent and Gaspésie regions, 42,337 km²) are generally similar in size and environmental conditions to northern New Brunswick. Aerial surveys of defoliation in New Brunswick detected only ~2500 ha of defoliation in 2017 and 500 ha in 2018, the third and fourth years of the outbreak, in comparison with 2,183,400 ha of defoliation in 2017 and 2,509,650 ha in 2018 [24] in the adjacent Bas-Saint-Laurent-Gaspésie regions of Québec (Figure 7B). Although there are some differences in methods, resolution, and mapping procedures between the defoliation aerial surveys in New Brunswick and Québec, there are orders of magnitude differences in observed defoliated area. Following 4 and 5 years of EIS treatments in New Brunswick, the Québec-New Brunswick border was evident from the air by defoliation on the Québec side versus none visible in New Brunswick. We should note that this is not a critique of the foliage protection approach being used in Québec, where ongoing foliage protection efforts have been successful in keeping treated areas below economic injury thresholds (50% defoliation) [28], but only a small proportion of the outbreak area is treated each year. However, with over ~8 million ha of defoliation in Québec, SBW populations are well beyond levels where an EIS approach would be feasible.

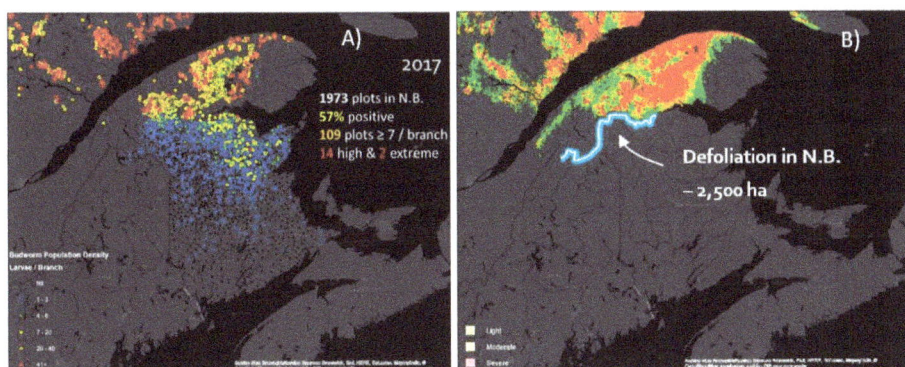

Figure 7. SBW L2 population (**A**) and aerial survey defoliation (**B**) in 2017, after 4 years of EIS treatments in northern New Brunswick and adjacent Québec (the border between the two provinces is indicated by the blue line in (B)).

4. Discussion

Over 5 years of development, testing, and refinement, the treatment priority and blocking tool has proven effective in directing insecticide treatments to reduce L2 populations and to facilitate determining costs and benefits of alternative size treatment programs. We tested a wide range of possible treatment program sizes (for example, in 2018 we tested five program sizes treating 100,000 to 300,000 ha, in 50,000 ha increments, which would equate to $4 million to $12 million treatment costs) and selected the 200,000 ha solution as covering nearly all L2 > 7 per branch and high spray priority areas. Advantages of using our blocking tool are (1) it is an objective, optimal solution, and (2) it can be rerun in hours to incorporate any desired changes, such as differing budgets and program sizes, alternative methods of L2 interpolation or scaling, changes to rules for including low spruce-fir areas, or updates of exclusion area.

How safe for the environment are the *Btk*, tebufenozide, and pheromone treatments? All products are federally registered and approved as safe for use by Health Canada. The research project is regulated by requirements of the Federal Pest Control Products Act and New Brunswick's Pesticides Control Act and Regulation. Any provincial permit and product label conditions were observed to ensure safe and responsible use. *Btk* is a naturally occurring soil bacteria and is not harmful to humans or other mammals, bees, birds, or fish when used according to label conditions. *Btk* has been used for the last 20 years to control SBW defoliation [47]. Tebufenozide is an insect growth regulator that larvae eat, which imitates a natural insect hormone that causes the developing caterpillars to molt prematurely as the larvae go through their growth stages. The caterpillars then quickly stop feeding and die. It is harmless to humans or other mammals, bees, birds, or fish when used according to strict label conditions. Pheromones occur naturally, are unique to each insect, and trigger behavioral changes in members of the same species. Pheromones pose no risk to humans or other animals. They are used to lure or attract insects to traps, and they can be used to disrupt mating cycles. SBW pheromones do not kill insects.

Tebufenozide is used on forestry, ornamentals, and a variety of crops. It is added to water and aerially applied at a rate of 1 to 2 liters per hectare, with the tebufenozide portion fixed at 290 mL per hectare. Nozzles on the aircraft break the liquid mixture into small droplets (atomize) so that when the drops land on foliage they are small enough to be eaten by SBW. Because the forest canopy acts as a filter, 90%–95% of the spray is deposited in the forest canopy [48]. The portion that reaches the ground stays in the upper 5 cm of the soil and leafy debris, does not leach away, and is broken down over time by soil microbes, sunlight, and moisture [49,50]. Tebufenozide deposited in the canopy is relatively rainfast and is not easily washed off by rainfall [51] and tebufenozide that reaches the ground is not harmful to soil invertebrates [52], Water bodies are identified on maps and are excluded from all treatment areas during the planning phase, so there is no targeting of visible water bodies. Tebufenozide that lands on water has no noticeable environmental impact; research showed that there were no significant harmful effects on most organisms at concentrations expected after aerial spraying, even if a water body were to be unintentionally sprayed [48,53,54]. The most recent review of tebufenozide states that 'No adverse effects on birds, mammals or aquatic species are likely to occur from exposure to tebufenozide' [55].

In the EIS trials conducted from 2014 to 2018, SBW populations in blocks treated with *Btk* or tebufenozide were consistently reduced and generally did not require treatment in the subsequent years. Based on intensive L2 sampling, SBW populations in northern New Brunswick continued to increase from 2015 to 2017, but at a relatively slow pace (Table 1). There are no other studies of similar area-wide insect population early intervention strategies for comparison.

In 2018, SBW populations declined such that only 10 L2 points with ≥ 7/branch were detected, compared to 112 such points in 2017 (0.5% versus 5.5%; Table 1). This SBW population decline was unexpected and is not currently understood, but thought to perhaps result from parasitoids, other natural enemies, or weather factors, possibly in combination with the EIS treatments. Analyses of parasitoids and diseases in collected SBW samples are ongoing. There are numerous examples of rapid year-to-year 10- or 100-fold magnitude increases or decreases of SBW populations, e.g., [56], so we view the 2018 SBW population decline as likely a temporary reprieve rather than a continuing population trend. However, in 2018, the area of SBW defoliation detected by aerial surveys in adjacent Bas St. Laurent-Gaspésie, Québec, increased from 2,257,000 ha in 2017 to 2,728,000 ha, while only 550–2500 ha of defoliation was detected under EIS treatments in New Brunswick in 2017–2018 (Figure 7B).

There are several requirements for conducting a study such as this. The first is buy-in from a large number of stakeholders: the provincial forest management agency, regional forest industry, private woodlot owners, researchers, and the community at large. To achieve this, a huge effort has gone into communication and outreach, under the auspices of the Healthy Forest Partnership (see www.healthyforestpartnership.ca). The second requirement is a large amount of funding, because area-wide application of insecticide treatments on a trial basis is very expensive given the large areas involved. Over $17 million was spent on project insecticide treatments from 2014–2018. Funding for the first 4 years was jointly provided through proposals to the Governments of Canada and New Brunswick, forest industry, and Natural Resources Canada. A third requirement is the involvement of a large research effort, including over 30 scientists from Natural Resources Canada and five universities. Our ability to assemble these project requirements was strengthened by the DSS tools to estimate impacts of the alternative, what would happen to timber supply, direct and indirect effects on the regional economy, and employment of a large-scale uncontrolled SBW outbreak, e.g., [7,9,10]. These impacts represent billions of dollars [7].

The positive and promising results from EIS from 2015 to 2017 resulted in the Healthy Forest Partnership submitting proposals to the Canadian federal government and all four Atlantic Canada provincial governments for funding to continue the EIS SBW project. This funding request was approved, with an additional $75 million of funding for continuation from 2018 to 2023. As a result, we are able to continue the trials, and to expand the research into several new areas including assessment of the ecological benefits of an EIS on watersheds, remote sensing of low-level defoliation, and assessment of climate change effects on SBW populations. Natural Resources Canada, all four Atlantic Canada provinces, and forest industry are supportive and contributing to the required investment. A strong coalition of researchers, landowners, forestry companies, governments, forest protection experts, communities, and citizens is committed to testing this strategy.

Recent timber supply projections conducted by NB ERD for Crown land in New Brunswick have indicated that at best, the foliage protection strategy would result in a 10%–15% long-term (in 50 years) reduction in spruce-fir harvest level, and would cost more than EIS. Timber supply projections were based on uncontrolled moderate and severe SBW outbreak scenarios as used in [21], foliage protection of 20%–40% of susceptible forest area, and economic impacts of the resulting harvest reductions [7]. Extrapolation of these results for a severe SBW outbreak scenario to all four Atlantic Canada provinces, based on the per hectare detailed New Brunswick estimates, projected that an uncontrolled outbreak would cause 96 million m^3 of timber harvest losses and economic cost of $15 billion over 50 years; that a foliage protection strategy on 20% of susceptible forest would reduce the harvest losses to 43 million m^3 and economic cost of $5 billion, but with a treatment cost of $2 billion; whereas EIS, if successful, was projected to cost $300 million from 2014 to 2026 and result in minimal harvest and economic impacts. Therefore, the Early Intervention Strategy, if it continues to work, has been termed a $300 million (potential) solution to a $15 billion problem. These values were based on the estimated cost of continued EIS treatments for all of Atlantic Canada to 2026, the projected end of the SBW outbreak, compared to timber supply and economic impacts if SBW was uncontrolled, extrapolated from previous studies in New Brunswick [7,21].

5. Conclusions

Following 5 years of EIS treatments of low but increasing SBW populations, L2 populations across northern New Brunswick are considerably lower than SBW populations across the border in adjacent Québec. SBW populations in blocks treated with *Btk* or tebufenozide were consistently reduced and generally did not require treatment in the subsequent year. The differences in defoliation and SBW outbreak patterns between New Brunswick and the immediately adjacent areas of Québec are probably due to EIS, as the forest types, weather, and site conditions are similar to northern New Brunswick. SBW defoliation detected from aerial surveys was less than 2500 ha in 2017 and 550 ha in 2018 in New Brunswick, compared to over 2.5 million ha in adjacent Bas St. Laurent-Gaspésie areas of Québec. Unexpectedly, SBW populations across northern New Brunswick, based on intensive L2 sampling, showed over 90% reductions in 2018. We expect that SBW populations may well rebound after this decline, but the first 5 years of EIS trials in New Brunswick are showing overall positive results. We do not know if the EIS approach will continue to work, but after 5 years of treatments, there are dramatic differences between New Brunswick and the SBW outbreak in adjacent Québec. Research into EIS and considerations that underlie the development of a pest containment program are being addressed in an ongoing manner in the overall EIS project, which is continuing to 2023 or beyond, and includes 10 projects conducted by over 30 scientists.

Author Contributions: Project conceptualization was led by D.A.M., J.R., and R.J.; SBW population sampling by D.C. and R.J.; data analyses by L.A.-B., C.H., and R.J.; project administration by D.A.M. and P.A.; SBW treatments by P.A., L.A.-B., D.C.; funding acquisition by D.A.M. and R.J.; initial draft manuscript preparation by D.A.M.; manuscript revisions and editing by D.A.M., R.J., C.H., P.A., L.A.-B., D.C. and J.R.

Funding: This research was funded by Atlantic Canada Opportunities Agency, Natural Resources Canada, Government of New Brunswick, and forest industry in New Brunswick.

Acknowledgments: This research was overseen by the Healthy Forest Partnership, a consortium of researchers, landowners, forestry companies, governments, and forest protection experts. Many scientists and staff of industry and government agencies have made important contributions, without which the project could not have proceeded. We thank the Québec Ministère des Forêts, de la Faune et des Parcs for permission to use their SBW L2 data.

Conflicts of Interest: The authors declare no conflict of interest.

References

1. Kettela, E.G. *A Cartographic History of Spruce Budworm Defoliation 1967 to 1981 in Eastern North America*; Canadian Forest Service: Ottawa, ON, Canada, 1983; p. 9.
2. Ostaff, D.P.; MacLean, D.A. Patterns of balsam fir foliar production and growth in relation to defoliation by spruce budworm. *Can. J. For. Res.* **1995**, *25*, 1128–1136. [CrossRef]

3. MacLean, D.A. Vulnerability of fir-spruce stands during uncontrolled spruce budworm outbreaks: A review and discussion. *For. Chron.* **1980**, *56*, 213–221. [CrossRef]

4. Ostaff, D.P.; MacLean, D.A. Patterns of balsam fir mortality caused by an uncontrolled spruce budworm outbreak. *Can. J. For Res.* **1989**, *19*, 1087–1095.

5. Virgin, G.V.; MacLean, D.A. Five decades of balsam fir stand development after spruce budworm-related mortality. *For. Ecol. Manag.* **2017**, *400*, 129–138. [CrossRef]

6. Baskerville, G.L. Spruce budworm: Super silviculturist. *For. Chron.* **1975**, *51*, 138–140. [CrossRef]

7. Chang, W.-Y.; Lantz, V.A.; Hennigar, C.R.; MacLean, D.A. Economic impacts of spruce budworm (*Choristoneura fumiferana* Clem.) outbreaks and control in New Brunswick, Canada. *Can. J. For. Res.* **2012**, *42*, 490–505.

8. James, P.M.A.; Robert, L.-E.; Wotton, B.M.; Martell, D.L.; Fleming, R.A.; Robert, L. Lagged cumulative spruce budworm defoliation affects the risk of fire ignition in Ontario, Canada. *Ecol. Appl.* **2017**, *27*, 532–544. [CrossRef] [PubMed]

9. Kneeshaw, D.; Sturtevant, B.R.; Cooke, B.; Work, T.; Pureswaran, D.; DeGrandpre, L.; MacLean, D.A. Insect disturbances in forest ecosystems. Chapter 7. In *Routledge Handbook of Forest Ecology*; Peh, K.S.-H., Corlett, R.T., Bergeron, Y., Eds.; Routledge: Oxon, UK, 2015; pp. 93–113.

10. MacLean, D.A. Impacts of insect outbreaks on tree mortality, productivity, and stand development. *Can. Entomol.* **2016**, *148*, S138–S159. [CrossRef]

11. Régnière, J.; Nealis, V.G. Ecological mechanisms of population change during outbreaks of the spruce budworm. *Ecol. Entomol.* **2007**, *32*, 461–477. [CrossRef]

12. Johns, R.C.; Flaherty, L.; Carleton, D.; Edwards, S.; Morrison, A.; Owens, E. Population studies of tree-defoliating insects in Canada: A century in review. *Can. Entomol.* **2016**, *148*, S58–S81. [CrossRef]

13. Royama, T.; Eveleigh, E.S.; Morin, J.R.B.; Pollock, S.J.; McCarthy, P.C.; McDougall, G.A.; Lucarotti, C.J. Mechanisms underlying spruce budworm outbreak processes as elucidated by a 14-year study in New Brunswick, Canada. *Ecol. Monogr.* **2017**, *87*, 600–631. [CrossRef]

14. Armstrong, J.A.; Cook, C.A. *Aerial Spray Applications on Canadian Forests: 1945 to 1990*; Forestry Canada: Ottawa, ON, Canada, 1993; p. 274.

15. Coulombe Commission (Commission d'étude sur la gestion de la forêt publique québécoise). Chapitre 4: Protection, conservation et gestion multiressource: Des axes de changement. In *Commission d'étude sur la gestion de la forêt publique québécoise, Rapport final*; Ministère des Forêts, de la Faune et des Parcs: Quebec, QC, Canada, 2004; pp. 47–92.

16. Lévesque, R.C.; Cusson, M.; Lucarotti, C. *BEGAB: Budworm Eco-Genomics: Applications & Biotechnology (BEGAB)*; Institut de biologie intégrative et des systèmes, Université Laval: Québec City, QC, Canada, 2010.

17. Hennigar, C.R.; MacLean, D.A.; Quiring, D.T.; Kershaw, J.A. Differences in spruce budworm defoliation among balsam fir and white, red, and black spruce. *For. Sci.* **2008**, *54*, 158–166.

18. LeDuc, A.; Joyal, C.; Bergeron, Y.; Morin, H. Balsam fir mortality following the last spruce budworm outbreak in northwestern Quebec. *Can. J. For. Res.* **1995**, *25*, 1375–1384.

19. Su, Q.; Needham, T.D.; MacLean, D.A. The influence of hardwood content on balsam fir defoliation by spruce budworm. *Can. J. For. Res.* **1996**, *26*, 1620–1628. [CrossRef]

20. Campbell, E.M.; MacLean, D.A.; Bergeron, Y. The severity of budworm-caused growth reductions in balsam fir/spruce stands varies with the hardwood content of surrounding forest landscapes. *For. Sci.* **2008**, *54*, 195–205.

21. Hennigar, C.; Erdle, T.; Gullison, J.; MacLean, D. Re-examining wood supply in light of future spruce budworm outbreaks: A case study in New Brunswick. *For. Chron.* **2013**, *89*, 42–53. [CrossRef]

22. Chang, W.-Y.; Lantz, V.A.; MacLean, D.A. Public attitudes about forest pest outbreaks and control: Case studies in two Canadian provinces. *For. Ecol. Manag.* **2009**, *257*, 1333–1343. [CrossRef]

23. Chang, W.-Y.; Lantz, V.A.; Hennigar, C.R.; MacLean, D.A. Benefit-cost analysis of spruce budworm (*Choristoneura fumiferana* Clem.) control: Incorporating market and non-market values. *J. Environ. Manag.* **2012**, *93*, 104–112. [CrossRef]

24. QMFFP: Québec Ministère des Forêts, de la Faune et des Parcs. *Aires infestées par la tordeuse des bourgeons de l'épinette au Québec en 2018*; Gouvernement du Québec, Direction de la protection des forêts: Québec City, QC, Canada, 2018; 20p.

25. Miller, C.A.; Kettela, E.G. Aerial control operations against the spruce budworm in New Brunswick, 1952–1973. In *Aerial Control of Forest Insects in Canada*; Prebble, M.L., Ed.; Department of the Environment: Ottawa, ON, Canada, 1975; pp. 94–112.

26. Irving, H.J.; Webb, F.E. Forest protection against spruce budworm in New Brunswick—An aerial applicator's perspective. *Pulp Pap. Can.* **1981**, *82*, 23–31.

27. Kline, A.W.; Lavigne, D.R.; MacLean, D.A. Effectiveness of spruce budworm spraying in New Brunswick in protecting the spruce component of spruce–fir stands. *Can. J. For. Res.* **1984**, *14*, 163–176.

28. Fuentealba, A.; Dupont, A.; Hébert, C.; Berthiaume, R.; Quezada-García, R.; Bauce, É. Comparing the efficacy of various aerial spraying scenarios using *Bacillus thuringiensis* to protect trees from spruce budworm defoliation. *For. Ecol. Manag.* **2019**, *432*, 1013–1021. [CrossRef]

29. Régnière, J.; Delisle, J.; Pureswaran, D.S.; Trudel, R. Mate-finding allee effect in spruce budworm population dynamics. *Entomol. Exp. Appl.* **2013**, *146*, 112–122.

30. Sharov, A.A.; Liebhold, A.M.; Roberts, A.E. Optimizing the use of barrier zones to slow the spread of gypsy moth (Lepidoptera: Lymantriidae) in North America. *J. Econ. Entomol.* **1998**, *91*, 165–174.

31. Sharov, A.A.; Leonard, D.; Liebhold, A.M.; Roberts, E.A.; Dickerson, W. "Slow the spread": A national program to contain the gypsy moth. *J. For.* **2002**, *100*, 30–36.

32. Eveleigh, E.S.; McCann, K.S.; McCarthy, P.C.; Pollock, S.J.; Lucarotti, C.J.; Morin, B.; McDougall, G.A.; Strongman, D.B.; Huber, J.T.; Umbanhowar, J.; et al. Fluctuations in density of an outbreak species drive diversity cascades in food webs. *Proc. Natl. Acad. Sci. USA* **2007**, *104*, 16976–16981. [CrossRef]

33. Huber, J.; Eveleigh, E.; Pollock, S.; McCarthy, P. The Chalcidoid parasitoids and hyperparasitoids (Hymenoptera: Chalcidoidea) of *Choristoneura* species (Lepidoptera: Tortricidae) in America north of Mexico. *Can. Entomol.* **1996**, *128*, 1167–1220.

34. Miller, J.D.; MacKenzie, S.; Foto, M.; Adams, G.W.; Findlay, J.A. Needles of white spruce inoculated with rugulosin-producing endophytes contain rugulosin reducing spruce budworm growth rate. *Mycol. Res.* **2002**, *106*, 471–479. [CrossRef]

35. Frasz, S.L.; Walker, A.K.; Nsiama, T.K.; Adams, G.W.; Miller, J.D. Distribution of the foliar fungal endophyte *Phialocephala scopiformis* and its toxin in the crown of a mature white spruce tree as revealed by chemical and qPCR analyses. *Can. J. For. Res.* **2014**, *44*, 1138–1143. [CrossRef]

36. MacLean, D.A.; Erdle, T.A.; MacKinnon, W.E.; Porter, K.B.; Beaton, K.P.; Cormier, G.; Morehouse, S.; Budd, M. The Spruce Budworm Decision Support System: Forest protection planning to sustain long-term wood supplies. *Can. J. For. Res.* **2001**, *31*, 1742–1757. [CrossRef]

37. Hennigar, C.R.; MacLean, D.A.; Porter, K.B.; Quiring, D.T. Optimized harvest planning under alternative foliage-protection scenarios to reduce volume losses to spruce budworm. *Can. J. For. Res.* **2007**, *37*, 1755–1769. [CrossRef]

38. Hennigar, C.R.; Wilson, J.S.; MacLean, D.A.; Wagner, R.G. Applying a spruce budworm decision support system to Maine: Projecting spruce-fir volume impacts under alternative management and outbreak scenarios. *J. For.* **2011**, *109*, 332–342.

39. McLeod, I.M.; Lucarotti, C.J.; Hennigar, C.R.; MacLean, D.A.; Holloway, A.G.L.; Cormier, G.A.; Davies, D.C. Advances in aerial application technologies and decision support for integrated pest management. In *Integrated Pest Management and Pest Control*; Soloneski, S., Larramendy, M.L., Eds.; InTech Open Access Publisher: Rijeka, Croatia, 2012; pp. 651–668. ISBN 978-953-307-926-4.

40. Miller, C.A.; Kettela, E.G.; McDougall, G.A. *A Sampling Technique for Overwintering Spruce Budworm and Its Applicability to Population Surveys*; Canadian Forestry Service: Fredericton, NB, Canada, 1971; p. 11

41. Zhang, B.; MacLean, D.A.; Johns, R.C.; Eveleigh, E.S. Effects of hardwood content on balsam fir defoliation during the building phase of a spruce budworm outbreak. *Forests* **2018**, *9*, 530. [CrossRef]

42. Bognounou, F.; De Grandpré, L.; Pureswaran, D.S.; Kneeshaw, D. Temporal variation in plant neighborhood effects on the defoliation of primary and secondary hosts by an insect pest. *Ecosphere* **2017**, *8*, e01759. [CrossRef]

43. Martel, V.; Johns, R.C.; Eveleigh, E.; McCann, K.; Pureswaran, D.; Sylvain, Z.; Morrison, A.; Morin, B.; Owens, E.; Hébert, C. Landscape level impacts of EIS on SBW, other herbivores and associated natural enemies (ACOA RD100 2.2.2). In Proceedings of the SERG-I Workshop, Fredericton, NB, Canada, 2–5 February 2017; pp. 112–118.

44. R Core Team. *R: A Language and Environment for Statistical Computing*; R Foundation for Statistical Computing: Vienna, Austria, 2017; Available online: https://www.R-project.org (accessed on 23 April 2019).

45. Mangiafico, S. *Rcompanion: Functions to Support. Extension Education Program. Evaluation*, Package Version 2.1.7; 2019. Available online: https://CRAN.R-project.org/package=rcompanion (accessed on 23 April 2019).

46. De Rosario-Martinez, H. *Phia: Post-Hoc Interaction Analysis*, R Package Version 0.2-1; 2015. Available online: https://CRAN.R-project.org/package=phia (accessed on 23 April 2019).

47. Van Frankenhuyzen, K. Development and current status of *Bacillus thuringiensis* for control of defoliating forest insects. In *Forest Insect Pests in Canada*; Armstrong, J.A., Ives, W.G.H., Eds.; Canadian Forest Service: Ottawa, ON, Canada, 1995; pp. 315–325.

48. Kreutzweiser, D.; Nicholson, C. A simple empirical model to predict forest insecticide ground-level deposition from a compendium of field data. *J. Environ. Sci. Health Part B* **2007**, *42*, 107–113. [CrossRef]

49. Sundaram, K.M.S. Persistence and mobility of tebufenozide in forest litter and soil ecosystems under field and laboratory conditions. *Pestic. Sci.* **1997**, *51*, 115–130. [CrossRef]

50. Thompson, D.; Kreutzweiser, D. A review of the environmental fate and effects of natural 'reduced risk' pesticides in Canada. In *Crop Protection Products for Organic Agriculture: Environmental, Health, and Efficacy Assessment*; Racke, K.D., Felsot, A., Eds.; ACS Books, American Chemical Society: Washington, DC, USA, 2007; pp. 245–274.

51. Sundaram, K.M.S. Photostability and rainfastness of tebufenozide deposits of fir foliage. In *Biorational Pest Control Agents*; American Chemical Society: Washington, DC, USA, 1995; pp. 134–152.

52. Addison, J.A. Safety testing of tebufenozide, a new molt-inducing insecticide for effects on non-target forest soil invertebrates. *Ecotoxicol. Environ. Saf.* **1996**, *33*, 55–61. [CrossRef]

53. Kreutzweiser, D.; Capell, S.; Wainio-Keizer, K.; Eichenberg, D. Toxicity of new molt inducing insecticide (RH-5992) to aquatic macroinvertebrates. *Ecotoxicol. Environ. Saf.* **1994**, *28*, 14–24. [CrossRef]

54. Kreutzweiser, D.P.; Gunn, J.M.; Thompson, D.G.; Pollard, H.G.; Faber, M.J. Zooplankton community responses to a novel forest insecticide, tebufenozide (RH-5992), in littoral lake enclosures. *Can. J. Fish. Aquat. Sci.* **1998**, *55*, 639–648. [CrossRef]

55. US Department of Agriculture, Forest Service. *Gypsy Moth Management in the United States: A Cooperative Approach: Final Supplementary Environmental Impact Statement*; US Department of Agriculture Forest Service: Newtown Square, PA, USA, 2012; Volume 1, p. 11.

56. Royama, T.; MacKinnon, W.E.; Kettela, E.G.; Carter, N.E.; Hartling, L.K. Analysis of spruce budworm outbreak cycles in New Brunswick, Canada, since 1952. *Ecology* **2005**, *86*, 1212–1224. [CrossRef]

forests

Article

Economics of Early Intervention to Suppress a Potential Spruce Budworm Outbreak on Crown Land in New Brunswick, Canada

Eric Ye Liu [1], Van A. Lantz [1], David A. MacLean [1,*] and Chris Hennigar [1,2]

[1] Faculty of Forestry and Environmental Management, University of New Brunswick, Fredericton, NB E3B 5A3, Canada; Eric.Liu@unb.ca (E.Y.L.); vlantz@unb.ca (V.A.L.); Chris.Hennigar@gnb.ca (C.H.)

[2] New Brunswick Department of Energy and Resource Development, Fredericton, NB E3B 5H1, Canada

* Correspondence: macleand@unb.ca; Tel.: +1-506-458-7552

Received: 1 May 2019; Accepted: 31 May 2019; Published: 1 June 2019

Abstract: We investigated the potential economic impacts of future spruce budworm (*Choristoneura fumiferana* Clem.) (SBW) outbreaks on 2.8 million ha of Crown land in New Brunswick, Canada and compared an early intervention strategy (EIS) with foliage protection approaches. We coupled the Spruce Budworm Decision Support System (SBW DSS) with a Computable General Equilibrium (CGE) model to assess the impacts of EIS and foliage protection on 0%, 5%, 10%, and 20% of susceptible Crown (publicly owned) forest, under moderate and severe SBW outbreak scenarios. Cumulative available harvest supply from 2017 to 2067 was projected to be reduced by 29 to 43 million m^3, depending upon SBW outbreak severity, and a successful EIS approach would prevent this loss. These harvest reductions were projected to reduce total economic output by $25 billion (CAD) to $35 billion. Scenarios using biological insecticide foliage protection over 20% of susceptible Crown forest area were projected to reduce losses to 6–17 million m^3 and $0.5–4.1 billion. Depending upon SBW outbreak severity, EIS was projected to have benefit/cost ratios of 3.8 to 6.4 and net present values of $186 million to $353 million, both higher than foliage protection strategies. Sensitivity analysis scenarios of 'what if' EIS partially works (80% or 90%) showed that these produced superior timber harvest savings than the best foliage protection scenario under severe SBW outbreak conditions and generally superior results under moderate outbreak scenarios. Overall, results support the continued use of EIS as the preferred strategy on economic grounds to protect against SBW outbreaks on Crown land in New Brunswick.

Keywords: insect population management; spruce budworm; early intervention; defoliation; economic losses; decision support system; computable general equilibrium model

1. Introduction

The forest sector is one of the most significant components of the Canadian economy, and contributed 8% to 10% of the manufacturing gross domestic product (GDP) from 2010 to 2015 [1]. Approximately $38 billion CAD of annual total economic output was generated by the forest sector in 2014, accounting for almost 6% of all Canadian exports, worth $30.7 billion CAD [1]. However, economic value of the forest sector is vulnerable to natural disturbances that reduce timber supply and cause large-scale economic losses. Pest outbreaks and forest fires are the dominant natural disturbances that affect the forest sector in Canada [2–5]. In eastern Canada, spruce budworm (*Choristoneura fumiferana* Clem.) (SBW) outbreaks have impacted millions of hectares of spruce-fir forests every 30–40 years, and the outbreaks that generally last about 10–15 years have destroyed hundreds of millions of cubic meters of timber [6–8].

To protect against forest pest outbreaks, various forest management/protection strategies have been used to reduce timber supply loss and tree mortality caused by forest pest outbreaks. These include (1) foliage protection using biological pesticides (i.e., *Bacillus thuringiensis* Kurstaki (*Btk*) or tebufenozide) on SBW infested stands to keep trees alive [4,9]; (2) salvage harvesting of dying or dead trees while they were still usable [10,11]; (3) re-planning harvest schedules to retain less vulnerable stand types or ages [12]; or (4) planting less vulnerable tree species [13]. Foliage protection using chemical pesticides in past decades [14] or biological pesticides today, salvage, or planting less susceptible spruce (*Picea* sp.) plantations to replace balsam fir (*Abies balsamea* (L.) Mill.) are the primary strategies used. During the last SBW outbreak in New Brunswick (NB), over 1.5 million ha of spruce-fir forest was treated with insecticides annually from 1970 to 1983 [15], which effectively prevented extensive spruce-fir mortality, unlike that which occurred in Cape Breton, Nova Scotia, where SBW-caused mortality exceeded 85% of the forest [8].

An alternative to a foliage protection strategy could be an 'early intervention strategy' (EIS) focused on altering SBW population dynamics before a full outbreak occurs [16–18]. The concept of a SBW EIS suggests that identification of low but rising SBW populations in 'hot spots' could be targeted for treatment with pesticide before defoliation occurs (i.e., much earlier in the SBW outbreak development than in past treatments [17]) and that this could either slow or prevent the progression of the SBW population rise in the treated area. Recent research in Québec has demonstrated low mating success of female SBW moths when populations are at low levels [19], and the EIS aims to use this feature by detecting low but rising SBW population hot spots and treating these with biological insecticides *Btk* or tebufenozide [16].

Chang et al. [2] evaluated market and non-market benefits and costs of controlling a future SBW outbreak on all Crown (publicly-owned) forest in NB. They used a timber supply model to estimate the potential timber harvest revenue benefits and the costs of pest control efforts [4], and a contingent valuation method [20] to estimate the non-market benefits of forest protection against SBW. Results showed that a pest control program protecting 10%–20% of the susceptible Crown forest had the highest benefit-cost ratio of 3.2–4.0 and net present value of $59–111 million, with the ranges of values for moderate and severe SBW outbreak scenarios [2].

Chang et al. [3] also used a recursive dynamic Computable General Equilibrium (CGE) model and timber supply losses projected using the SBW Decision Support System (DSS; [21]) to investigate the potential economic impacts of foliage protection, salvage and re-planning harvest scheduling. A total of 16 forest protection scenarios were examined: two SBW outbreak severities (moderate vs. severe), four SBW control program levels (0%, 10%, 20%, and 40% protection of susceptible Crown forest area), and two pest management strategies (with or without re-planning scheduling and salvage). They found that a total of $3.3 and $4.7 billion (CAD) in present value lost output would result from uncontrolled moderate and severe SBW outbreak scenarios, respectively, over the 2012–2041 period in NB [3]. Total output value loss was projected to decrease by 40%, 56%, and 66% when spraying biological pesticides on 10%, 20%, and 40% of susceptible Crown forest areas, respectively [3]. Combining SBW control with re-planning scheduling and salvage strategy was projected to reduce the negative impacts of SBW outbreak by a further 1%–18% of output value depending on the level of control implemented.

In this paper, we used methods generally similar to those of Chang et al. [2,3] to evaluate the potential economic costs, benefits and economy-wide impacts of EIS compared to foliage protection strategies on NB Crown forest. Since EIS effectiveness is still being investigated [17], our analysis used past SBW outbreak scenarios (e.g., [4]) and results to date of EIS trials conducted on over 425,000 ha of NB forest from 2014 to 2018 [18]. Chang et al. [3] used a relatively simple CGE model, and we developed a more detailed CGE model to evaluate effects of SBW outbreak, foliage protection, and EIS, on the regional economy in NB. Objectives of this study were to (1) assess the potential timber volume savings associated with implementing a SBW EIS (assuming that EIS works), under moderate and severe SBW outbreak scenarios, compared to impacts under foliage protection on 0%, 10%, 20%, and 40% of susceptible Crown forest; (2) evaluate costs, benefits, and economy-wide impacts of EIS and foliage protection strategies on Crown forest under moderate and severe outbreak scenarios; and (3) test two sensitivity analyses, of costs/benefits if EIS partially works, resulting in saving 80% or 90% of the harvest volume reductions resulting from an uncontrolled outbreak scenario.

2. Methods

2.1. Study Area

The study area was all Crown land in NB, one of three Maritime Provinces in eastern Canada. NB is 85% forested [22] and approximately 50% (3.4 million ha) is Crown land owned by the provincial government and managed by forest industry. Generally, the NB Crown forest is spread across the province, but much of it lies in large, consolidated blocks in the central, northcentral, and northwest regions. For management purposes, the NB Crown land is divided into 10 Crown Licenses and leased to six forest companies [22].

The forest sector in NB includes more than 200 company locations and directly employs approximately 10,600 workforce [23]. Another 3400 people are indirectly employed by the forest sector, which builds on close to 500 forestry and logging operation locations (2011). The forest sector directly contributes about 4.4% of annual gross domestic production (i.e., $969 million CAD during 2011), and accounts for over 10% of total merchandise exports (i.e., $1.522 billion) [23]. Moreover, there is a significant indirect economic impact given the industry's strong linkages with other sectors of the economy such as metal fabrication, transportation and distribution, and professional and technical services [23]. Overall, NB is the most forest-dependent economy of any province in Canada [24].

Historically, SBW outbreaks have been the most devastating factor that causes large-scale defoliation and mortality on balsam fir and spruce stands in NB. For example, approximately 3.5 and 2.0 million ha of forestlands were moderately-severely defoliated (30%–100% of current year foliage) in the peak years of 1975 and 1983 during the 1970s–1980s SBW outbreak in the province [25]. SBW outbreaks in NB have recurred every 30–40 years [26], and a severe SBW outbreak is underway in Québec and beginning in the adjacent areas of northern NB [27]. With the first defoliation detected in NB in 2015, it is widely believed that a new provincial-scale SBW outbreak is beginning [18].

To protect against a potential SBW outbreak and test the impact of EIS, a SBW EIS research program was conducted in NB from 2014 to 2018 in an $18 million project [16–18]. As part of this, the ongoing SBW monitoring program was approximately doubled, with much of the additional annual SBW overwintering larvae sampling on mid-crown branches contributed by forest industry crews. Based on positive results to date, the EIS research trial is being continued from 2018 to 2022, with an additional $75 million of funding from the Government of Canada. The EIS trials on increasing SBW populations in northern NB resulted in insecticide treatment of hotspots covering 15,300, 56,800, 150,000, and 200,000 ha in 2015, 2016, 2017, and 2018, respectively [18]. Continuation of these SBW EIS test results will determine the effectiveness of EIS, but results have been positive to date [18].

2.2. SBW DSS Model

The timber supply impacts of potential SBW outbreak scenarios on Crown forest in NB were assessed using *Accuair ForPRO*, the most recent version of the SBW DSS [4,28]. This system consists of three components: (1) a SBW pre-defined tree defoliation-damage multiplier file [29], which alters stand dynamics in terms of tree growth and survival; (2) a stand impact matrix [21], which estimates relative stand volume reductions over time in different forest types as a function of stand dynamics with and without defoliation and insecticide treatment scenarios; and (3) a timber supply model [30], specifically a model developed by NB Department of Energy and Resource Development (NB ERD) for NB Crown land, which optimizes the schedule of stand interventions and harvest treatments over time to maximize the sustainable supply of forest values (e.g., timber, habitat, water quality) from 2017 to 2097. Overall, the *ForPRO* system links these three components to quantify the timber harvest volume impacts of SBW outbreak scenarios (defined as annual defoliation over each year of an outbreak) and benefits of insecticide treatments during the simulation period.

Each Crown License was assumed to re-plan their original timber harvest schedule and to implement salvage activities to harvest dead and dying trees, when feasible, in response to the potential SBW outbreak, since re-planning harvest scheduling and salvage can substantially reduce SBW impacts [3,31]. Salvage harvesting to capture defoliation-caused mortality and re-planning to modify the harvest schedule in light of defoliation-caused stand development changes can partially mitigate wood supply impact of an SBW outbreak [4]. For Crown land in NB, Hennigar et al. [4] showed that salvage and re-planning reduced wood supply impacts that accumulated over the 15 years after outbreak onset by about 20% for moderate and severe outbreak scenarios. Additionally, linear programming was used to minimize the SBW impacts on spruce-fir-jack pine (*Pinus banksiana*) harvest and total timber volume harvest and also to optimally prioritize areas for foliage protection treatments. More details of *Accuair ForPRO* DSS implementation are given in [4,28].

2.3. SBW Outbreak and Control Scenarios

To simulate SBW impacts in the *Accuair ForPRO* system, effects of 5-year cumulative defoliation levels were calculated and analyzed. Current-year defoliation patterns for balsam fir under moderate and severe outbreak scenarios [21] were used to estimate the annual SBW population (overwintering second-instar larvae (L2) per branch) using a linear relationship between balsam fir defoliation and SBW L2 population calculated by NB ERD. The SBW L2 population patterns were input into the *Accuair ForPRO* System, which calculated 5-year cumulative defoliation patterns for each host tree species. To reflect differing defoliation patterns among SBW host tree species, defoliation of white spruce (*Picea glauca* (Moench) Voss), red spruce (*Picea rubens* Sarg.), and black spruce (*Picea mariana* (Mill.) B.S.P.) were scaled to 72%, 41%, and 28% that of balsam fir [32]. However, these relative differences between fir and other host tree species were assumed to not occur at very high SBW population levels, when balsam fir defoliation was 90%–100%; i.e., if balsam fir defoliation was very high, spruce defoliation also would be higher.

To evaluate the impacts of future SBW outbreaks and effectiveness of alternative forest protection strategies, a scenario planning approach [33] was used. Overall, a total of 12 future SBW outbreak and forest protection scenarios were analyzed using the *Accuair ForPRO* system over an 80-year (2012–2092) period. Scenarios included two future SBW outbreak patterns (moderate and severe), each for EIS, no protection, re-planning only, and three foliage protection strategies (5%, 10%, or 20% of the area protected each year) scenarios.

The no protection scenario assumed no insecticide treatments to control a future SBW outbreak, and the resulting projected available harvest supply losses were used as the baseline for comparison with the defoliation control scenarios, to determine the marginal available harvest supply benefits of protection. The EIS scenario assumed that biological insecticide (i.e., *Btk* or tebufenozide) would be applied in all years in SBW hotspots [18]. The insecticide spray treatment varied based on the number of detected overwintering SBW L2 larvae per branch that were detected: (1) no protection with ≤6 L2/branch; (2) applying one *Btk* or tebufenozide application in all SBW hotspots with 7–20 L2/branch; and (3) applying two *Btk* applications in SBW hotspots when >20 L2/branch were detected. The objective of these EIS treatments was to keep SBW population levels low and maintain SBW defoliation below 10%. Additionally, the L2 sample point data (1500–2000 points sampled each year from 2014 to 2018 [18]) was converted to a spatial layer using spatial interpolation (80 × 80 m cells). Then EIS treatments were prioritized based upon the interpolated L2 raster and the percentage of spruce-fir. Overall, higher spray priority values occurred with larger number of L2/branch and with higher percentage of host species located in those areas. More details of the actual pest control treatment blocking and treatments are described in [18].

The foliage protection scenarios tested assumed that biological insecticides were applied to protect foliage in all years of the simulated moderate or severe SBW outbreaks. Insecticide treatment was simulated to be applied to 5%, 10%, or 20% of susceptible Crown forest area by spraying *Btk* or tebufenozide in all years when balsam fir defoliation exceeds 40%. In addition, re-planning of harvesting was combined with insecticide spray treatments in simulations, to minimize the defoliation-caused impact on NB timber volume harvest. The overall objective of foliage protection was to reduce defoliation to the NB ERD foliage protection target of 50% for balsam fir.

In previous analyses [34], we tested a combined EIS and foliage protection scenario, which assumed that the initial EIS treatments failed to slow or prevent the progression of SBW outbreak rise, and a moderate or severe SBW outbreak would occur. Results of these combined EIS + foliage protection scenarios were largely indistinguishable, in terms of pattern of timber supply reductions over time, from the comparable foliage protection scenario, slightly higher and somewhat later in time [34]. Therefore, we decided not to include such analyses in the current paper, partly because after 5 years of trials EIS appears to be working [18] and because results were so similar to the comparable foliage protection scenario [34]. Instead, we tested two 'what if' sensitivity analysis scenarios, assuming that EIS worked partially, resulting in saving 80% or 90% of the harvest volume reductions resulting from an uncontrolled outbreak scenario. Our rationale here was to test 'what if' EIS generally works in keeping SBW population levels low, but that some defoliation occurs, which results in low impacts (10% or 20% of the full outbreak scenario).

There are divergent views about the processes involved in the early stages of an SBW outbreak, as described by Johns et al. [17]. Areas to be treated with insecticide under the foliage protection scenarios were calculated based on the moderate or severe SBW outbreak scenario duration [9], area of susceptible forest in NB [4], and 5%, 10%, or 20% of susceptible forest to be protected. Areas to be treated annually under the EIS scenario were estimated based on several assumptions: (1) the SBW hotspot dynamic determining treatment areas would follow a similar pattern to the SBW population dynamic, spreading from existing hotspots to a larger geographical area; (2) actual measured growth rate of SBW hotspot areas was used from 2015 to 2017; (3) SBW population increases would last 10 years, 2016 to 2026; (4) SBW populations would begin to decrease in northern NB in 2021–2024; (5) there is less spruce-fir area to treat in southern NB than in northern NB; and (6) if EIS treatments are effective and moth in-flights from Québec decline after 2020, the growth rate and duration of SBW hotspot area also would decline. Overall, the total SBW hotspot area requiring EIS treatments was predicted to increase from 15,900 ha in 2015 to 562,500 ha from 2020 to 2022, and then decline to 161,370 ha by 2026. MacLean et al. [18] describe positive results of the first 5 years of EIS treatments in NB, corroborating the initial years of the simulation scenarios.

2.4. CGE Model

A CGE model was used to estimate the regional economic impacts of SBW outbreak and forest protection scenarios on the NB economy. This CGE model, which comprised a series of simultaneous linear and non-linear equations (described in Appendix A Tables A1–A3), was similar in structure to those of [35–38].

In the model, the NB economy was defined as a single region that is recursive dynamic and deterministic in nature, with the assumption of a small-open-economy and constant returns to scale technology. The model was calibrated to the regional economy of NB in 2010 (the latest year for which data were available) using NB input-output data from Statistics Canada [39]. Following the Canadian System of National Accounts [40], the NB economy consisted of 32 sectors, but for ease of presentation, we aggregated 12 sectors that were least impacted by a potential SBW outbreak. Overall, 20 sectors were defined: crop and animal; forestry and logging; fishing, hunting, and trapping; support activities for agriculture and forestry; mining and oil and gas extraction; utilities; construction; manufacturing; wholesale trade; retail trade; transportation and warehousing; information and cultural industries; finance, insurance, real estate, rental and leasing, and holding companies; professional, scientific, and technical services; administrative and support, waste management, and remediation services; educational services; health care and social assistance; arts, entertainment and recreation; accommodation and food services; and other services (except public administration) sectors. In addition, we assumed that producers from each sector produce intermediate and primary goods or provide services.

Three primary factors were considered in this study: (1) labor, (2) capital, and (3) stumpage, which is the market value of standing trees that must be paid by firms for the right to harvest timber, measured in $/cubic meter. Several assumptions were made in association with these primary factors. First, we assumed that labor was mobile and employed across all sectors, and the labor supply was fixed without the effect of interprovincial/international labor movement. Thus, the labor market followed the neoclassical economic theory of labor market, suggesting that labor was only affected by adjustment of the wage rate [41,42]. As a result, labor input could be calculated by aggregating wages and salaries, supplementary labor income, and mixed income from the input-output table [39]. Capital was assumed to be mobile and used in production across all sectors. Therefore, capital could freely move from one industry to another in the economy, but with a different rental rate [42]. Capital input was defined as the summation of taxes on products and production, subsidies on products and production, and gross operating surplus (except stumpage expenditures, as described below) from the 2010 NB input-output table [39]. Stumpage, the third primary factor, was defined only for the forestry and logging sector. While stumpage expenditures were included in gross operating surplus in the Statistics Canada input-output table, Statistics Canada does not report separately on this expenditure. Therefore, we followed Chang et al. [3] by using stumpage revenue data from the National Forestry Database [43] as input in the forestry and logging sector. We subtracted this value from gross operating surplus to appropriately adjust the capital input described above.

With regard to model structure, simultaneous linear and non-linear equations (Appendix A Table A3) were used to describe: (1) the behavior of economic agents; (2) market conditions; (3) macroeconomic balances; and (4) growth projections for primary inputs between periods. Overall, these equations were designed to operate recursively for each year over 50 years.

Regarding the behavior of economic agents, producers were assumed to simultaneously maximize their profit and minimize their production cost. To explain the behavior of producers, a constant elasticity of substitution (CES) production technology was assumed. Through the CES function, producers could make their choice between primary factors and thereby could substitute primary factors freely in response to the change of factor prices to produce a most efficient final value-added composite. Once the fraction of primary factors was determined, a Leontief-type technology was adopted to combine value-added composite with a fixed-share of intermediate demands. Specifically, the fixed-share of intermediate demands was assumed since the proportion of intermediate demands was only determined by existing technology rather than producers' decisions on primary factors [35]. Overall, the final output price would be derived from the combined costs of value-added composite and intermediate demands.

A CES Armington function [44] was established to explain producers' behavior in selecting intermediate inputs; producers were assumed to have choice to purchase intermediate inputs from either domestic or international markets. Since the relative prices of domestic goods and imports (including the rate of tariff) were essential to producers' production cost, the final ratio of imports to domestic goods was determined by producers' cost minimizing decision-making on domestic or foreign goods. Moreover, a constant elasticity of transformation (CET) function was used to distinguish producers' behavior for selling their products; to maximize profit, producers were assumed to have choice to sell their products to the market with the highest returns (where these returns were determined by multiplying the world price to the exchange rate adjusted for any taxes or subsidies). Overall, the final composite goods, including imported and domestic goods, were assumed to fulfill demands for both intermediate goods and final goods. Specifically, the intermediate goods, which were consumed by producers, were determined by technology and the composition of sectoral production; while the final goods, which were consumed by households, were determined by household income and the composite of aggregate demand.

Households, as the second economic agent, were assumed to maximize their utilities subject to their received income, including returns from supplying factors of production (capital, labor, and stumpage), and domestic government transfers (i.e., unemployment benefits, pensions, and other transfers from their domestic governments). The supplies of capital and labor were assumed to be fixed within a given time and mobile among all 20 sectors. Households were assumed to spend a proportion of their total income, and invest/save the remainder. All households were assumed to have identical consumption preferences, with consumption of each commodity affected by prices and incomes. Households make their consumption decisions by maximizing their utility subject to their household budget constraint. A Stone–Geary utility function was specified to model households' consumption preferences.

Government, as the final economic agent, was assumed to maximize its utility subject to the received tax revenue, levied on (1) capital and labor use (i.e., capital and labor tax revenues); (2) income of households (i.e., income tax revenue); (3) consumption by households (i.e., consumption tax revenue); and (4) importing (i.e., tariff revenue). A proportion of tax revenue was transferred from government to households (i.e., unemployment payments, pension, etc.), and the rest was spent on government's expenditures on capital, labor, and other commodities. A Cobb–Douglas utility function [45] was specified when the government was maximizing its utility.

We assumed that market equilibriums were achieved in the CGE model in the goods market and in the factor market. Specifically, equilibrium in the goods market consisted of aggregate demand for each commodity (from household and government consumption) equaling aggregate supply from domestic production and imported goods. To achieve equilibrium in the current account, the value of imports was required to equal the sum of value of exports and government savings. Since government savings and exchange rates were defined to be fixed, the change in value of imports would equal that of exports. Regarding the factor market, it was assumed that there was an equilibrium between the factors' demand and supply. Factor prices on production would simultaneously adjust to ensure that factor demand was equal to supply. Additionally, an equilibrium was also defined between household saving and investment; since the household saving rate was assumed to be fixed, household saving would passively adjust until it was equivalent to the investment spending.

To estimate the growth projection of primary inputs over time, it was assumed that, without any SBW outbreaks, the NB economy would operate on the historical, steady-state growth path over the next 50 years (2010–2060). Specifically, the annual growth rate of the labor force was exogenously set at 2% to approximately sustain a 2% annual GDP growth rate. Following Chang et al. [3] and Ochuodho and Lantz [46], the capital stock was endogenously determined through a capital accumulation equation that was influenced by the previous period's capital stock and household total savings (see Appendix A Tables A1–A3 for details).

To assess the economic impacts of SBW outbreak and forest protection scenarios, the CGE model was designed to re-run annually with the additional exogenous stumpage input changes based on the timber volume harvest results from the SBW DSS. To incorporate the SBW DSS output into the CGE model, stumpage was implemented as a primary input in the forestry and logging sector. Timber volume harvest results from each SBW outbreak and forest protection scenario estimated from the SBW DSS were converted into stumpage values by multiplying the 5-year periodic growth rate of total timber volume harvest from Crown land to the initial stumpage value in 2010 [43]. These stumpage values were used to shock the stumpage input in the CGE model. General Algebraic Modeling System (GAMS) software [47] and the CONOPT 3 solver were used to solve the CGE model. Economic impact assessment was performed in the CGE model to evaluate the impacts of each SBW outbreak and protection scenario on several major economic variables: total stumpage revenue; value of regional outputs; and value of provincial trade (i.e., net exports). Additionally, following Chang et al. [3], a 4% discount rate was adopted to estimate the present-value of these economic variables.

2.5. Benefit-Cost Analysis

We used a benefit-cost analysis (BCA) to compare market and non-market benefits and costs associated with each SBW defoliation control scenario. Market benefits included the value of timber harvest volume, and market costs included the aerial spraying costs per hectare multiplied by the number of hectares treated. Non-market benefits of each SBW defoliation control scenario were evaluated by aggregating use and non-use values including recreational, wildlife habitat, and other environmental service values. Market benefits, derived from the SBW DSS, were measured by comparing differences between the per-period (i.e., 5 year) values of timber harvest volume reduction (i.e., all harvest tree species) for each defoliation control scenario compared to the no protection scenario. Per-period values of timber harvest volume reduction were calculated by multiplying the wood product harvest volume losses (i.e., pulpwood, sawlogs) with wood product values determined based on: stumpage values (estimated using prices of pulp and sawlog materials on NB private land (NB ERD 2016 Crown model)), producer surplus in timber and wood products markets (assumed to be approximately 20% of stumpage prices [2]), a Crown timber License management fee (i.e., estimated at $3/m^3 (NB ERD 2016 Crown model)), and a social adjustment cost (i.e., costs of job searching, assumed to be 10% of stumpage prices (from [2], based on Van Kooten and Wang [48])). Overall, wood product values were estimated at $10/m^3 for pulpwood and $27/m^3 for sawlogs.

Market costs of each SBW defoliation control scenario were calculated by multiplying treatment area per year by spray program costs: one application of *Btk* or tebufenozide costs $40/ha; and two applications of *Btk* costs $80/ha (personal communication, Forest Protection Limited, 2018). Average costs for logistical, monitoring, and other costs specific to each strategy were estimated through consultation with experts in the field.

Non-market benefits of each SBW control effort were adopted from the Chang et al. [20] willingness to pay (WTP) estimation of the public's benefit from controlling future SBW outbreaks in NB. A contingent valuation method (CVM) analysis determined that average WTP was $86.19 (CAD 2007) per household per year for 5 years (2007–2011), including timber value, which was considered a market benefit, or $53.87 for the non-market benefit [20]. The associated value for a moderate SBW outbreak was estimated in proportion to the volume lost. All values were assessed in current value (undiscounted) and present value (discounted) terms, with the latter using a 4% discount rate.

Benefit-cost ratio (BCR; present value benefits divided by present value costs) and net present value (NPV; present value benefits minus present value costs) measures were compared among forest protection strategies to determine the preferred strategy on economic grounds. The strategy with the highest BCR would be selected if the objective was to maximize protection benefits per unit of control cost, as a cost-effectiveness measure. The strategy with the highest NPV would be preferred if the primary goal was to maximize net returns of SBW control efforts, as an economic efficiency measure.

3. Results

3.1. Projected Cumulative Timber Harvest Volume Impacts

The projected relative timber volume harvest levels for Crown land in NB from 2012 to 2066 under different SBW outbreak patterns and forest protection scenarios are presented in Figure 1. In this analysis, we assumed that the EIS works scenario was the same as the no SBW outbreak scenario, for both moderate and severe outbreak analyses (Figure 1). Without any forest protection, future SBW outbreaks were forecast to significantly reduce NB available harvest supply from Crown land over the next 50 years. The NB available harvest supply was projected to substantially decline until it reached the nadir during the 2032–2046 period, and then to start to recover after defoliation was simulated to end in 2026. SBW defoliation impact on NB available harvest supply under a moderate outbreak scenario (Figure 1a) was projected to be more gradual than for a severe outbreak (Figure 1b). Relative to the baseline no outbreak scenario, projected timber volume harvest levels without any forest protection were projected to be reduced by up to 13% and 27% under moderate and severe outbreak patterns, respectively. Overall, the NB projected available harvest supply (or harvest level) was projected to be reduced by 28 and 43 million m^3 under moderate and severe outbreaks, respectively, over the 50-year period (Table 1).

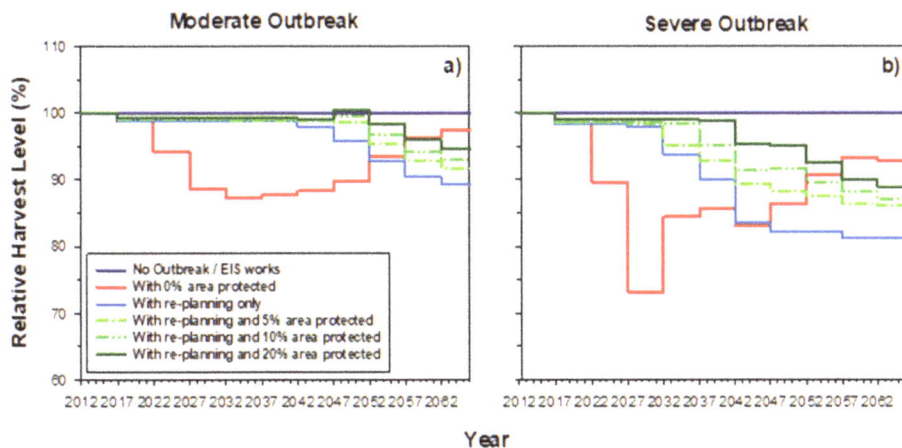

Figure 1. Projected relative total timber volume harvest for all Crown land in New Brunswick from 2012 to 2066 under (**a**) moderate and (**b**) severe spruce budworm (*Choristoneura fumiferana* Clem.) (SBW) outbreak and protection scenarios.

Since we assumed that a future SBW outbreak would be prevented if the EIS treatment worked, the projected NB available harvest supply was assumed to be not affected over the next 50 years. Consequently, the projected relative timber volume harvest level under the EIS scenario would remain at the baseline level of 100% (Figure 1). As a result, the 28 and 44 million m^3 of cumulative available harvest volume losses estimated above were anticipated to be saved by EIS treatments under moderate and severe outbreak patterns, respectively (Table 1). Empirical results from EIS trials from 2014 to 2018 support the assumption that, thus far, EIS works [18].

Under the foliage protection scenarios, available harvest supply savings were also substantial (Figure 1a,b). Re-planning of harvest scheduling alone was projected to save 12.5 million m^3 of cumulative available harvest volume from loss over the next 50 years (Table 1). By combining foliage protection with re-planning of harvesting, SBW outbreak impacts on NB available harvest supply were projected to be reduced further, increasing with the area of Crown forest protected (Figure 1a,b). Under the moderate outbreak scenario, foliage protection with re-planning of harvesting was projected to save 17, 19, and 22 million m^3 of cumulative available harvest volume from loss when protecting 5%, 10%, and 20% of total susceptible Crown forest, respectively (Table 1). Under a severe outbreak scenario, a projected 27 million m^3 of NB available harvest supply loss was saved when 20% of susceptible Crown forest was protected (Table 1). We also tested two sensitivity analysis scenarios considering that EIS partly worked, assuming that it would reduce 80% or 90% of the harvest volume impact. This amounted to saving 22–25 or 35–39 million m^3 under moderate and severe SBW outbreaks, respectively (Table 1). Both of these would exceed the cumulative timber volume saving of the re-planning and 20% area protected scenarios (Table 1).

Table 1. Summary of available harvest volume impact (cumulative from 2017 to 2067) on Crown land in New Brunswick from moderate and severe spruce budworm outbreak scenarios.

Impact on:	Timber Harvest Volume (million m^3) by Outbreak and Protection Strategy Scenarios	
	Moderate outbreak	Severe Outbreak
Cumulative timber volume harvest loss (−)	−27.86	−43.54
With 0% area protected		
Cumulative timber volume harvest saving (+)		
With re-planning only	12.54	1.20
With re-planning and 5% area protected	16.90	13.36
With re-planning and 10% area protected	19.15	19.42
With re-planning and 20% area protected	21.81	26.71
With Early Intervention Strategy	27.86	43.54
If Early Intervention Strategy works 90%	25.07	39.19
If Early Intervention Strategy works 80%	**22.29**	**34.83**

3.2. CGE Model Results

3.2.1. Current Value Stumpage Revenue Impacts

Figure 2 presents the projected relative stumpage revenues for Crown land in NB under different SBW outbreak and forest protection scenarios over the 50-year period. Since stumpage revenue values were calculated based on the timber volume harvest results, percentage reductions in stumpage revenue values were similar to reductions in available harvest supply. Relative stumpage revenues without SBW control declined by at most by 13% (moderate outbreak) and 27% (severe outbreak). Overall, cumulative current value stumpage revenue losses without forest protection were estimated at $178 million and $273 million under moderate and severe outbreak patterns, respectively. Under the EIS scenario, stumpage revenue savings would equal potential outbreak reductions, because timber supplies were assumed to remain at their baseline levels over the 50-year period, i.e., no outbreak (Figure 2). The foliage protection scenarios were projected to save $124–150 million (moderate outbreak) and $115–188 million (severe outbreak) of stumpage revenue over 50 years, when protecting 5%–20% of total susceptible Crown forest area, respectively.

3.2.2. Current Value Domestic Output Impacts

Simulations showed that uncontrolled moderate and severe outbreaks would significantly impact NB's domestic economic output over the next 50 years, with losses projected to be $24.6 billion and $35.3 billion with no SBW protection (Table 2). Under the EIS scenario (assuming EIS treatments worked), the negative impacts of SBW outbreaks on NB domestic output were eliminated over the 50-year period, so cumulative current value total domestic output savings would also be $24.6 billion (moderate) and $35.3 billion (severe) (Table 2). The foliage protection scenarios were projected to reduce the impacts of future SBW outbreaks on NB domestic output over the 50 year period by $23–24 billion (moderate outbreak) and $18–31 billion (severe outbreak) when protecting 5%–20% of total susceptible Crown forest, respectively (Table 2).

Moderate Outbreak / Severe Outbreak

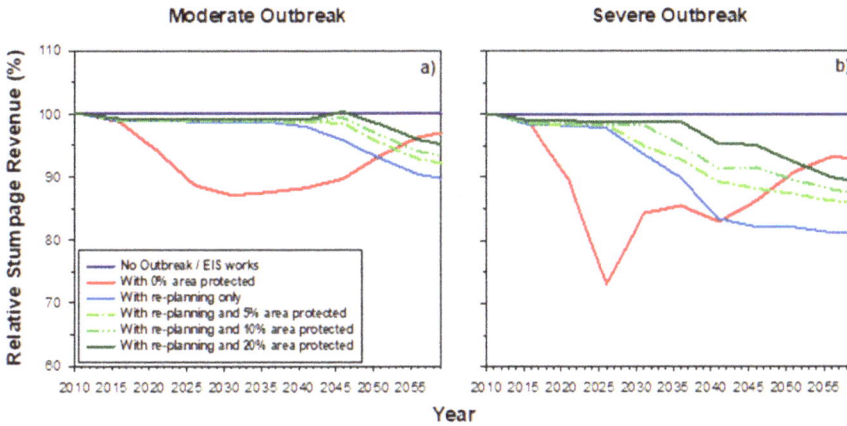

Figure 2. Projected relative stumpage revenue from all Crown land harvest in New Brunswick from 2010 to 2060 under (**a**) moderate and (**b**) severe SBW outbreak and protection scenarios.

Table 2. Current value [a] output and net export impacts, cumulative from 2010 to 2060, for moderate and severe spruce budworm outbreak scenarios on Crown land in New Brunswick.

Impact on:	Economic Output and Net Export ($ billion) by Outbreak and Protection Strategy Scenarios	
	Moderate Outbreak	Severe Outbreak
Output ($ billion):		
Output loss (−) With 0% area protected	−24.63	−35.31
Output saving (+) With re-planning only	21.71	9.58
With re-planning and 5% area protected	23.14	18.10
With re-planning and 10% area protected	23.65	22.88
With re-planning and 20% area protected	24.10	31.24
With Early Intervention Strategy	24.63	35.31
Net Export ($ billion):		
Net Export loss (−) With 0% area protected	−19.57	−27.79
Net Export saving (+) With re-planning only	17.80	7.75
With re-planning and 5% area protected	18.80	14.45
With re-planning and 10% area protected	19.14	18.22
With re-planning and 20% area protected	19.42	25.04
With Early Intervention Strategy	19.57	27.79

[a] Values are presented in current-value Canadian dollar terms.

3.2.3. Current Value Net Export Impacts

Projected current value net export losses in NB, compared to the EIS works/no outbreak baseline scenario, were predicted to be $19.6 billion and $27.8 billion over the next 50 years under uncontrolled moderate and severe SBW outbreak scenarios (Table 2). Under the EIS works assumption of no future outbreak, all of this loss would be prevented. Re-planning alone was projected to reduce losses by $17.8 billion (moderate outbreak scenario) or $7.8 billion (severe outbreak scenario). Foliage protection applied over 20% of susceptible Crown forest would reduce net export losses up to $19.4 billion (moderate outbreak scenario) or $25.0 billion (severe outbreak) (Table 2). The large differences of $6.7–8.5 billion in output and net export savings between the re-planning only versus re-planning with 5% area protected (severe SBW outbreak—Table 2) occur because the SBW DSS is very effective in selecting stands that will sustain the highest timber harvest volume losses and directing insecticide protection to prevent those losses [21,31]. Therefore, the model initially selects stands with the largest marginal timber supply benefit of protection, and the biggest benefit occurs in the initial 5%–10% of protection [21,31]. In contrast, the re-planning only scenario has no insecticide protection of stands.

3.2.4. Present Value Stumpage Revenue, Domestic Output, and Net Export Impacts

Table 3 describes the present value stumpage revenue, domestic output, and net export impacts of future SBW outbreaks and forest protection scenarios on the NB economy and forest-based industries over the 2010–2060 period. Present values were smaller than the current values presented in Tables 1 and 2 because they were discounted at a 4% market rate of interest. An uncontrolled future

SBW outbreak was forecast to cause $63.6 and $101.2 million of present value stumpage revenue losses under the moderate and severe outbreak scenarios, respectively (Table 3). The available harvest supply reductions reflected in these present value stumpage revenue reductions were forecast to cause $6.1–9.6 billion (moderate and severe outbreak scenarios) of domestic output reductions and $4.7–7.3 billion (moderate and severe outbreak) of net export losses (Table 3). If the EIS is successful, it would prevent these reductions.

An uncontrolled SBW outbreak was projected to cause large-scale sectoral output and net export reductions in NB (Table 3). The manufacturing, forestry and logging, and support activities for agriculture and forestry sectors were the sectors projected to be most affected by a future SBW outbreak (Table 3). Under a severe SBW outbreak scenario, total economic output was projected to be reduced from $9.6 billion over 50 years with no outbreak or EIS, to $4.0 billion with no re-planning or foliage protection to reduce losses (Table 3). This would result in a reduction of net exports from $7.3 billion to $3.0 billion (Table 3).

3.3. Benefit-Cost Analysis Results

Table 4 presents the present value market and non-market benefits, market costs, BCR and NPV results under SBW outbreak and forest protection scenarios. Market benefits, based on the NB Crown SBW DSS model (for timber supply volume savings and market prices), were projected to increase significantly with more forest protection. Under the EIS scenario (Works 100% column in Table 4), the cumulative present value market benefits were projected to be $162 million and $319 million under moderate and severe outbreak patterns, respectively (Table 4). The cumulative present value market benefits under the foliage protection strategy scenarios were projected to be $118–131 million (moderate outbreak) and $147–210 million (severe outbreak) when protecting 5%–20% of susceptible Crown forest (Table 4).

Present value cumulative market costs increased with more Crown forest protected (Table 4). Present value market cost for the EIS scenarios was $65.5 million, based on treating a cumulative total of 2,040,000 ha (Figure 3). The foliage protection scenario costs ranged from $35 million to $133 million (moderate) and $43 million to $164 million (severe) when protecting 5%–20% of total susceptible Crown forest (Table 4). These costs were from treating 1,136,000 ha (5% of susceptible Crown forest scenario), 2,273,000 ha (10% scenario), or 4,546,000 ha (20% scenario) for moderate outbreak scenarios (Table 4, Figure 3). These were cumulative amounts over the period from 2017 to 2028 (Figure 3). Areas treated under the severe outbreak scenario were 280,000–1,136,000 ha higher than for the moderate outbreak, because defoliation under the severe scenario lasted 2 years longer (Figure 3). This resulted in about 23% higher costs for protection under the severe outbreak scenario, ranging from $7.9 million to $30.7 million higher depending upon the area protected (Table 4).

For non-market benefits, the public's WTP in NB was highest under the EIS scenario, where NB households were estimated to be willing to pay up to $90 million and $99 million (present value) over the next 50 years to prevent future moderate or severe outbreaks, respectively (Table 4). However, the public's WTP was significantly lower if future SBW outbreaks were not be prevented. Under the foliage protection strategy scenarios, present value non-market benefits ranged from $4.5–18 million (moderate outbreak) and $4.9–20 million (severe outbreak), with 5%–20% of area protected (Table 4).

Table 3. Present value [a] cumulative (2010–2060) stumpage revenue, output ($ million CAD) and net export ($ billion CAD) impacts, by sector, of spruce budworm outbreak and protection scenarios on Crown Land in New Brunswick.

Outbreak Scenarios and Impacts by Sector:	Re-Planning and Foliage Protection Strategy, by % area protected				Early Intervention Strategy [b] (2040) [d]
	0% [c]	5% [c] (1136) [d]	10% [c] (2273) [d]	20% [c] (4546) [d]	
Moderate Spruce Budworm Outbreak Scenario					
Stumpage revenue loss ($ million)	−44.49	−49.58	−52.73	−55.53	−63.61
Output loss ($ million):					
Forestry and logging	−790.64	−882.15	−939.26	−989.95	−1137.09
Support activities for A&F [e]	−52.48	−60.32	−64.97	−69.44	−81.15
Manufacturing	−6624.32	−6989.64	−7127.82	−7237.06	−7398.01
Rest of economy	2024.59	2186.10	2265.80	2334.43	2493.33
Total	−5442.85	−5746.01	−5866.25	−5962.02	−6122.93
Net export loss ($ million):					
Forestry and logging	−174.77	−200.93	−217.78	−233.52	−278.89
Support activities for A&F	−21.76	−25.41	−27.53	−29.64	−34.94
Manufacturing	−5860.56	−6183.30	−6305.26	−6401.66	−6543.45
Rest of economy	1790.36	1929.61	1997.54	2055.71	2188.49
Total	−4266.73	−4480.03	−4553.03	−4609.10	−4668.79
Severe Spruce Budworm Outbreak Scenario					
Stumpage revenue loss ($ million)	−39.74	−57.16	−67.88	−80.07	−101.19
Output loss ($ million):					
Forestry and logging	−709.46	−1,018.19	−1,208.84	−1,425.36	−1,807.11
Support activities for A&F	−41.29	−65.92	−80.39	−96.72	−128.27
Manufacturing	−4609.69	−6990.15	−8396.04	−10,469.37	−11,563.92
Rest of economy	1396.76	2185.22	2647.49	3295.06	3864.69
Total	−3963.68	−5889.03	−7037.77	−8696.39	−9634.61
Net export loss ($ million):					
Forestry and logging	−152.40	−227.26	−273.33	−325.46	−437.66
Support activities for A&F	−15.90	−27.03	−33.44	−40.63	−55.12
Manufacturing	−4089.39	−6194.64	−7438.60	−9269.10	−10,235.98
Rest of economy	1267.10	1958.66	2363.30	2925.33	3412.06
Total	−2990.58	−4490.28	−5382.06	−6709.86	−7316.71

[a] All values are presented in present value (2011) Canadian dollar terms using a 4% discount rate. [b] Stumpage revenue, output, and net exports under the EIS are equal to projected losses under a no protection scenario. [c] Percent of area protected. [d] Values in parentheses represent 000's of hectares protected for the moderate outbreak scenario. Values for the severe outbreak scenario were 1420, 2841, and 5682 ha, respectively. [e] Agriculture and Forestry.

Regarding efficiency of forest protection scenarios, the NPV analysis indicated that the EIS scenario was the most economically efficient (i.e., highest NPV) strategy for both market and total (i.e., market and non-market) values (Table 4). The market NPV of the EIS scenario was projected to be $96 million (moderate outbreak) and $254 million (severe outbreak). Including non-market values, total NPV of the EIS scenario increased to $186 million (moderate outbreak) and $353 million (severe outbreak) (Table 4). The foliage protection scenarios were forecast to provide the least efficient forest protection. The market and total NPV of the foliage protection strategy under a moderate outbreak scenario were estimated to range from $82 million and $87 million with 5% of Crown forest protected, to $–2 million and $16 million with 20% of Crown forest protected. The NPV estimates were substantially higher under a severe outbreak scenario, with values of $104–109 million (market and total NPV, respectively) with 5% area protected and $46–66 million with 20% of total susceptible Crown forest protected (Table 4).

Table 4. Projected present value[a] benefits and costs of spruce budworm control programs on Crown land in New Brunswick over 50 years from 2010 to 2060.

Outbreak Scenarios and Values:	Foliage Protection Strategy, by % Susceptible Crown Forest Protected ('000 ha) [b]			Early Intervention Strategy [c] (2040)		
	5% (1136)	10% (2273)	20% (4546)	Works 100%	Works 90%	Works 80%
Moderate Spruce Budworm Outbreak Scenario						
PV [f] Market Benefit ($ million) [d]	117.55	123.94	130.91	161.79	145.61	129.43
PV Market Costs ($ million) [e]	−35.35	−67.87	−132.94	−65.50	−65.50	−65.50
PV Non-Market Benefits ($ million)	4.48	8.96	17.92	89.63	80.66	71.70
PV BCR [g] of Protection [Market value] ($/$)	3.33	1.83	0.98	2.47	2.22	1.98
PV BCR of Protection [Market + Non-Market] ($/$)	3.45	1.96	1.12	3.84	3.45	3.07
NPV [h] of Protection [Market value] ($ million)	82.20	56.07	−2.03	96.29	80.11	63.94
NPV of Protection [Market + Non-Market] ($ million)	86.68	65.03	15.90	185.92	160.78	135.64
Severe Spruce Budworm Outbreak Scenario						
PV [f] Market Benefit ($ million) [d]	147.07	173.77	209.84	319.33	287.40	255.47
PV Market Costs (−$ million) [e]	−43.26	−83.41	−163.67	−65.50	−65.50	−65.50
PV Non-Market Benefits ($ million)	4.94	9.89	19.77	98.87	88.98	79.09
PV BCR [g] of Protection [Market value] ($/$)	3.40	2.08	1.28	4.88	4.39	3.90
PV BCR of Protection [Market + Non-Market] ($/$)	3.51	2.20	1.40	6.39	5.75	5.11
NPV [h] of Protection [Market value] ($ million)	103.81	90.35	46.17	253.84	221.90	189.97
NPV of Protection [Market + Non-Market] ($ million)	108.75	100.24	65.94	352.71	310.89	269.07

[a] All values are presented in present value (2011) Canadian dollar terms using a 4% discount rate. [b] Foliage protection strategy assuming a potential moderate or severe outbreak scenario starts in 2015, forest insecticide spraying starts in 2018, and covers 5%, 10%, or 20% of susceptible Crown forest. Areas in ha treated for the moderate outbreak scenario are in parentheses, and similar areas for the severe outbreak scenario were 1420, 2841, and 5682 ha, respectively. The total SBW susceptible area was estimated at 2,840,860 ha in New Brunswick Crown forest [3]. [c] Assuming that the Early Intervention Strategy works completely (100%) or works partially (90% or 80%, see Table 1) as a sensitivity analysis. [d] Market benefits were estimated using the Crown stumpage revenue net of license management fees from the 2016 New Brunswick Crown timber supply model. [e] Market costs contain forest insecticide spraying program treatment costs and monitoring costs. [f] PV = present value. [g] BCR = benefit-cost ratio (i.e., BCR of protection = PV benefits/PV costs). [h] NPV = net present value (i.e., NPV of protection = (PV benefits-PV costs)/(1 + discount rate) ^{n}).

With regard to cost-effectiveness, BCR analysis revealed that the EIS was the most cost-effective (i.e., highest BCR) forest protection strategy, with a BCR of 3.8 (moderate outbreak) and 6.4 (severe outbreak), including both market and non-market values. Only considering market values, the EIS scenario had the highest BCR of 4.9 under a severe outbreak, but the foliage protection strategy with 5% of Crown forest protected scenario had the highest BCR of 3.3 under a moderate outbreak. Similar to the NPV estimates, higher forest protection levels generally led to lower BCRs. For example, market BCRs for the foliage protection strategy scenarios declined from 3.3 to 1.0 when forest protection area increased from 5% to 20% of total susceptible Crown forest (Table 4).

The sensitivity analysis ('what if' EIS works 90% or 80%) showed that under a severe SBW outbreak scenario, EIS works 90% and EIS works 80% always ranked second and third, after EIS works 100% and with the foliage protection strategies ranked numbers 4–6 (Table 4). Under a moderate SBW outbreak, the EIS works 90% scenario ranked second for market and non-market benefits, BCR and NPV (market + non-market), but third for BCR and NPV (market benefits only) (Table 4). The EIS works 80% resulted in third or fourth ranks, behind foliage protection 20% for market benefits and behind the foliage protection 5% for BCR and NPV (market) (Table 4). Therefore, it is likely that even if the EIS only partially works (saving 80%–90% of the uncontrolled outbreak timber harvest losses), it will be economically superior to foliage protection scenarios.

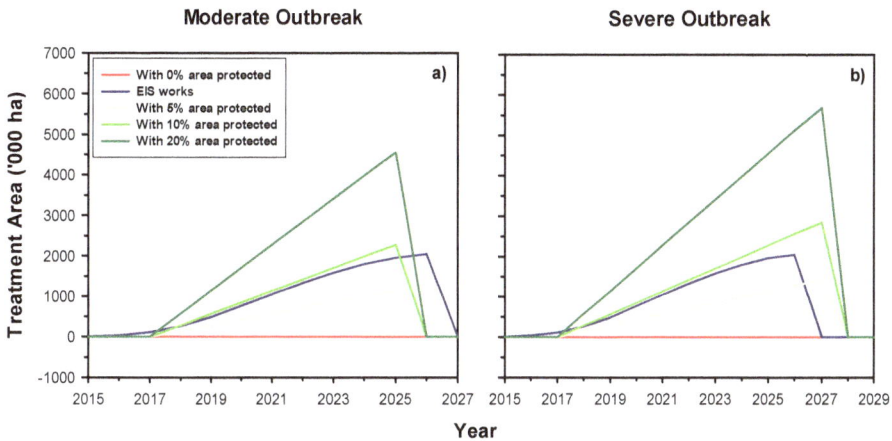

Figure 3. Projected total treatment area for Crown land in New Brunswick under (**a**) moderate and (**b**) severe SBW outbreak and protection scenarios.

4. Discussion

We coupled an advanced SBW DSS model with a dynamic CGE model to assess the impacts of future SBW outbreaks and forest protection strategies on available timber harvest from Crown land in NB and the provincial economy. Several key findings resulted from this research. First, the SBW DSS and CGE model results showed that unprotected future moderate or severe SBW outbreaks would have very large impacts on the available timber harvest from Crown land in NB and the regional economy. Sectoral economic outputs and net exports from forest-related industries in NB would be severely reduced. Approximately 50% of the NB provincial wood supply is from Crown land [22], so overall impacts from all land in NB would be roughly double what is reported here, but developing a timber supply model for private land was beyond the scope of this study. Our results were consistent with, but relatively higher than, previous economic impact research on SBW outbreaks in NB conducted by Chang et al. [3]. We utilized a revised SBW DSS model [4,28], applied to more recent (2017) NB forest inventory than in [3], as well as more refined modelling of the economic structure in CGE model. Our results were also for a longer study horizon, 50 years instead of 30 years, which is the main explanation of the larger impacts relative to previous results from [3].

Results from this study are applicable to other locations, countries, or different insect pests in several ways. First, SBW outbreaks occur across Canada and north-eastern United States, and the general level of timber harvest reductions and resulting economic and cost/benefit effects are applicable in other regions. Secondly, the SBW DSS tool has already been applied in several other jurisdictions: Maine, USA [49], Quebec [50], and four other provinces in Canada [28]. Thus, existing SBW DSS installations in these areas could be used with regional CGE or other economic models to estimate regional economic effects of SBW outbreaks. Third, the general approach of using a scenario planning approach to evaluate effects of defoliation on timber harvest volumes and resulting economic effects is applicable for other insects and countries, but would require local calibration of models. Finally, if the EIS approach continues to work, our cost/benefit results are useful in providing rationale for conducting other early intervention trials.

The EIS and foliage protection strategy both significantly mitigated SBW outbreak impacts on the regional economy. By definition, because it was assumed to result in no outbreak, the EIS scenario saved the most total domestic output and net exports over the 50-year period. Sensitivity analysis if EIS works 80% or 90% scenarios were superior to the best foliage protection scenario for all economic indicators under a severe outbreak, but were less cost effective than the 5% area protected for BCR and NPV of market values. It is impossible to know without a longer-term test whether EIS will continue to work, as the EIS research is less than one-half of the way through the current trial in NB, but thus far, results have been very positive [18]. Following 5 years of EIS treatments of low but increasing SBW populations, L2 populations across northern NB were much lower than SBW populations in adjacent Québec [18]. SBW populations in areas treated with *Btk* or tebufenozide were consistently reduced, and defoliation detected from aerial surveys in 2017–2018 in NB was only 550–2500 ha, compared to over 2.5 million ha in adjacent Bas St. Laurent-Gaspésie areas of Québec [50]. The foliage protection strategy also substantially reduced impacts of SBW outbreaks on the NB economy, but under the severe outbreak scenario, would still result in $4.1 billion lower economic output over 50 years than no outbreak or the EIS works scenario, at 2.5 times higher cost for the 20% area protected, severe SBW outbreak scenario.

Although larger foliage protection areas were consistently associated with higher available harvest supply and larger sectoral/economy-wide savings, they were generally not more cost-effective or efficient. BCA and NPV analyses revealed that treating 5% of susceptible Crown forest with insecticides was often the most cost effective and efficient due largely to the rate at which treatment costs increased with the area treated. This finding is consistent with previous results [2] except the market costs in our study were comparatively higher than those estimated by Chang et al. [2], because they assumed that foliage protection was applied only in peak defoliation years, when current balsam fir defoliation was higher than 70%, whereas we assumed foliage protection whenever defoliation exceeded 40%. The Chang et al. [2] approach would apply foliage protection only when required to keep trees alive, whereas our approach would limit defoliation sufficiently to prevent growth reduction. Overall, this finding suggests that it is essential for forest policy makers to consider BCR and NPV economic criteria when making informed foliage protection strategy.

Results also showed that re-planning of harvesting alone was forecast to significantly reduce SBW outbreak impacts on NB gross domestic output and net exports, particularly if the outbreak severity was only moderate. Combining re-planning of harvesting with foliage protection led to further substantial output and net export savings, similar to results of Chang et al. [3].

A final, and possibly most important, finding was that the EIS scenario was the most cost-effective and efficient forest protection strategy with the highest BCR of protection of 3.8–6.4 (moderate-severe outbreak scenarios). This largely had to do with the significant available harvest supply and market/non-market savings associated with the assumption that EIS treatments would be effective in preventing an SBW outbreak. Even the most aggressive foliage protection strategy still resulted in billions of dollars of economic impact over a 50-year period. Foliage protection on 20% of susceptible Crown forest cost 2.5 times more than the EIS scenario. Although market benefits and available harvest supply increased with area of foliage protection, due to high treatment cost, foliage protection of 20% of susceptible Crown forest was neither the most cost-effective nor economically efficient strategy, and in fact resulted in a negative market NPV of $–2.03 million of protection over 50 years.

While interesting findings emerged from this research, there were still several limitations. We used a single regional CGE model, whereas a multi-regional CGE model would better reflect reality by allowing for more detailed accounting of import/export substitution by region of origin and destination [35]. This was beyond the scope of the current study, but it is the authors' intent to attempt a multi-regional CGE model for more in-depth economic impact analysis in future. Our projection only addressed available harvest from Crown land, about 50% of timber supply in NB [22]. Another limitation, particularly with regard to the BCA and NPV analysis, was the lack of data on costs associated with the re-planning harvests. Better estimates might change the cost-effectiveness and efficiency ranking of forest protection strategies. Further, the non-market benefit estimates in this study, from Chang et al. [20], only included recreation and wildlife values. These values may have changed over time, and other non-market values such as carbon and water quality, which were not valued, may be important to the NB population. Including updated and more robust non-market values might lead to even stronger economic support for forest protection scenarios with high forest protection levels.

5. Conclusions

Our results have shed important light on potential biophysical and economic consequences of future SBW outbreaks and forest protection strategies. The findings confirmed that an EIS, assuming that it continues to work, was predicted to best mitigate negative economic impacts of SBW and was the most cost-effective and economically efficient forest protection strategy. This indicates that it is justified on economics grounds to continue recent efforts to utilize EIS to control future SBW outbreaks in NB. Following 5 years of EIS treatments of low but increasing SBW populations, SBW across northern NB are considerably lower than populations across the border in adjacent Québec [18]. These positive results from 2014 to 2018 trials have resulted in the Healthy Forest Partnership EIS team being approved for $75 million of additional funding from Natural Resources Canada to continue EIS trials from 2018 to 2022.

Author Contributions: Project conceptualization was led by D.A.M. and V.A.L.; timber supply analyses by C.H.; CGE modeling by E.Y.L. and V.A.L.; data analyses by E.Y.L.; project administration by D.A.M. and V.A.L.; funding acquisition by D.A.M.; initial draft manuscript preparation by E.Y.L.; manuscript revisions and editing by D.A.M., V.A.L., C.H., and E.Y.L.

Funding: This research was funded by Atlantic Canada Opportunities Agency, Natural Resources Canada, Government of New Brunswick, and forest industry in New Brunswick.

Acknowledgments: This research was overseen by the Healthy Forest Partnership, a consortium of researchers, landowners, forestry companies, governments, and forest protection experts. Many scientists and staff of industry and government agencies have made important contributions to the overall EIS research, without which the project could not have proceeded.

Conflicts of Interest: The authors declare no conflict of interest.

Appendix A

Table A1. Computable General Equilibrium (CGE) model variables.

Variables	Description	Variables	Description
Production Block		*Household Block*	
FAD$_{if}$	Factor input demand	INC	Household total gross income
FAS$_{if}$	Factor supply	SAH	Household savings
VAD$_i$	Value-added input demand	CBUD	Household disposable income (budget) after saving
IDE$_i$	Composite intermediate input demand	SBUD	Household discretionary (supernumerary budget)
PVA$_i$	Value-added input price	CON$_i$	Household consumption demand of commodities
PID$_i$	Intermediate input price	SAT	Household savings
PF$_f$	Factor price	INV$_i$	Household investment demand for commodities
P$_i$	Price of composite commodities demand (input)		
PD$_i$	Price of composite domestic production supply (output)	*Government Block*	
PDD$_i$	Price of domestic output delivered to home markets	KG	Government capital demand
X$_i$	Domestic sales of composite commodities	LG	Government labor demand
XD$_i$	Domestic production (output)	CG$_i$	Public demand for commodities
XDD$_i$	Domestic output delivered to home markets	SAG	Government savings
		TAXR	Total tax revenues
		TRMT	Total import tariff revenues
Foreign Trade Block		TRF	Total government transfer
M$_i$	Composite import	TRO	Other government transfer
E$_i$	Composite export	UNEMP	Unemployment level (Philips curve)
PM$_i$	Domestic import price	CPI	Consumer price index
PE$_i$	Domestic export price	tc$_i$	Tax rate on consumer commodities
SAF	Foreign savings	tk$_i$	Tax rate on capital use
ER	Exchange rate	tl$_i$	Tax rate on labor use
OBJ	Dummy objective variable	ty$_i$	Tax rate on income

* Subscript i is a set that denote sectors of the economy (i.e., 1, 2, 3, ... , 23); subscript f is a set that denotes input factors (i.e., capital, labor, and stumpage).

Table A2. CGE model parameters.

Parameters	Description
Elasticities of substitution	
σV_i	Substitution in the composite value-added function
σP_i	Substitution between the composite value-added input and the composite intermediate input
σA_i	Armington substitution between imports and domestic commodities
σT_i	CET substitution between domestic and export markets
σY_i	Income elasticities of demand for commodities
Share parameters	
γV_{if}	Share parameter in composite value-added input function
γp_i	Share parameter in total cost (production) function
γA_i	CES share parameter in Armington function
γT_i	CET share parameter in transformation function
Efficiency (shift) parameters	
$\emptyset V_i$	Shift parameter in the composite value-added input function
$\emptyset P_i$	Shift parameter in total cost (production) function
$\emptyset A_i$	Shift parameter in Armington function
$\emptyset T_i$	Shift parameter in transformation function
Other parameters	
αCG_i	Cobb–Douglas power of commodities bought by government
αKG	Cobb–Douglas power of capital use by government
αLG	Cobb–Douglas power of labor use by government
αI_i	Cobb–Douglas power share parameter for investment goods
trep	Replacement rate
IO_i	Technical coefficients of intermediate input
η	Philips curve parameter
Ψ_i	Budget shares in nested-LES household utility function
μH_i	Household subsistence consumption level
λ_i	Marginal propensity to save
Dynamic Growth Path	
GRW	Initial steady-state labor growth rate
RRR	Real rate of return on capital
$Time_t$	Time period into future from base year 2010
$GrowthTS_t$	Annual stumpage revenue growth rate

* Subscript i is a set that denote sectors of the economy (i.e., 1, 2, 3, ... , 23); subscript f is a set that denotes input factors (i.e., capital, labor, and stumpage); subscript t is a set that denotes time period in years from base year 2010 (i.e., 1, 2, 3, ... , 50).

Table A3. CGE model equations.

Equation	Description
Production Block	
$FAD_{if} = \left[VAD_i (\gamma V_i)^{\sigma V_i}\right] / \{OV_i (PF_f)^{1-\sigma V} \left[\sum_{f=1}^{3} (\gamma V_i)^{\sigma V_i} (PF_f)^{1-\sigma V}\right]^{\frac{\sigma V_i}{1-\sigma V_i}}\}$ where f denotes labor, capital for all sectors, and stumpage for forestry sector only. Eq.(A.1)	Factor demand by firm
$VAD_i = \left(\frac{XD_i}{OP_i}\right) \left(\frac{\gamma P_i}{PVA_i}\right)^{\sigma P_i} \left[\gamma P_i^{\sigma P_i} PVA_i^{1-\sigma P_i} + (1-\gamma P_i)^{\sigma P_i} PID_i^{1-\sigma P_i}\right]^{\frac{\sigma P_i}{1-\sigma P_i}}$ Eq.(A.2)	Value-added demand
$PID_i = \left(\frac{XD_i}{OP_i}\right) \left(\frac{1-\gamma P_i}{PID_i}\right)^{\sigma P_i} \left[\gamma P_i^{\sigma P_i} PVA_i^{1-\sigma P_i} + (1-\gamma P_i)^{\sigma P_i} PID_i^{1-\sigma P_i}\right]^{\frac{\sigma P_i}{1-\sigma P_i}}$ Eq.(A.3)	Composite intermediate input
$PD_i XD_i = PVA_i VAD_i + PID_i IDE_i$ Eq.(A.4)	Zero profit condition for the firm
Household Block	
$INC = \sum_{f=1}^{3} (PF_f FAS_f) + TRF$ Eq.(A.5)	Household total gross income
$SAH = \lambda_t INC$ Eq.(A.6)	Household savings
$CBUD = INC - SAH$ Eq.(A.7)	Household disposable income after tax and savings
$SBUD = CBUD - \sum_{i=1}^{20} P_i \mu H_i$ Eq.(A.8)	Household discretionary budget
$(1+tc_i) P_i CCN_i = (1+tc_i) P_i \mu H_i + \Psi_i \left[CBUD - \sum_{i=1}^{20} \mu H_i (1+tc_i) P_i\right]$ Eq.(A.9)	Household consumption demand of commodities
$SAT = SH + SG\ CPI + SF\ ER$ Eq.(A.10)	Household total savings
$P_i INV = \alpha I_i SAT$ Eq.(A.11)	Investment demand for commodities
$\left(\frac{\frac{PF_f}{CPI}}{\frac{PF_f^0}{CPI^0}} - 1\right) = \eta \left(\frac{\frac{UNEMP}{FAS_f}}{\frac{UNEMP^0}{FAS_f^0}} - 1\right)$ where subscript f denotes labor Eq.(A.12)	Unemployment level (Philips curve)
$CPI = \left(\sum_{i=1}^{20} P_i CON_t\right) / \left(\sum_{i=1}^{20} P_i^0 CON_t^0\right)$ Eq.(A.13)	Consumer price index

Table A3. *Cont.*

Equation		Description
Government Block		
$P_i CG_i = \alpha CG_i(TAXR - TRF - SG * CPI).$	Eq(A.14)	Government demand for commodities
$PF_f KG = \alpha KG(TAXR - TRF - SG * CPI)$ where f denotes capital	Eq(A.15)	Government capital demand function
$PF_f LG = \alpha LG(TAXR - TRF - SG * CPI)$ where f denotes labor	Eq(A.16)	Government labor demand function
$TAXR = ty_i INC + \sum_{i=1}^{20}\left(P_i tc_i CON_i + FAD_f tk_i PF_f + FAD_f tl_i PF_f + M_i tm_i PM_i ER\right).$	Eq(A.17)	Total tax revenues
$TRF = trep\, PF_f * UNEMP + TROCPI$ where f denotes labor	Eq(A.18)	Total transfers
$TRMT = \sum_{i=1}^{20} tm_i M_i PM_i ER$	Eq(A.19)	Total tariff revenue
Market Clearing Block		
$\sum_{i=1}^{20} FAD_{if} + KG = FAS_f - UNEMP$ where f denotes labor	Eq(A.20)	Market clearing for labor
$\sum_{i=1}^{20} FAD_{if} + LG = FAS_f$ where f denotes capital	Eq(A.21)	Market clearing for capital
$FAD_{if} = FAS_f$ where f denotes stumpage and I denotes forestry and logging sector	Eq(A.22)	Market clearing for stumpage
$X_i = CON_i + INV_i + CG_i + \sum_{i=1}^{20} IO_i XD_i$	Eq(A.23)	Market clearing for commodities
$\sum_{i=1}^{20} M_i PM_i = \sum_{i=1}^{20} E_i PE_i + SAF$	Eq(A.24)	Trade Balance of payments

Table A3. *Cont.*

Equation		Description
Trade Block		
a) Export side		
$XDD_i = (XD_i/OT_i)\left(\frac{1-\gamma T_i}{PDD_i}\right)^{\sigma T_i}\left[\gamma T_i^{\sigma T_i}PE_i^{1-\sigma T_i} + (1-\gamma T_i)^{\sigma T_i}PDD_i^{1-\sigma T_i}\right]^{\frac{\sigma T_i}{1-\sigma T_i}}$	Eq(A.25)	Domestic supply of domestic output (supply side)
$E_i = (XD_i/OT_i)\left(\frac{\gamma T_i}{PE_i}\right)^{\sigma T_i}\left[\gamma T_i^{\sigma T_i}PE_i^{1-\sigma T_i} + (1-\gamma T_i)^{\sigma T_i}PDD_i^{1-\sigma T_i}\right]^{\frac{\sigma T_i}{1-\sigma T_i}}$	Eq(A.26)	Export demand for domestic output
b) Import side		
$PD_iXD_i = PE_iE_i + PDD_iXDD_i$	Eq(A.27)	CET zero profit condition (profit maximization)
$XDD_i = (XD_i/OA_i)\left(\frac{1-\gamma A_i}{PDD_i}\right)^{\sigma A_i}\left[\gamma A_i^{\sigma A_i}PM_i^{1-\sigma A_i} + (1-\gamma A_i)^{\sigma A_i}PDD_i^{1-\sigma A_i}\right]^{\frac{\sigma A_i}{1-\sigma A_i}}$	Eq(A.28)	Domestic demand for domestically produced goods (demand side)
$M_i = (XD_i/OA_i)\left(\frac{\gamma A_i}{PM_i}\right)^{\sigma A_i}\left[\gamma A_i^{\sigma A_i}PM_i^{1-\sigma A_i} + (1-\gamma A_i)^{\sigma A_i}PDD_i^{1-\sigma A_i}\right]^{\frac{\sigma A_i}{1-\sigma A_i}}$	Eq(A.29)	Domestic demand for composite imported goods
$PD_iXD_i = PM_iM_i + PDD_iXDD_i$	Eq(A.30)	Armington CES zero profit condition (cost minimization)
Macroeconomic Closures		
$PD_iXD_i = PM_iM_i + PDD_iXDD_i$	Eq(A.31)	Exogenously fix factor endowments
$\overline{SAF} = SAF^0$	Eq(A.32)	Exogenously fix foreign savings
$\overline{SAG} = SAG^0$	Eq(A.33)	Exogenously fix government savings
$\overline{TRO} = TRO^0$	Eq(A.34)	Exogenously fix government other transfer
Artificial Objective Function		
$OBJ = 1$	Eq(A.35)	Dummy objective variable
Dynamic Growth Path		
$RRR = PF_f^0 FAS_f^0\left(\frac{GRW}{SAT}\right)$ where f denotes capital factor	Eq(A.36)	Real rate of return on capital
$GRW_t = (SATRRR)\left(PF_fFAS_f\right)$ where f denotes capital factor input	Eq(A.37)	Growth path for each time period recursive loop run
$\overline{FAS_f} = (1+GRW_t)FAS_f$ where f denotes capital factor input	Eq(A.38)	Exogenously fixing capital growth path dynamic loop
$\overline{FAS_f} = (1+GSW)FAS_f$ where f denotes labor factor input	Eq(A.39)	Exogenously fixing labor growth path dynamic loop
$\overline{FAS_f} = (1+GrowthTS)FAS_f^0$ where f denotes stumpage in forestry and logging sector	Eq(A.40)	Exogenously fixing stumpage growth path dynamic loop

* Subscript i is a set that denotes sectors of the economy (i.e., 1, 2, 3,..., 23); subscript f is a set that denotes input factors (i.e., capital, labor, and stumpage); subscript t is a set that denotes time period in years from base year 2010 (i.e., 1, 2, 3,..., 50).

References

1. Natural Resources Canada. Overview of Canada's Forest Industry. The official website of the Government of Canada, 2015. Available online: http://www.nrcan.gc.ca/forests/industry/overview/13311 (accessed on 19 January 2016).
2. Chang, W.-Y.; Lantz, V.A.; Hennigar, C.R.; MacLean, D.A. Benefit-cost analysis of spruce budworm (*Choristoneura fumiferana* Clem.) control: Incorporating market and non-market values. *J. Environ. Manag.* **2012**, *93*, 104–112. [CrossRef] [PubMed]
3. Chang, W.-Y.; Lantz, V.A.; Hennigar, C.R.; MacLean, D.A. Economic impacts of spruce budworm (*Choristoneura fumiferana* Clem.) outbreaks and control in New Brunswick, Canada. *Can. J. For. Res.* **2012**, *42*, 490–505. [CrossRef]
4. Hennigar, C.R.; Erdle, T.A.; Gullison, J.J.; MacLean, D.A. Reexamining wood supply in light of future spruce budworm outbreaks: A case study in New Brunswick. *For. Chron.* **2013**, *89*, 42–53. [CrossRef]
5. Niquidet, K.; Tang, J.; Peter, B. Economic analysis of forest insect pests in Canada. *Can. Ent.* **2016**, *148*, 357–366. [CrossRef]
6. Sterner, T.E.; Davidson, A.G. *Forest Insect and Disease Conditions in Canada 1981*; Canadian Forest Service: Ottawa, ON, Canada, 1982.
7. Blais, J.R. Trends in the frequency, extent, and severity of spruce budworm outbreaks in eastern Canada. *Can. J. For. Res.* **1983**, *13*, 539–547. [CrossRef]
8. MacLean, D.A.; Ostaff, D.P. Patterns of balsam fir mortality caused by an uncontrolled spruce budworm outbreak. *Can. J. For. Res.* **1989**, *19*, 1087–1095. [CrossRef]
9. MacLean, D.A.; Beaton, K.P.; Porter, K.B.; MacKinnon, W.E.; Budd, M.G. Potential wood supply losses to spruce budworm in New Brunswick estimated using the spruce budworm decision support system. *For. Chron.* **2002**, *78*, 739–750. [CrossRef]
10. Blum, B.M.; MacLean, D.A. Potential silviculture, harvesting and salvage practices in eastern North America. In *Recent Advances in Spruce Budworms Research: Proceedings of the CANUSA Spruce Budworms Research Symposium*; Canadian Forestry Service: Ottawa, ON, Canada, 1985; pp. 264–280.
11. Spence, C.E.; MacLean, D.A. Regeneration and stand development following a spruce budworm outbreak, spruce budworm inspired harvest, and salvage harvest. *Can. J. For. Res.* **2012**, *42*, 1759–1770. [CrossRef]
12. Sainte-Marie, G.B.; Kneeshaw, D.D.; MacLean, D.A.; Hennigar, C.R. Estimating forest vulnerability to the next spruce budworm outbreak: Will past silvicultural efforts pay dividends? *Can. J. For. Res.* **2014**, *45*, 314–324. [CrossRef]
13. Etheridge, D.A.; MacLean, D.A.; Wagner, R.G.; Wilson, J.S. Effects of intensive forest management on stand and landscape characteristics in northern New Brunswick, Canada (1945–2027). *Landsc. Ecol.* **2006**, *21*, 509–524. [CrossRef]
14. Kettela, E.G. Insect control in New Brunswick, 1974–1989. In *Forest Insect Pests in Canada*; Armstrong, J.A., Ives, W.G.H., Eds.; Natural Resources Canada, Canadian Forest Service: Ottawa, ON, Canada, 1995; pp. 655–665.
15. Miller, C.A.; Kettela, E.G. Aerial control operations against the spruce budworm in New Brunswick, 1952–1973. In *Aerial Control of Forest Insects in Canada*; Prebble, M.L., Ed.; Department of the Environment: Ottawa, ON, Canada, 1975; pp. 94–112.
16. Healthy Forest Partnership. Research Area Map. Healthy Forest Partnership, 2017. Available online: http://www.healthyforestpartnership.ca/en/research/what-where-and-when (accessed on 12 June 2017).
17. Johns, R.C.; Régnière, J.; MacLean, D.A.; James, P.; Martel, V.; Pureswaran, D.; Stastny, M. A conceptual framework for the spruce budworm Early Intervention Strategy: Can outbreaks be contained? *Forests* **2019**. submitted for publication.
18. MacLean, D.A.; Amirault, P.; Amos-Binks, L.; Carleton, D.; Hennigar, C.; Johns, R.; Régnière, J. Positive results of an early intervention strategy to suppress a spruce budworm outbreak after five years of trials. *Forests* **2019**, *10*, 448. [CrossRef]
19. Régnière, J.; Delisle, J.; Pureswaran, D.S.; Trudel, R. Mate-finding allee effect in spruce budworm population dynamics. *Entomol. Exper. Applic.* **2013**, *146*, 112–122. [CrossRef]
20. Chang, W.Y.; Lantz, V.A.; MacLean, D.A. Social benefits of controlling forest insect outbreaks: A contingent valuation analysis in two Canadian provinces. *Can. J. Agric. Econ.* **2011**, *59*, 383–404. [CrossRef]

21. MacLean, D.A.; Erdle, T.A.; MacKinnon, W.E.; Porter, K.B.; Beaton, K.P.; Cormier, G.; Morehouse, S.; Budd, M. The Spruce Budworm Decision Support System: Forest protection planning to sustain long-term wood supplies. *Can. J. For. Res.* **2001**, *31*, 1742–1757. [CrossRef]

22. Erdle, T.A.; Ward, C. *Management alternatives for New Brunswick's public forest: Report of the New Brunswick Task Force on Forest Diversity and Wood Supply: Summary*; New Brunswick Department of Natural Resources: Fredericton, NB, Canada, 2008.

23. New Brunswick Department of Economic Development. Rebuilding New Brunswick: New Brunswick Value-Added Wood Sector Strategy 2012–2016. Government of New Brunswick. Available online: http://www2.gnb.ca/content/dam/gnb/Corporate/pdf/EcDevEc/Wood.pdf (accessed on 20 August 2017).

24. Atlantic Provinces Economic Council. *The New Brunswick Forest Industry: The Potential Economic Impact of Proposals to Increase the Wood Supply*; Atlantic Provinces Economic Council: Halifax, NS, Canada, 2003.

25. National Forestry Database. Forest Insects - Quick Facts. Canadian Council of Forest Ministers, 2016. Available online: http://www.nfdp.ccfm.org/en/data/insects.php (accessed on 10 January 2016).

26. Royama, T.; MacKinnon, W.E.; Kettela, E.G.; Carter, N.E.; Hartling, L.K. Analysis of spruce budworm outbreak cycles in New Brunswick, Canada, since 1952. *Ecology* **2005**, *86*, 1212–1224. [CrossRef]

27. Healthy Forest Partnership. *Spruce Budworm L2 and Defoliation, 2018*; Healthy Forest Partnership, 2018. Available online: http://forestprotectionlimited.maps.arcgis.com/apps/Viewer/index.html?appid=1c6e488b56864964a917b06861382929 (accessed on 15 January 2018).

28. McLeod, I.M.; Lucarotti, C.J.; Hennigar, C.R.; MacLean, D.A.; Holloway, A.G.L.; Cormier, G.A.; Davies, D.C. Advances in aerial application technologies and decision support for integrated pest management. In *Integrated Pest Management and Pest Control*; Soloneski, S., Larramendy, M.L., Eds.; InTech Open Access Publisher: Rijeka, Croatia, 2012; pp. 651–668, ISBN 978-953-307-926-4.

29. Erdle, T.A.; MacLean, D.A. Stand growth model calibration for use in forest pest impact assessment. *For. Chron.* **1999**, *75*, 141–152. [CrossRef]

30. Remsoft Inc. *Spatial Woodstock 2008.12 User Guide*; Remsoft Inc.: Fredericton, NB, Canada, 2008.

31. Hennigar, C.R.; MacLean, D.A.; Porter, K.B.; Quiring, D.T. Optimized harvest planning under alternative foliage-protection scenarios to reduce volume losses to spruce budworm. *Can. J. For. Res.* **2007**, *37*, 1755–1769. [CrossRef]

32. Hennigar, C.R.; MacLean, D.A.; Quiring, D.T.; Kershaw, J.A. Differences in spruce budworm defoliation among balsam fir and white, red, and black spruce. *For. Sci.* **2008**, *54*, 158–166.

33. Schoemaker, P.J.H. Scenario planning: A tool for strategic thinking. *Sloan Manag. Rev.* **1995**, *36*, 25–40.

34. Liu, E.Y. Economics of early intervention to suppress a potential spruce budworm outbreak in New Brunswick, Canada. Master's Thesis, University of New Brunswick, Fredericton, NB, Canada, 2018.

35. Ochuodho, T.O.; Lantz, V.A.; Lloyd-Smith, P.; Benitez, P. Regional economic impacts of climate change and adaptation in Canadian forests: A CGE modeling analysis. *For. Pol. Econ.* **2014**, *25*, 100–112. [CrossRef]

36. Das, G.G.; Alavalapati, J.R.R.; Carter, D.R.; Tsigas, M.E. Regional Impacts of environmental regulations and technical change in the US forestry sector: A multiregional CGE analysis. *For. Pol. Econ.* **2005**, *7*, 25–38. [CrossRef]

37. Lofgren, H.; Harris, R.L.; Robinson, S. *A Standard Computable General Equilibrium (CGE) Model in GAMS*; International Food Policy Research Institute: Washington, DC, USA, 2002.

38. Zhang, J.; Alavalapati, J.R.R.; Shrestha, R.K.; Hodges, A.W. Economic impacts of closing national forests for commercial timber production in Florida and Liberty County. *J. For. Econ.* **2005**, *10*, 207–223. [CrossRef]

39. Statistics Canada. Provincial Input-Output Tables, Inputs and Outputs, Summary Level, Basic Price, Annual (Dollars), CANSIM (Database). Statistics Canada, 2014. Available online: http://www5.statcan.gc.ca/cansim/a26?lang=eng&retrLang=eng&id=3810028&&pattern=&stByVal=1&p1=1&p2=-1&tabMode=dataTable&csid= (accessed on 18 August 2016).

40. Statistics Canada. Canadian System of National Accounts (CSNA) 2012-Canada. 2012. Available online: https://www.statcan.gc.ca/eng/nea/classification/io_ind/cat (accessed on 5 August 2017).

41. Alavalapati, J.; White, W.; Jagger, P.; Wellstead, A. Effect of land use restrictions on the economy of Alberta: A computable general equilibrium analysis. *Can. J. Reg. Sci.* **1996**, *19*, 349–365.

42. Alavalapati, J.; Adamowicz, W.; White, W. A comparison of economic impact assessment methods: The case of forestry developments in Alberta. *Can. J. For. Res.* **1998**, *28*, 711–719. [CrossRef]

43. Canadian Council of Forest Ministers. *Statement of Revenues from the Sale of Timber from Provincial Crown Land*; National Forestry Database, 2016. Available online: http://www.nfdp.ccfm.org/en/data/revenues.php (accessed on 2 October 2016).

44. Armington, P.S. A theory of demand for products distinguished by place of production. In *International Monetary Fund Staff Papers*; Palgrave Macmillan: London, UK, 1969; Volume 16, pp. 244–252.

45. Cobb, C.W.; Douglas, P.H. A theory of production. *Amer. Econ. Rev.* **1928**, *18*, 139–165.

46. Ochuodho, T.O.; Lantz, V. Economic impacts of climate change in the forest sector: A comparison of single-region and multiregional CGE modeling frameworks. *Can. J. For. Res.* **2014**, *44*, 449–464. [CrossRef]

47. Rosenthal, R. *GAMS: A User's Guide*; GAMS Development Corporation: Washington, DC, USA, 2010.

48. Van Kooten, G.C.; Wang, S. Estimating economic costs of nature protection: British Columbia's forest regulations. *Can. Public Pol.* **1998**, *24*, 63–71. [CrossRef]

49. Hennigar, C.R.; Wilson, J.S.; MacLean, D.A.; Wagner, R.G. Applying a spruce budworm decision support system to Maine: Projecting spruce-fir volume impacts under alternative management and outbreak scenarios. *J. Forestry* **2011**, *109*, 332–342.

50. Québec Ministère des Forêts de la Faune et des Parcs. *Aires Infestées Par la Tordeuse des Bourgeons de L'épinette au Québec en 2018-Version 1.0*; Gouvernement du Québec, Direction de la Protection des Forêts: Québec, QC, Canada, 2018; pp. 1–17, ISBN 978-2-550-82373-5.

forests

Article

Dynamics and Management of Rising Outbreak Spruce Budworm Populations

Jacques Régnière [1,*], Barry J. Cooke [2], Ariane Béchard [1], Alain Dupont [3] and Pierre Therrien [4]

1   Natural Resources Canada, Canadian Forest Service, P.O. Box 10380 Stn Ste-Foy, Quebec,
    QC G1V-4C7, Canada
2   Natural Resources Canada, Canadian Forest Service, 1219 Queen St E, Sault Ste Marie, ON P6A-2E6, Canada
3   Société de Protection des Forêts contre les Insectes et Maladies, 1780 Rue Semple, Quebec,
    QC G1N 4B8, Canada
4   Ministère des Forêts, de la Faune et des Parcs du Québec, 2700 rue Einstein, Quebec, QC G1P 3W8, Canada
*   Correspondence: Jacques.Regniere@Canada.ca; Tel.: +1-418-648-5257

Received: 12 August 2019; Accepted: 30 August 2019; Published: 1 September 2019

Abstract: Management of spruce budworm, *Choristoneura fumiferana* (Clem.), outbreak spread requires understanding the demographic processes occurring in low, but rising populations. For the first time, detailed observations were made in the early stages of outbreak development. We sampled populations over a three-year period in both treated and untreated populations in the Lower St-Lawrence region of Quebec, Canada, and measured the density-dependence of survival and population growth rates, and the impact of natural enemies and insecticides. Insecticides tested were *Bacillus thuringiensis* (Berliner 1915) and tebufenozide. We recorded strong density-dependence of survival between early larval stages and adult emergence, explained largely by the variation of natural enemy impacts and overcrowding. We also observed inverse density-dependence of apparent fecundity: net immigration into lower-density populations and net emigration from the higher, linked to a threshold of ~25% defoliation. Because of high migration rates, none of the 2013 treatments reduced egg populations at the end of summer. However lower migration activity in 2014 allowed population growth to be reduced in treated plots. This evidence lends support to the conclusion that, for a budworm population to increase to outbreak density, it must be elevated via external perturbations, such as immigration, above a threshold density of ~4 larvae per branch tip (L_4). Once a population has increased beyond this threshold, it can continue growing and itself become a source of further spread by moth migration. These findings imply that populations can be brought down by insecticide applications to a density where mortality from natural enemies can keep the reduced population in check, barring subsequent immigration. While we recognize that other factors may occasionally cause a population to exceed the Allee threshold and reach outbreak level, the preponderance of immigration implies that if all potential sources of significant numbers of moths are reduced on a regional scale by insecticide applications, a widespread outbreak can be prevented, stopped or slowed down by reducing the supply of migrating moths.

Keywords: spruce budworm; *Choristoneura fumiferana*; forest protection; early intervention strategy; survival; apparent fecundity; immigration; growth rate; treatment threshold; insecticides

---

1. Introduction

Knowing how an insect outbreak starts is central to the development of any preventative management strategy. The spruce budworm (SBW), *Choristoneura fumiferana* (Clem.), is the most destructive forest insect pest of eastern North America [1]. Millions of dollars were spent during the last outbreak cycle (1975–1995) trying to predict what would happen, and to mitigate the damage [2,3]. A new cycle of spruce budworm outbreak has recently emerged in the province of Quebec [4] so the time is right to ask how best to manage this emerging threat. However, despite the considerable investment in population ecology research over the last cycle, we still have only a meagre understanding of the dynamics of rising outbreak SBW populations. This is due to the fact that funding for pest research typically lags a few years behind the development of an outbreak. Consequently, entomologists frequently have little opportunity to study the early phase of outbreak development.

There are two distinct views on how spruce budworm outbreaks start, and what controls their pattern of termination and recurrence (Figure 1; for a good description of these two contrasting theories, see Table SI1 in [5]). One theory dates back to the 1960's and was formalized in the 1970's as "Double-equilibrium theory" [6]. It was based on life-table studies done in Green River, NB (Canada) in the 1950s and 1960s [7]. The second, which is dubbed the "oscillatory theory" here, was formulated in the early 1980's after a thorough re-analysis of the same Green River life tables that formed the basis of the earlier double-equilibrium theory [8]. The Green River data, rather crucially, lacked any observations relating to the start of the outbreak, so both these theories suffer from a lack of empirical foundation.

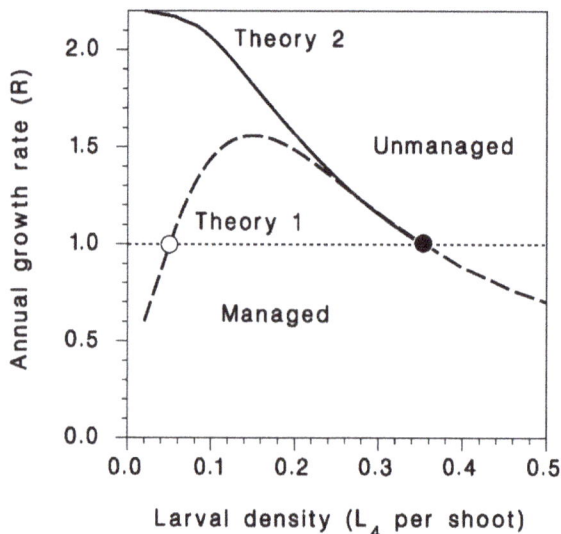

Figure 1. Population growth patterns hypothesized for managed and unmanaged incipient populations of spruce budworm, according to two contrasting theories of cycle initiation. Under Theory 1, management aims at maintaining populations below the threshold density (open circle) below which growth rates are <1 and populations are kept in check by natural factors.

The distinction between the two theories of periodic outbreaks matters for its practical as well as theoretical significance. If recurring spruce budworm outbreaks are the result of a triggered eruption (Theory 1) then there are significant opportunities for delaying, or even preventing, outbreaks—either through population management (i.e., direct control) or forest management (i.e., indirect control). If, instead, spruce budworm outbreaks are the result of a periodic predator-prey cycling process (Theory 2), then there is little opportunity to alter the timing or the trajectory that populations will follow; all that is possible is to reduce peak populations somewhat, to limit forest damage below some modest threshold. In the former case, the earlier the intervention, the more likely it is to maintain populations below the eruption threshold. In the latter case, the best one can hope for with early intervention is to modulate cycle amplitude through a continual series of population reductions that are suspended once the population cycle peaks [9,10]. If the first is true, then management can aim at preventing or retarding the development of an outbreak [11].

In sum, a great deal hinges on our understanding of the early phase of outbreak initiation, which is why so much effort has been invested recently in understanding outbreak initiation from incipience. The results of population dynamics studies done since the 1980s in Ontario and Quebec have brought important new data to bear on our understanding of the outbreak process, particularly the terminal stage of outbreak collapse. First, the "gradual" nature of changes in larval survival from natural enemies at the end of outbreaks is questionable. Evidence suggests that mortality increases in sufficiently old outbreaks only after an initial drop in the budworm population, caused by disease, starvation, emigration, low survival among dispersing larvae or theoretically pesticide applications. Natural enemies then take control and an endemic period starts. This was well documented during the collapse of the 1975–1995 spruce budworm outbreak in Ontario and Quebec (see Figure 7 in [12]). Second, over nearly three decades of observations in endemic populations starting in the mid 1980's in Quebec indicate that mortality from natural enemies remains very high for long periods after collapse. And there are no signs yet of any natural enemy rarefication that would allow the budworm to begin rising from endemic levels. Clearly, the outbreak process is not strictly sinusoidal, or populations in this part of Quebec would have risen a decade or more ago. Finally, a strong Allee effect [13,14] has recently been documented in spruce budworm mating success, whereby females in low-density populations have a very low probability of attracting a male and mating [15]. This density-dependent mating success, combined with very high mortality among larvae in endemic populations raises the possibility that budworm actually goes extinct locally or regionally between outbreaks—a result that is more consistent with the older eruptive theory of budworm population dynamics than with the newer oscillatory theory.

But how do these observations on population collapse and endemicity through the 1990s and 2000s relate to what happens during the rising phase of the outbreak cycle? We sought to characterize the nature of population growth in incipient populations of spruce budworm in the Lower St. Lawrence (LSL) region of Quebec in 2012–2014 by examining detailed patterns of population change in a set of monitoring plots where moderately low, but rising populations were divided into two groups: (a) Those to be treated with a suppression tactic, and (b) those to be left alone. Our expectations are expressed in Figure 1. We predicted that spruce budworm would not exhibit the monotonic growth functions consistent with Theory 2 (solid line), but would instead exhibit positive density dependence and non-monotonic growth more consistent with Theory 1 (dashed line). In other words, we predicted eruptive, rather than cyclic patterns of growth. We also predicted that our suppression tactics would lead to consistently low rates of recruitment across all density classes (below the dotted horizontal line), leading to the overall conclusion that there may be circumstances where early intervention could lead to the prolonged suppression of budworm populations.

The research reported here had two main objectives: (1) To observe the dynamics of early rising outbreak populations, over as wide a range of densities as possible, from endemic to full outbreak; (2) to conduct efficacy trials of three common suppression tools: (i) A bacterial insecticide, (ii) a molting hormone analog, and (iii) a pheromone-based mating disruptant. Here, we report measurements of population levels, defoliation, survival and recruitment recorded in treated and untreated locations in the LSL between 2012 and 2014, as a regional outbreak was rising. We were particularly interested in whether the three suppression treatments were equally efficacious across the population density spectrum. If the treatments varied in patterns of efficacy then the existence of eruptive patterns of early population growth would lead to a complex interaction that could serve as a specific guide for optimal suppression. The main objective of pheromone mating disruption as a population management tool is to reduce egg recruitment by preventing local females from mating. The disappointing results of those trials are presented elsewhere [16]. Suffice it to state that mating disruption had no measurable effect of SBW demographics, and those treatments will not be further discussed here.

2. Materials and Methods

2.1. Plots

Our observations were made between 2012 and 2015, in a series of plots located in a broad triangle on the south shore of the St-Lawrence River in Quebec, Canada, between Rimouski to the west, Matane to the east, and Causapscal to the south (Figure 2). This region was chosen because it harboured spruce budworm populations that were in the early stages of outbreak development starting in 2011. Three criteria were used to select those plots: (1) Spruce budworm population density, (2) stand composition (sufficient conifer content, tree size and age), and (3) accessibility. Plots were between 30 and 100 ha. In 2012, we used 6 plots, three untreated and three treated with *Bacillus thuringiensis* (Berliner 1915) var kurstaki (Btk), a naturally-occurring bacterial pathogen of insects. In 2013, a total of 12 plots were sampled, four untreated, four treated with tebufenozide, a synthetic insecticide that mimics the insect hormone ecdysone, and four treated with a synthetic pheromone as a mating disruptant ("Disrupt Bioflake"). In 2014 our observations were made in 13 plots, nine of which were untreated and four treated with Btk. The Société de Protection des Forêts contre les Insectes et Maladies (SOPFIM) sampled an additional ten sites, located further to the south west at the edge of the developing outbreaks. In 2015, we used 13 plots, this time with no treatments.

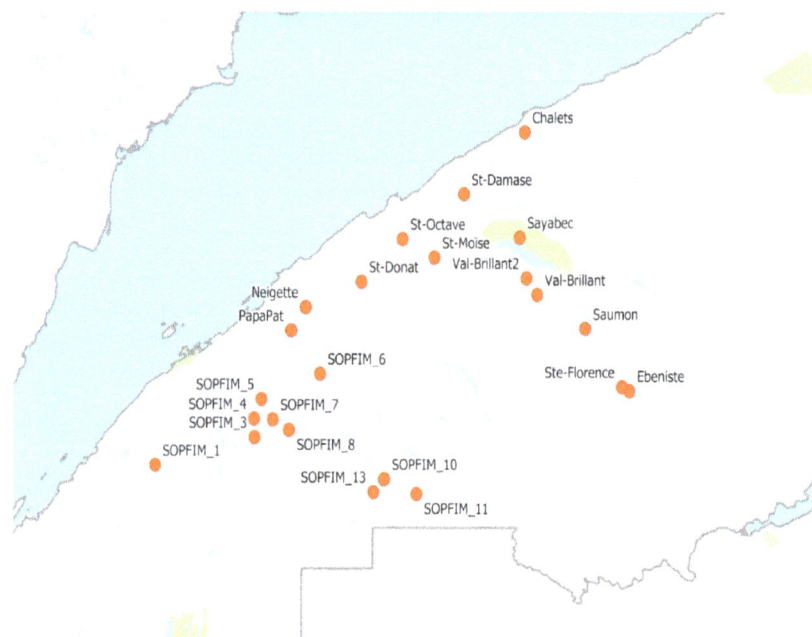

Figure 2. Map of the Lower St-Lawrence region of Quebec, with location of all plots in this study.

2.2. Treatments

In 2012, a standard aerial application of Btk (Foray 76B, Valent BioSciences) was applied on 7 June 2012 between 5:00 and 7:00 AM. The application was made under calm, stable air conditions, at the rate of 30 BIU/ha in 1.5 L/ha by a Cessna 188 fixed-wing aircraft equipped with four Micronair AU4000 atomizers (Micron Sprayers Limited, Bromyard, United Kingdom). The aircraft was guided by on-board GPS through the Accuair Aerial Management System, using deposit-optimization software based on meteorological information collected in flight by the AIMMS 20 system (McLeod et al. 2012). There was no assessment of insecticide deposit.

In 2013, tebufenozide (Mimic®, Valent BioSciences) was applied aerially in the second half of June, when larvae had reached the 5th instar (see [17] for details). Disrupt Bioflakes® (Hercon Environmental) were applied in four plots (see [16] for details).

In 2014, a standard aerial application of Btk (Foray 76B; 30 BIU/ha in 1.5 L/ha). Two plots were treated with an AT-502 (4 ASC-A10 atomizers at 9000 RPM) on 10 June (AM) at peak L_4 with on-board AG-NAV GUIA system. The other two were treated on 16 June, slightly later at average instar 4.5, with an AT-802F (10 Micronair AU4000 atomizers at 8000 RPM) was guided by on-board GPS through the Accuair Aerial Management System, using deposit-optimization software. Deposit was assessed using an ELISA-based assay. In each treated plot, 15 mid-crown branches were collected from balsam fir or white spruce trees hours after spray application. From each branch 5 new-growth shoots were collected for analysis using Agdia's Btk Pathoscreen.

In 2014, SOPFIM conducted a pheromone mating disruption trial in an additional set of 10 plots (5 treated, 5 untreated) located near the south-western edge of the developing SBW outbreak in the LSL region (SOPFIM sites in Figure 2; see [16] for details).

2.3. Foliage Sampling

The main objective of sampling was to determine the relationship between population density and (1) survival between the early larval stages (L_4) and the adult stage, and (2) reproduction, or recruitment from the adult to the egg stage. To achieve this, 30–100 mid-crown branch tips (45–60 cm) were taken in each site at the peak of the 4th instar, at the end of the pupal stage (around 50% moth emergence), and at the end of egg hatch, from dominant and co-dominant balsam fir and white spruce trees. To provide information on the natural enemies (parasitoids and pathogens) in late larval stage, an additional sample was taken at the 6th instar. Sample sizes were varied to reduce sampling error as much as possible given the resources available, using well-established mean-variance relationships [18]. All live spruce budworms found on the branches, and all current year shoots were counted. Density was expressed as spruce budworm per current shoot. The density of emerged adults was estimated from the number of pupal exuviae found on foliage in the pupal stage sample, plus live larvae and pupae in these samples that subsequently survived in rearing to emerge as adults (see Section 2.4). The sex ratio of emerged adults was very near 1:1 overall and is not presented here.

Eggs in each mass collected from the egg sample were counted. In 2014, pupal exuviae found on the foliage of the egg sample (taken several weeks after the end of adult emergence) were counted in all plots. This was done to test the hypothesis that both adults and eggs could be counted from the same sample to calculate egg recruitment (under the assumption that pupal cases remain on the foliage for at least that long). Based on the results, we used a single sample, taken after egg hatch, to estimate egg recruitment in 2015. In all years, defoliation was measured using the Fettes method [19] from branches in the egg-mass (end-of-season) branch samples in all plots (except in the SOPFIM plots of 2014, where defoliation was estimated from the L_6 sample, too early for use here).

2.4. Survival in Rearing

Random subsamples of up to 100 live budworm larvae and pupae recovered from the foliage of each plot in the L_4, L_6, and pupal samples were placed individually in rearing on artificial diet to monitor survival to the adult stage and identify the natural enemies causing mortality (parasitoids and pathogens). In 2012, all larvae from individual branches were placed in rearing until the maximum of 100 was reached. In all other years, the subsamples were selected at random from all individuals recovered from each plot. Very few pathogens caused mortality in rearing, and those are not presented. Parasitoids were identified to species as much as possible or to family by staff of the Ministère de la Faune, des Forêts et des Parcs du Québec. Parasitoids of these low-density SBW populations are very diverse, and sample sizes are too small to allow an analysis species by species, except for the most abundant. We do not attempt analysis by species here.

2.5. Data Analysis

Population growth rates are the combined result of survival and reproduction. We are interested in determining the pattern of density dependence of both, as well as the relationship that exists between survival and the frequency of parasitoids in larval and pupal samples. We also examine the relationship between population density (at the early larval stage) and the resulting defoliation. Throughout, log() is base 10 logarithm while ln() is natural logarithm.

2.5.1. Field Survival

The density of emerged pupae (exuviae) recovered from the egg samples in 2014 was compared by regression analysis to the density of adults estimated from the pupal sample, to verify they were equal. Their equality is important to confirm the validity of the emerging adult density estimate obtained from the egg sample in 2015.

Survival from early larvae to emerged adults was calculated as the ratio $S = A/N$, where A is the density of emerging adults estimated from the pupal sample and N is the density of early-instar larvae estimated from the L_4 sample. We used a non-linear regression model based on the Weibull function to relate S to N:

$$S = \frac{p_0 + p_{0,S}}{N}\left(1 - \exp\left\{-\left[\left(p_1 + p_{1|2013} + p_{1|2014} + p_{1|S} + p_{2|2012} + p_{2|2013} + p_{2|2014}\right)N\right]^{p_3}\right\}\right) \quad (1)$$

where p_0 is maximum adult density (equivalent to a "carrying capacity"), and $p_{0,S}$ represents a distinct value of this maximum for the SOPFIM (2014) dataset, where adult densities seemed considerably higher than in other datasets; p_1 is a common slope parameter for untreated populations that applies to all datasets (2012, 2013, 2014 and SOPFIM), and each additional subscripted p_1 parameter tests for differences in slope between datasets, with $p_{1|S}$ applying specifically to the SOPFIM data of 2014; p_2 parameters test for the effect of insecticides in each dataset (there was no insecticide treatment in the SOPFIM dataset); p_3 is a common shape parameter. For illustration and further computations, this model was then reduced by removing the least significant terms one at a time until all remaining parameters were significant at $\alpha = 0.05$. Multiplying Equation (1) by the density of larvae N, we obtain a sigmoid relationship between adult and larval density.

2.5.2. Defoliation

We tested the effects of larval population density and insecticide applications on defoliation using a General Linear Model with a logit transformation of defoliation:

$$\ln\left(\frac{D/100}{1 - D/100}\right) = a + b_y + c_T + d_{yT}\left(e + f_y + g_T + h_{yT}\right)N \quad (2)$$

where D is defoliation (%) and N is L_4 density (larvae per shoot). Parameters a and e are common intercept and slope, while b_y, c_T, d_{yT}, f_y, g_T and h_{yT} apply to specific years (index y), insecticide treatments (index T) and interactions (index yT). This model was reduced by removing least-significant terms one at a time until all remaining terms were significant at $\alpha = 0.05$.

We also examined the relationship between defoliation and adult density (A) to illustrate the probable role of competition for food in the determination of maximum adult density (parameter p_0 in Equation (1)).

2.5.3. Survival in Rearing

A detailed analysis of the contributions of the various parasitoid species to mortality in rearing is beyond the scope of this paper and will be reported separately. For the purposes of the present analysis, we limit ourselves to relating field survival and survival in rearing. Because survival in successive samples is affected by very different, successive natural enemies that change from stage to stage, their combined effect must be calculated. Survival estimates from insects reared from the L_4 (S_4), L_6 (S_6) and pupal samples (S_P) were combined (multiplied) to provide an estimate of overall rearing survival. Because emerged pupae in the pupal sample represent survivors, rearing survival at the pupal stage was corrected for adult emergence with $S'_p = P_E + S_P \times (1-P_E)$ where P_E is the proportion of emerged pupae in the sample:

$$S_R = S_4 \, S_6 \, S'_P \tag{3}$$

Field survival from L_4 to adult (S) was then related to rearing survival S_R, year y and insecticide treatment T (0 or 1) by a general linear model:

$$\log\left(\frac{S}{1-S}\right) = a_y + b_T + \left(c_y + d_T\right)\log\left(\frac{S_R}{1-S_R}\right) \tag{4}$$

where parameters a_y, b_T, c_y and d_T are year- and treatment-specific intercepts and slopes. This model was reduced by removing the least significant terms one at a time until all remaining terms were significant at $\alpha = 0.05$.

2.5.4. Recruitment to the Egg Stage

A theoretical framework for the analysis and interpretation of egg recruitment in spruce budworm population dynamics was recently developed [20]. The analysis of egg recruitment is done in accordance to that work. We use the following equation to relate egg density E to the density of surviving adults A by non-linear regression using:

$$\log(E) = \log(I_Y + F_Y A) \tag{5}$$

where I_Y represents the dataset-specific immigration rate (eggs/shoot), F_Y is the realized fecundity of local moths prior to emigration (eggs/adult) and index Y distinguishes datasets (year: 2012, 2013, 2014, 2015 or SOPFIM in 2014). From these regressions the apparent fecundity can be calculated as:

$$E/A = I_Y/E + F_Y \tag{6}$$

We noted an inverse relationship between apparent fecundity and defoliation D (in % of current-year foliage), which was analyzed with a general linear model:

$$\log(E/A) = p_{0,Y} + \frac{p_{1,Y}}{D} \tag{7}$$

where $p_{0,Y}$ and $p_{1,Y}$ represent the year (dataset) effects on the intercept and slope. We also tested the hypothesis that egg mass size, an indication of the reproductive status of females (the reproductively older, the smaller the egg mass; see [21]), with the model:

$$M = p_{0,Y} + p_1 \log(E/A) \tag{8}$$

where $p_{0,Y}$ represents year effects on the intercept, and p_1 is a common slope. For this analysis, one sample (3 egg masses) from LSL site 10 in 2012 was excluded as an outlier (average 34 eggs/mass).

2.5.5. Annual Population Growth Rate

Annual population growth rates were calculated from L_4 densities in successive years (N_t and N_{t+1}) as $R_t = N_{t+1}/N_t$ in each site and year where data were available. Estimates of N were available for all sites in 2012–2014. No L_4 samples were taken in 2015. However, estimates of 2015 L_4 density were obtained from 2014 egg density, using a regression equation developed from egg and L_4 data collected in 2012–2014 (pooling years). This regression was:

$$\log(N_{t+1}) = 0.598 \log(E_t) - 0.374, \; \left(R^2 = 0.64\right) \tag{9}$$

Observations from the three years were compared with expected annual growth rates, calculated as the product of Equations (1) and (6).

3. Results

3.1. Population Density and Survival

Several population density expressions have been used: Per branch, per m^2, per 10 ft^2 or 10 m^2 or per kg of foliage are common in the literature. Insects per kg of branch reduces branch to branch variance, is easy to measure, and eliminates differences in SBW density between balsam fir, white spruce and black spruce [18]. While counting growing shoots is more tedious, we have found it has the same virtues of variance reduction and homogenization between host plants. We believe that, because the number of shoots per unit foliage weight can vary in response to defoliation, and because it is shoots that are the main budworm food resource, whenever possible this is the ideal unit to use as denominator in density calculations. On balsam fir, the transformations between density expressions can be done with the following factors: 1052 shoots/m^2 and 722 shoots/kg, and 0.686 m^2/kg.

There was no significant difference in adult density estimated from the pupal samples or from pupal exuviae recovered in the eggs samples in 2014 (intercept -0.001 ± 0.003, not significantly different from 0: $t = -0.46$, $p = 0.65$; slope 1.065 ± 0.083, not significantly different from 1: $t = 0.78$, $p = 0.45$). Thus, estimating adult density using the exuviae recovered from the egg sample is an adequate cost-cutting strategy, provided that the egg sample is not delayed any longer than necessary once egg hatch is complete.

The most important result of this study is the density-dependent shape of the relationship between survival from early feeding larvae to adults S, and larval density N, well described with Equation (1) ($R^2 = 0.71$; Table 1; Figure 3), excluding the observation from plot LSL-24, an outlier where an unexpectedly high survival to adults was observed (see Figure 3e). Two terms of this model were not significant (after stepwise reduction). In particular, there was no significant difference in survival between untreated populations in 2012 and 2013 (parameter $p_{1|2013}$ not significantly different from 0, Table 1). Survival in the LSL as well as the SOPFIM untreated plots was significantly higher in 2014 than in 2012 (parameters $p_{1|2014}$ and $p_{1|SOPFIM}$ both significantly higher than 0, Table 1).

Table 1. Parameter estimates of Equation (1) relating survival from L$_4$ to adult. Full model. $R^2 = 0.87$.

Parameter	Estimate	SE	t	p
p_0	0.06206	0.01378	4.50269	0.000028
$p_{0,S}$	0.05220	0.03037	1.71885	0.046687
p_1	1.98880	1.18228	1.68218	0.050163
$p_{1|2013}$	1.63544	1.22133	1.33907	0.094054
$p_{1|2014}$	3.81979	1.36986	2.78846	0.004034
$p_{1|S}$	2.66913	1.38153	1.93201	0.030232
$p_{2|2012}$	1.26555	1.39848	0.90495	0.185456
$p_{2|2013}$	-2.35263	1.24347	1.89198	0.032874
$p_{2|2014}$	-4.79843	1.20321	3.98801	0.000138
p_3	1.94932	0.20292	9.60643	0.000000

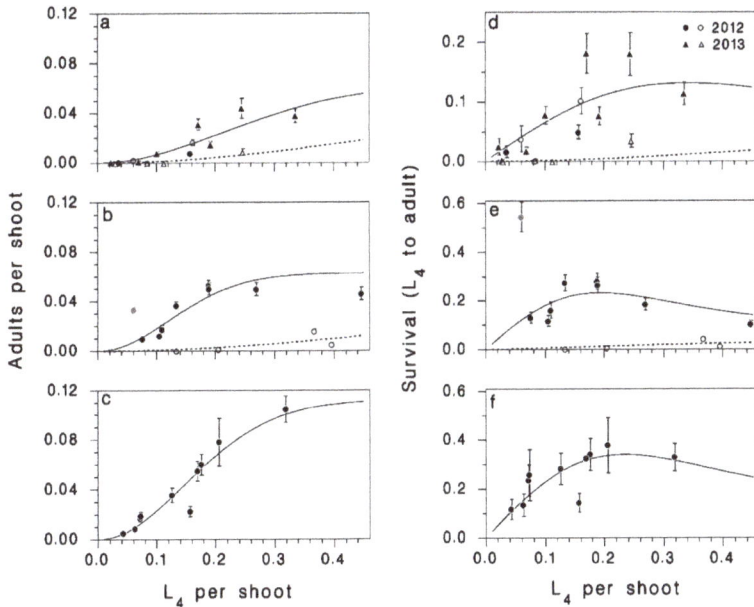

Figure 3. Left column: Relationship between larval and adult density in (**a**) 2012 (untreated and Bt-treated populations included) and 2013; (**b**) 2014; (**c**) 2014 among the SOPFIM plots (no insecticide treatments). Lines: Equation (1) (reduced, × N). Right column: Relationship between larval density and survival to the adult stage (**d**) in 2012 and among 2013; (**e**) 2014; (**f**) SOPFIM plots in 2014. Lines: Equation (1) (reduced; solid: Untreated; dotted: Insecticide). Vertical bars: SEM. Open symbol in (**b,e**): Outlier from plot LSL-24 in 2014 not used in model fitting.

The relationship between larval and adult density (given by the product $A = N \times S$ with S predicted by Equation (1)) is sigmoidal, with a rapidly increasing slope as larval density increases, and levelling off near 0.05 adults per shoot at larval densities exceeding 0.25–0.3 L_4/shoot, suggesting increased competition for food (Figure 3a–c). Survival S followed a non-monotonic pattern of rise and decline in each dataset (compare to Figure 1, Theory 1 curve): Low survival at low density, high survival at medium density, and decreasing survival at still higher density (Figure 3d–f).

These data confirm that even in rising outbreaks, low-density incipient populations face heavy mortality. Rising spruce budworm populations that are high compared to endemic populations, but still low compared to full outbreak are still facing growth challenges due to density-dependent mortality, consistent with double-equilibrium theory (Figure 1, Theory 1) but not with oscillatory theory (Figure 1, Theory 2).

The treatment parameter $p_{2|2012}$ for the effect of Btk in 2012 was not significantly different from zero (Table 1), and there was no detectable mortality attributable to Btk in 2012 (Figure 3a,d). Because deposit was not measured in 2012, there are no data to support the hypothesis that poor deposit was the cause for this poor product performance. Yet, because the application was done with a small Cessna aircraft with weak wake, it is possible that little product ended up on the target foliage. The tebufenozide applications in 2013 and the Btk applications in 2014 both generated very high mortality rates (parameters $p_{2|3013}$ and $p_{2|2014}$ both significantly lower than zero, Table 1; Figure 4). We estimated the mortality inflicted by the insecticide treatments π from the observed survival in treated plots S and the survival expected to occur in untreated populations at the same larval population density calculated with Equation (1), $S(N)$:

$$\pi = 1 - S/S(N) \tag{10}$$

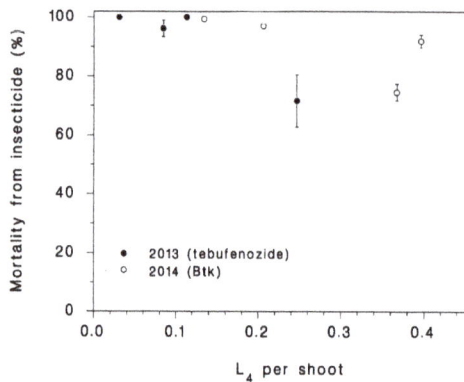

Figure 4. Mortality directly attributable to insecticides, as calculated with Equation (10). Vertical bars: SEM.

The resulting estimates range from 70% to 100% mortality from insecticides (Figure 4). It is interesting that mortality attributable to insecticides was higher in lower density populations both in 2013 and in 2014, despite the fact that two products with very different modes of action were used. This could very well be an indication of compensation (see [5]): Some of the mortality due to competition was relaxed by the insecticide applications.

3.2. Defoliation

The relationship between defoliation and larval population density (at the L_4) varied by year and insecticide treatment (Table 2; Figure 5). From other sources, one would expect a saturating shape with a monotonically-declining slope [22]. High natural mortality rates in lower density rising populations are probably responsible for the sigmoid shape of this relationship, depressing defoliation at lower densities. Among untreated populations, defoliation increased significantly from year to year at lower population densities, reflecting the increasing survival trend noted in the previous section. The Btk applications of 2012 had no effect on defoliation. The tebufenozide treatment in 2013 produced a significant but modest reduction in defoliation (dotted line in Figure 5a). In 2014, the Btk applications had a very pronounced effect on defoliation (bold dotted line in Figure 5a). We suspect that the limited impact of tebufenozide on defoliation in 2013 was caused by the late application (5th instar) compared to the 4th instar application of Btk in 2014.

Table 2. Analysis of Variance table for defoliation against L_4 larval density (N), year, and insecticide treatment (Equation (2)).

Source	DF	Adj SS	Adj MS	F-Value	p-Value
N	1	22.27	22.277	104.57	0.000
Year	2	0.228	0.114	0.53	0.595
Insecticide	1	2.061	2.061	9.68	0.006
Year × Insecticide	2	1.092	0.546	2.56	0.103
N × Year	2	2.997	1.499	7.04	0.005
N × Insecticide	1	0.325	0.325	1.52	0.232
N × Year × Insecticide	2	0.256	0.128	0.60	0.559
Error	19	4.047	0.213		
Total	30	73.30			

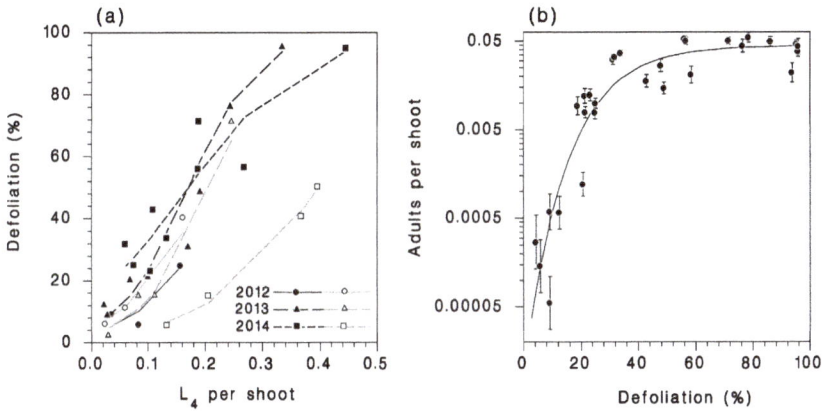

Figure 5. (a) Relationship between defoliation of current-year shoots and larval density at the L_4 (lines: Equation (2)), reduced (black lines and closed symbols: Untreated; grey lines and open symbols: Treated with insecticides). (b) Relationship between adult density and defoliation. Vertical bars: SEM.

There was a strong relationship between emerging adult density at the end of the season and the defoliation inflicted by the larval populations in untreated plots (Figure 5b). This relationship suggests that adult density is limited by competition for food, and that this competition reduces survival once defoliation exceeds 40%, corresponding to larval populations of 0.2 L_4 per shoot.

3.3. Survival in Rearing

There was strong and significant relationship between survival in rearing S_R (defined by Equation (3)) and field survival S from L_4 to adult, well described by Equation (4) ($R^2 = 0.83$; Table 3; Figure 6). This relationship varied slightly from year to year but was unaffected by insecticide applications. Thus, rearing survival reflects field survival, although field survival varied over a much wider range than did rearing survival. The analysis of the contributions of various parasitoid species to mortality in rearing is beyond the scope of this paper, and will be addressed in a subsequent contribution.

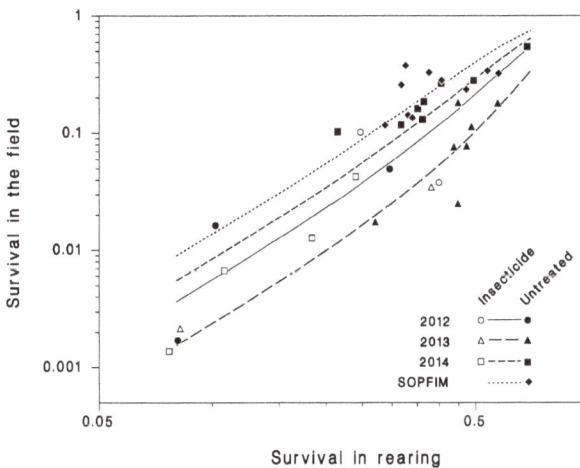

Figure 6. Relationship between field survival from L_4 to adult S and combined survival in rearing S_R (lines: Equation (4)), reduced.

Table 3. Analysis of Variance table for log odds of survival in the field against log odds of combined survival in rearing, year, and insecticide treatment (Equation (4)).

Source	DF	Adj SS	Adj MS	F-Value	p-Value
Year	3	1.6136	0.53787	6.70	0.002
Insecticide	1	0.0201	0.02006	0.25	0.621
Logit(S_R)	1	3.8566	3.85659	48.06	0.000
Logit(S_R) × Year	3	0.3090	0.10301	1.28	0.300
Logit(S_R) × Insecticide	1	0.0114	0.01138	0.14	0.709
Error	27	2.1665	0.08024		
Total	36	19.2402			

3.4. Recruitment to the Egg Stage

There was a clear relationship between egg and surviving adult density in each year (or dataset) except 2013. This relationship was very well described by Equation (5), fitted by non-linear regression ($R^2 = 0.92$) (Table 4; Figure 7). The intercept parameter *I*, representing the immigration rate (eggs/shoot), varied between 0.015 ± 0.006 in 2012 to 0.248 ± 0.072 eggs per shoot in 2014, and was always significantly different from zero implying that there was always detectable immigration in those sites. The realized fecundity parameter *F* ranged from 1.61 ± 1.61 in 2013 (not significantly different from 0), and 45.7 ± 7.7 in 2015, a level of variation that indicates a wide range of emigration rates from year to year.

Table 4. Parameter estimates of Equation (5) describing the relationship between egg and surviving adult density in the 5 datasets available in this study. $R^2 = 0.92$. The *t* statistics tests the hypothesis that the term is zero.

Parameter	Estimate	SE	t	p
I_{2012}	0.015	0.006	2.621	0.006
I_{2013}	0.157	0.028	5.507	0.000
I_{2014}	0.248	0.072	3.450	0.001
I_{2015}	0.112	0.050	2.221	0.016
I_S	0.073	0.035	2.072	0.023
F_{2012}	12.356	3.759	3.287	0.001
F_{2013}	1.617	1.608	1.006	0.161
F_{2014}	33.777	6.362	5.309	0.000
F_{2015}	45.713	7.7495	5.899	0.000
F_S	3.321	1.147	2.897	0.003

In 2012, when the developing outbreak in the LSL was just beginning and populations were generally low in the region, the immigration rate and the realized fecundity of resident moths (parameters *I* and *F* of Equation (5)) were low (Figure 7a). As a result, the relationship between apparent fecundity and resident adult density had a slope near 0, an indication of limited migration activity (Figure 7b) (see [20] for a thorough discussion of this topic). The same was true in 2014 among SOPFIM sites (Figure 7e,f). Although we suspect the resident adult density in the SOPFIM sites was overestimated in 2014 due to an early pupal sample (as evidenced by the high upper asymptote of adult density in Figure 3c), these parameter values suggest net emigration of gravid females with little immigration, but may also have been the result of low mating success [15]. The exact contribution of those factors to the low reproduction in 2012 and again in 2014 among SOPFIM sites is not clear, but we speculate that intensive emigration with little immigration is the most likely explanation. Those populations were located at the edge of the developing outbreak in the corresponding years. In both situations, spruce budworm populations were sparse in the study area and sources of regional immigrants would have been few or distant.

This is in sharp contrast with 2013, when there was no significant relationship between egg and resident adult density (Figure 7c). In that year, the realized fecundity of moths (F_{2013}) was so low that it was not significantly different from zero, which suggests that many moths emigrated before laying eggs. While this is biologically unlikely, it does indicate that 2013 was a particularly intensive migration year (high immigration, low realized fecundity of resident moths). As a result, the slope of the relationship between apparent fecundity and resident moth density in 2013 was very near the extreme of −1 (Figure 7d), which is a telltale sign of regionally random redistribution of eggs through extensive migration of ovipositing female moths (defined as panmixis and explained in detail in [20]). In 2014, the immigration rate was highest among the LSL sites ($I_{2013} = 0.25 \pm 0.07$ eggs per shoot), and realized fecundity was high ($F_{2013} = 33.8 \pm 9.4$ eggs per moth) (Figure 7e). In 2015, realized fecundity of resident moths was high ($F_{2015} = 45.7 \pm 7.7$ eggs per moth), and immigration rate was near average ($I_{2015} = 0.112 \pm 0.05$ eggs per shoot), suggesting a year with low migration activity (Figure 7g). Because of the relatively high adult populations in the LSL in both 2014 and 2015, the slopes of the apparent fecundity relationships with resident adult density was closer to 0 than in 2013, which indicates that those were not years of extensive moth migration (Figure 7f,h).

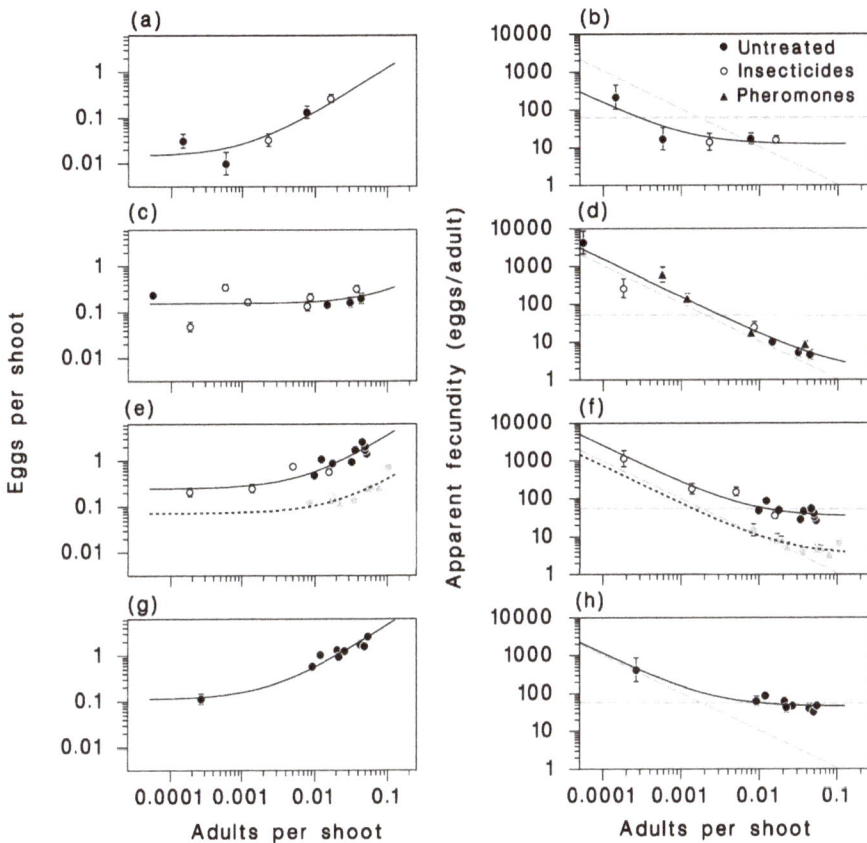

Figure 7. Relationship between resident adult density, egg density field (left column) and apparent fecundity (right column. Lines: Equations (5) (left column) or (6) (right column). (**a,b**): 2012. (**c,d**): 2013. (**e,f**): 2014, dotted lines and grey symbols: SOPFIM sites). (**g,h**) 2015. Grey dashed lines in the right panels represent the extremes of no migration (slope 0) and random distribution of eggs (slope −1).

The relationship between apparent fecundity and current-year defoliation was highly significant (Table 5) and was well described by Equation (7) (Figure 8a). Its slope and intercept varied between years (datasets), but generally apparent fecundity dropped to its minimum at defoliation in the range of 20%–30%. This pattern supports the hypothesis that spruce budworm females tend to emigrate from defoliated stands, and that they can sense a fairly low amount of defoliation.

Table 5. Analysis of Variance table for log odds of survival in the field against log odds of combined survival in rearing, year, and insecticide treatment (Equation (7)). $R^2 = 0.85$.

Source	DF	Adj SS	Adj MS	F-Value	p-Value
Year	3	3.531	1.177	16.08	0.000
1/D	1	0.656	0.656	8.96	0.006
Year × 1/D	3	4.562	1.521	20.77	0.000
Error	29	2.123	0.073		
Total	36	14.005			

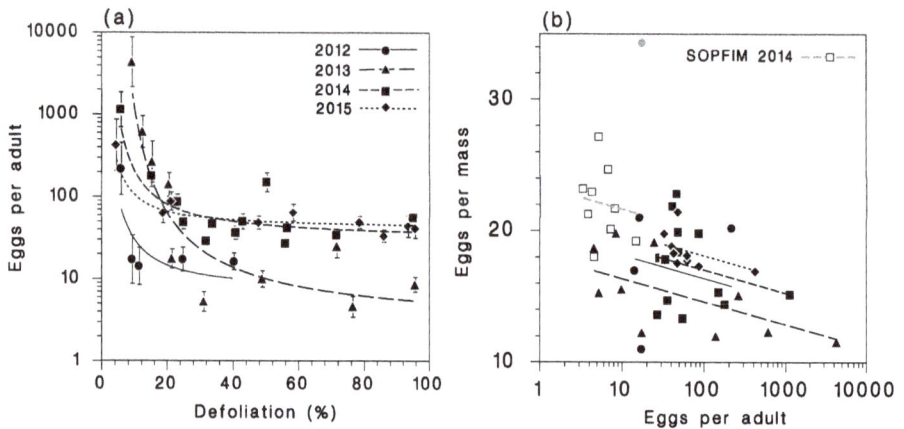

Figure 8. Relationships between (**a**) apparent fecundity and defoliation (Equation (7)) and (**b**) egg mass size and apparent fecundity (Equation (8)). Greyed symbol in (**b**): Outlier omitted from analysis.

Additional support for this hypothesis comes from the significant reduction of egg mass size in populations with clear net immigration, as represented by Equation (8) (Table 6, Figure 8b). While egg mass size variability was very high (due in large part to small sample sizes), the relationship explained 47% of this variation. The mechanism for this relationship seems simple: Immigrant moths are often mostly "spent" (have few eggs left in their oviducts), and there is a strong correlation between egg mass size and remaining fecundity in spruce budworm [21]. Thus, small egg masses indicate oviposition by immigrants, while large egg masses indicate oviposition by residents.

Table 6. Analysis of Variance table for log odds of survival in the field against log odds of combined survival in rearing, year, and insecticide treatment (Equation (8)). $R^2 = 0.47$.

Source	DF	Adj SS	Adj MS	F-Value	p-Value
Year	4	121.24	30.310	3.83	0.010
LogEM	1	41.80	41.804	5.29	0.027
Error	39	308.41	7.908		
Total	44	587.99			

3.5. Annual Population Growth Rate

Population growth rates in 2012 were near the replacement line ($R = 1$), and were not clearly affected by applications of Btk (Figure 9a). In 2013, population growth rates were very high in low density populations, and well below replacement in higher density sites. Although the tebufenozide applications of 2013 were highly efficacious, they did not result in a clear reduction of population growth rates (Figure 9b). Moth migration activity was extremely pronounced in 2013, blending populations over the entire study area. In 2014, among the 13 LSL sites, annual growth rates were very high in untreated populations, and at or below replacement in sites treated with Btk (Figure 9c). Among the 10 SOPFIM sites in 2014, located at the western periphery of the expanding outbreak of the LSL, growth rates were close to replacement, and had a very similar density dependence to that observed in 2012 in the LSL (Figure 9d). Clearly, insecticide applications can have a strong impact on population growth rates, but not in years when extensive moth dispersal occurs, such as in 2013 in the LSL region.

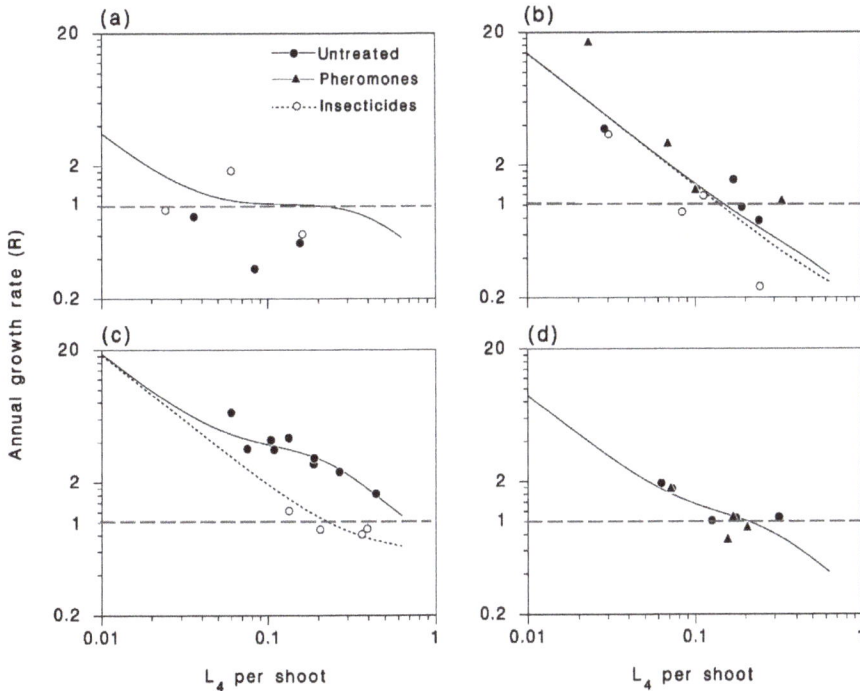

Figure 9. Relationships between larval population density (at the L_4) and annual population growth rate (**a**) 2012–2013; (**b**) 2013–2014; (**c**) 2014–2015 in the Lower St-Lawrence; (**d**) 2014–2015 in SOPFIM sites. Symbols: Observed. Lines: Expected from Equations (1) and (5). Horizontal dashed lines: Replacement level where $R = 1$.

4. Discussion

The non-monotonic density-dependent pattern of survival during the feeding larval and pupal stages of spruce budworm observed in this study (Figure 3) indicates that early outbreak populations are subject to a demographic Allee threshold (Figure 1, Theory 1), and are not increasing from endemic density because of high survival rates, as the oscillatory theory would predict (Figure 1, Theory 2). This suggests either a multiple equilibrium system (with an unobserved endemic equilibrium and a crowding-related upper equilibrium), or a system with a single outbreak equilibrium and an Allee threshold, where low-density populations actually go locally extinct unless supplemented by immigration, a possibility that was discussed by other workers [23]. In the short three-years of this study, we noted a general increase in survival in the third year (2014). We cannot determine whether this was part of a trend, or if it was simple annual variation in the impact of various mortality factors.

However, this increase was not accompanied by lower mortality in rearing (Figure 6) and so was probably not linked to a change in the impact of parasitoids. However, other natural enemies such as predators, could have been involved.

It therefore appears that endemic spruce budworm populations are kept in check by generalist parasitoids. Contrary to Theory 2 (Figure 1), our observations do not support the idea of a gradual relaxation of their impact over time, and thus populations do not rise after a gradual drop of mortality from natural enemies (JR, unpublished data; this study; [24]). But if budworm outbreaks do not occur as a result of a cyclical (regular) change in the impact of natural enemies, but rather in response to a sudden change in survival or apparent fecundity that shifts populations beyond the Allee threshold, how can a more-or-less regular outbreak cycle materialize? We do not have an answer to this question. Fluctuations in the impact of natural enemies may be involved under some circumstances, but not generally [24]. It is possible that the likelihood of such a sudden shift in survival increases over time as forest stands recovering from a previous outbreak age and host-tree masting becomes more prevalent over a landscape [25]; masting events occur synchronously over rather large areas. Survival of spruce budworm larvae is known to be favored when balsam fir produces large cone crops [26,27]. This bottom-up effect may be sufficient to generate initial epicenters on its own. It is also possible that improved survival over large areas also increases the density of migrating moths being concentrated by convergent airflow, and landing in sufficiently large numbers in particular areas of the landscape such as deep valleys [28]. This would lead to highly patterned spatial distributions of epicenters [4]. Once these initial outbreak epicenters appear in the landscape, they can become emitters of migrating moths that spread the outbreak by increasing the apparent fecundity of the surrounding low-density populations [20].

We observed, in each year of this study, but most strikingly in 2013, net immigration into lower-density populations, and net emigration away from higher-density populations (Figure 7). This pattern of inverse density-dependence of apparent fecundity is ubiquitous in spruce budworms [20], and its regional average and slope vary from year to year in response to factors that affect the migratory behavior of moths as well as to the variability of population density in the source region. We noted that apparent fecundity dropped sharply in populations where defoliation exceeds 20% (Figure 8a), a level at which the density of emerging adults starts to level off (Figure 5b), an indication of intraspecific competition for food. This evidence suggests a mechanism through which crowding triggers increased emigration of moths. It was recently discovered that moth emigration is density dependent in the closely-related western spruce budworm [20].

These results suggest that there is a threshold density below which spruce budworm populations are readily kept in check by natural mortality, especially parasitoids. To calculate an approximate value for such a threshold, we multiplied the survival rates provided by Equation (1) by the geometric mean of apparent fecundity among LSL budworm populations in 2012 and 2013 (27.3 and 49.6 eggs per adult, respectively). We did not use the 2014 survival because at that time the regional outbreak in our study area had become generalized. The resulting annual growth rates (Figure 10) show a range of densities between 2.3 and 4.3 L_4 per branch tip (assuming 80 shoots per branch) as threshold between decreasing and increasing populations. Of course, there is additional mortality from egg to L_4 that is not taken into account in this calculation. Therefore, these are conservative estimates, and actual thresholds are probably somewhat higher.

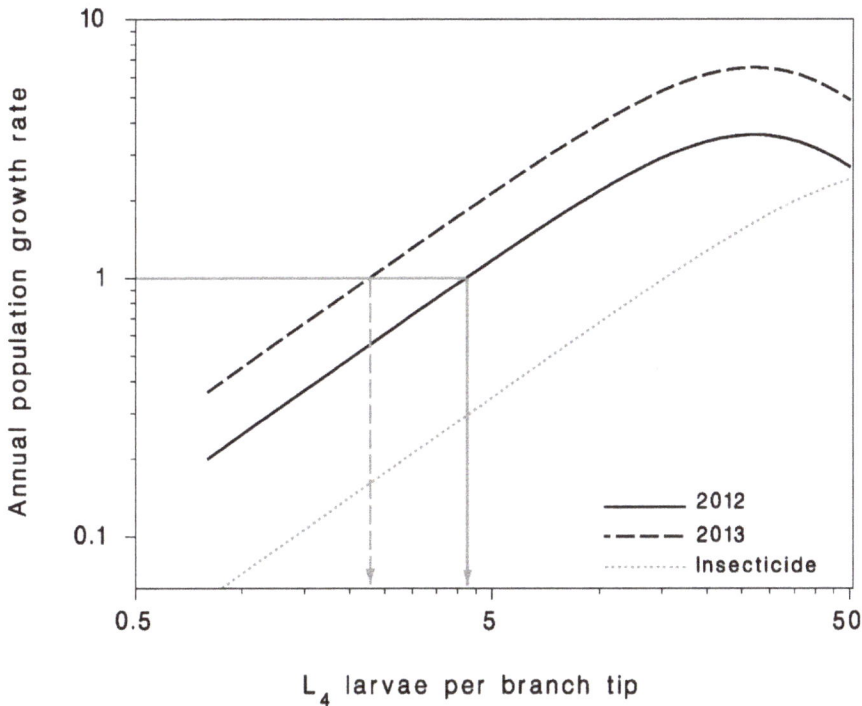

Figure 10. Relationship between expected population growth rate and density of feeding larvae (L$_4$) early in the development of a new spruce budworm outbreak (solid line: 2012, dashed line: 2013). Grey dotted line represents growth rate of populations treated with an insecticide (here, Btk in 2013). Grey downward arrows point to threshold densities for population increase, where $R = 1$ (4.3 and 2.3 L$_4$ larvae per branch tip in 2012 and 2013, respectively).

5. Conclusions

All of this evidence lends support to the conclusion that, for a budworm population to increase to outbreak density, it must be "propped-up" somehow above a certain threshold density (the Allee threshold). While other triggers or mechanisms may be involved at times, immigration from a nearby outbreak population seems to be the main way this can happen. Once a population has increased beyond the Allee threshold, it can continue growing and then become a hot spot itself, the source of further spread by moth migration.

We also demonstrated the high efficacy of Btk and tebufenozide as management tactics applicable to incipient populations of spruce budworm, while mating disruption with pheromones has been shown to be ineffective [16]. Taken together, these results indicate that insecticide applications, when efficacious, can reduce populations that happen to increase above the unstable eruption threshold. This is the essence of what is termed an "early intervention" strategy, where treatment in one year may have benefits lasting well beyond the treatment year [5].

These findings have significance with respect to the development of an early intervention strategy. First, they imply that populations can be brought down to a density where mortality from natural enemies and mating failure can keep the reduced population under check, barring massive immigration. This dependence on immigration implies that if all potential sources of significant numbers of moths are reduced on a regional scale, an outbreak can be stopped (prevented) by drying-up the supply of migrating moths. It also implies that moth migration poses a threat to an early intervention strategy. In years of extensive migration activity, the area being managed may be overwhelmed by immigrant, egg-laying moths, and the area to be treated may become prohibitively large. An attempt at implementing such a management strategy has been undertaken in New Brunswick, Canada, since 2015, and shows promising results [29].

Author Contributions: Conceptualization, J.R.; methodology, J.R., A.B., A.D.; formal analysis, J.R.; investigation, A.B., A.D., J.R.; resources, J.R., A.D.; data curation, A.B.; writing—original draft preparation, J.R., B.J.C.; writing—review & editing, B.J.C., A.B., A.D. and P.T.; project administration, A.B., A.D., P.T.; funding acquisition, J.R., A.D., P.T.

Funding: This research was funded by the natural resources and forest departments of Newfoundland, Nova Scotia, New Brunswick, Quebec, Ontario, Manitoba, Saskatchewan, Alberta and British Columbia, as well as the U.S.D.A. Forest Service through SERG-International. Considerable funds and in-kind contributions were also provided by the Atlantic Canada Opportunities Agency, Forest Protection Limited, and the Société de Protection des Forêts contre les Insectes et Maladies du Québec (SOPFIM).

Acknowledgments: Thanks to the Ministère des Forêts, de la Faune et des Parcs du Québec whose personnel contributed by rearing and diagnostic of mortality in rearing. Special thanks to the owners of the 13 woodlots in which much of this work was carried out. Thanks also to Kees van Frankenhuyzen, Rob Johns, Deepa Pureswaran, Véronique Martel and Michel Cusson for their help with early planning and fieldwork.

Conflicts of Interest: The authors declare no conflict of interest.

References

1. MacLean, D.A. Vulnerability of fir spruce stands during uncontrolled spruce budworm outbreaks: A review and discussion. *For. Chron.* **1980**, *56*, 213–221. [CrossRef]
2. Armstrong, J.A.; Cook, C.A. *Aerial Spray Applications on Canadian Forests: 1945 to 1990*; Forestry Canada: Ottawa, ON, Canada, 1993.
3. Chang, W.Y.; Lantz, V.A.; Hennigar, C.R.; MacLean, D.A. Economic impacts of spruce budworm (*Choristoneura fumiferana* Clem.) outbreaks and control in New Brunswick, Canada. *Can. J. For. Res.* **2012**, *42*, 490–505. [CrossRef]
4. Bouchard, M.; Auger, I. Influence of environmental factors and spatio-temporal covariates during the initial development of a spruce budworm outbreak. *Lands. Ecol.* **2014**, *29*, 111–126. [CrossRef]
5. Johns, R.; Bowden, J.; Carleton, R.D.; Cooke, B.J.; Edwards, S.; Emilson, E.; James, P.M.A.; Kneeshaw, D.; MacLean, D.A.; Martel, V.; et al. A conceptual framework for the spruce budworm early intervention strategy: Can outbreaks be stopped? *Forests* **2019**, in review.
6. Ludwig, D.; Jones, D.D.; Holling, C.S. Qualitative analysis of insect outbreak systems: The spruce budworm and forest. *J. Anim. Ecol.* **1978**, *47*, 315–332. [CrossRef]
7. Morris, R.F. The dynamics of epidemic spruce budworm population. *Mem. Entomol. Soc. Can.* **1963**, *31*, 30–32. [CrossRef]
8. Royama, T. Population dynamics of the spruce budworm *Choristoneura Fumiferana*. *Ecol. Monogr.* **1984**, *54*, 429–462. [CrossRef]
9. Berryman, A.A. The theory and classification of outbreaks. In *Insect Outbreaks*; Barbosa, P., Schultz, J.C., Eds.; Academic Press: New York, NY, USA, 1987; pp. 3–30. ISBN 13: 9780120781485.
10. Régnière, J.; Delisle, J.; Bauce, E.; Therrien, P.; Kettela, E.; Cadogan, L.; Retnakaran, A.; van Frankenhuyzen, K. Understanding of spruce budworm population dynamics: Development of early intervention strategies. In *Boreal Odyssey, Proceedings to the North American Forest Insect Work Conference, Edmonton, AB, Canada, 14–18 May 2001*; Volney, W.J.A., Spence, J.R., Lefebvre, E.M., Eds.; Canadian Forest Service: Victoria, BC, Canada, 2001; pp. 57–68.
11. Clark, W.C.; Jones, D.D.; Holling, C.S. Lessons for ecological policy design: A case study of ecosystem management. *Ecol. Model.* **1979**, *7*, 1–53. [CrossRef]
12. Régnière, J.; Nealis, V. Ecological mechanisms of population change during outbreaks of the spruce budworm. *Ecol. Entomol.* **2007**, *32*, 461–477. [CrossRef]
13. Allee, W.C.; Bowen, E. Studies in animal aggregations: Mass protection against colloidal silver among goldfishes. *J. Exp. Zool.* **1932**, *61*, 185–207. [CrossRef]
14. Stephens, P.A.; Sutherland, W.J.; Freckleton, R.P. What is the Allee effect? *Oikos* **1999**, *87*, 185–190. [CrossRef]
15. Régnière, J.; Delisle, J.; Pureswaran, D.; Trudel, R. Mate-finding Allee effect in spruce budworm population dynamics. *Entomol. Exp. Appl.* **2012**, *146*, 112–122. [CrossRef]

16. Régnière, J.; Delisle, J.; Dupont, A.; Trudel, R. Results of three mating disruption trials for population management of the spruce budworm (Lepidoptera: Tortricidae). *Forest* **2019**, in review.

17. Van Frankenhuyzen, K.; Régnière, J. Multiple effect of tebufenozide on the survival and performance of the spruce budworm (Lepidoptera: Tortricidae). *Can. Entomol.* **2016**, *149*, 227–240. [CrossRef]

18. Régnière, J.; Sanders, C.J. Optimal sample size for the estimation of spruce budworm (Lepidoptera: Tortricidae) populations on balsam fir and white spruce. *Can. Entomol.* **1983**, *115*, 1621–1626. [CrossRef]

19. Sanders, C.J. *A Summary of Current Techniques Used for Sampling Spruce Budworm Populations and Estimating Defoliation in Eastern Canada*; Canadian Forest Service: Victoria, BC, Canada, 1980.

20. Régnière, J.; Nealis, V.G. Moth dispersal, egg recruitment and spruce budworms: Measurement and interpretation. *Forests* **2019**, *10*, 706. [CrossRef]

21. Régnière, J. An oviposition model for the spruce budworm, *Choristoneura* fumiferana (Lepidoptera: Tortricidae). *Can. Entomol.* **1983**, *115*, 1371–1382. [CrossRef]

22. Régnière, J.; You, M. A simulation model of spruce budworm (Lepidoptera: Tortricidae) feeding on balsam fir and white spruce. *Ecol. Model.* **1991**, *54*, 277–297. [CrossRef]

23. Stedinger, J.R. A spruce budworm-forest model and its implications for suppression programs. *For. Sci.* **1984**, *30*, 597–615. [CrossRef]

24. Bouchard, M.; Martel, V.; Régnière, J.; Therrien, P.; Correia, D.L.P. Do natural enemies explain fluctuations in low-density spruce budworm populations? *Ecology* **2018**, *99*, 2047–2057. [CrossRef]

25. Bouchard, M.; Régnière, J.; Therrien, P. Bottom-up factors contribute to large-scale synchrony in spruce budworm populations. *Can. J. For. Res.* **2017**, *48*, 277–284. [CrossRef]

26. Blais, J.R. The relationship of the spruce budworm (*Choristoneura fumiferana* Clem.) to the flowering condition of balsam fir (*Abies balsamea* (L.) Mill.). *Can. J. Zool.* **1952**, *30*, 1–19. [CrossRef]

27. Bauce, E.; Carisey, N. Larval feeding behaviour affects the impact of staminate flower production on the suitability of balsam fir trees for spruce budworm. *Oecologia* **1996**, *105*, 126–131. [CrossRef] [PubMed]

28. Alerstam, T.; Chapman, J.W.; Bäckman, J.; Smith, A.D.; Karlsson, H.; Nilsson, C.; Reynolds, D.R.; Klaassen, H.G.; Hill, J.K. Convergent patterns of long-distance nocturnal migration in noctuid moths and passerine birds. *Proc. R. Soc. B Biol. Sci.* **2011**, *278*, 3074–3080. [CrossRef] [PubMed]

29. MacLean, D.; Amirault, P.; Amos-Binks, L.; Cerleton, D.; Hennigar, C.; Johns, R.; Régnière, J. Positive results of an early intervention strategy to suppress a spruce budworm outbreak after five years of trials. *Forests* **2019**, *10*, 448. [CrossRef]

Article

Density Dependence of Egg Recruitment and Moth Dispersal in Spruce Budworms

Jacques Régnière [1,*] and Vincent G. Nealis [2]

[1]   Natural Resources Canada, Canadian Forest Service, 1055 PEPS Street, Quebec, QC G1V 4C7, Canada
[2]   Natural Resources Canada, Canadian Forest Service, 506 Burnside Road West, Victoria, BC V8Z 1M5, Canada
*   Correspondence: Jacques.Regniere@Canada.ca; Tel.: +1-418-648-5257

Received: 30 July 2019; Accepted: 15 August 2019; Published: 20 August 2019

Abstract: Egg recruitment quantifies the relative importance of realized fecundity and migration rates in the population dynamics of highly mobile insects. We develop here a formal context upon which to base the measurement and interpretation of egg recruitment in population dynamics of eastern and western spruce budworms, two geographically separated species that share a very similar ecology. Under most circumstances, per capita egg recruitment rates in these budworms are higher in low-density populations and lower in high-density populations, relative to the regional mean: Low-density populations are nearly always migration sinks for gravid moths, and dense populations nearly always sources. The slope of this relationship, measured on a log scale, is negatively correlated with migration rate, and ranges between 0 and −1. The steeper the slope, the more marked net migration. Using our western spruce budworm observations, we found strong evidence of density-dependent emigration in budworms, so migration is not simply a random perturbation in the lagged, density-dependent stochastic process leading to budworm outbreaks. It is itself statistically and biologically density-dependent. Therefore, moth migration is a synchronizing factor and a spread mechanism that is essential to understanding the development and expansion of spruce budworm outbreaks at regional scales in the boreal forests of North America.

Keywords: spruce budworm; forest protection; early intervention strategy; egg recruitment; apparent fecundity; growth rate

1. Introduction

Recruitment of progeny is a fundamental ecological process determining inter-generation rates of change in the density of insect populations. When expressed per capita or as an apparent fecundity, recruitment is a survival rate. It can be combined with other stage-specific survival rates to connect the density of adults in one generation to that of eggs of the next generation across a time series of population measures. In many species, the adult is also a dispersal stage. If dispersing adults carry some or all of their progeny to new locations, then egg recruitment becomes a more complex phenomenon, particularly for spatial dynamics. Displacement of eggs resulting from dispersal of adult females can range from local redistribution of individuals within a habitat and gradual diffusion at the periphery, to long-range transport of winged, gravid females on air currents over hundreds of kms [1]. Even when it can be observed, the significance of dispersal may remain obscure as the source and destination of the dispersing adults, their survival en route, and reproductive status, are uncertain.

Forest insects display some of the most spectacular examples of eruptive population behavior, with periodic changes in population densities over several orders of magnitude. These outbreaks are often extensive as populations increase and decrease in apparent synchrony over vast areas [2–6]. Massive migrations of adult insects during outbreaks capture popular attention, but their importance in population dynamics is less obvious. In some cases, there is evidence migrations have initiated outbreaks de novo and even resulted in significant range expansions (e.g., [7]). However, more usually it is difficult to assess the net effect of dispersal on population patterns such as synchrony, as the practical problem of measuring dispersal is compounded by the close association between flight behavior and migration with meteorological and forest conditions at multiple spatio-temporal scales.

Such is the case with the spruce budworms, *Choristoneura fumiferana* (Clem.) and *C. occidentalis* Free., two of the most studied eruptive defoliators in the conifer forests of North America. In the eastern spruce budworm, synchrony of outbreaks over millions of hectares is documented in dendrochronologies [6,8], defoliation surveys [9,10] and direct measurement of populations [11,12]. Innovative field research has established the normative, detailed flight behavior of budworm moths, including gravid females, and their displacement under particular meteorological conditions, sometimes over great distances [1,13]. This research has enabled modeling flight behavior interfaced with high-resolution atmospheric models to simulate historic observations of spruce budworm dispersal over a large forested landscape [14]. Despite these insights, the extent to which outbreaks are initiated, as opposed to accelerated or supplemented, by immigration of egg-bearing moths, remains debatable [15,16]. Further, we are uncertain whether the periodic nature and large-scale spatial synchrony of spruce budworm outbreaks are dominated by common, intrinsic responses to widespread ecological conditions through the Moran effect [4,16] or if dispersal plays a distinct role in these large-scale features of their population ecology [17]. This has significant implications for management of outbreaks. If dispersal and egg recruitment is an entirely stochastic process, then suppression of spruce budworm outbreaks can be implemented at the local, forest stand level with minimal regard for area-wide outbreak conditions. If, however, recruitment is dependent on area-wide densities and directional in the sense that these populations become synchronized, then increases in populations anywhere can trigger increases everywhere.

The per capita egg recruitment rate, or simply recruitment rate, was first defined as a survival rate for spruce budworm [18] and analyzed as the "female survival ratio" [19]. Royama [16] modified this to the "E/M ratio", the number of eggs recruited to the next generation per the number of moths in the previous generation. We prefer the more general term "apparent fecundity" for this measured per capita rate of recruitment. Apparent fecundity, in the absence of migration, is expected to be somewhat less than half the maximum lifetime fecundity of females, or ≈ 100 eggs/moth, assuming a 1:1 sex ratio. Significant positive deviations from this expectation indicate net immigration of gravid moths, actually their eggs, and negative deviations indicate emigration. Royama 16] argued that apparent fecundity was density-independent and stochastic. His re-analysis of the Green River data concluded migration of gravid moths was a source of external noise that perturbs the details of the intrinsic temporal pattern during an outbreak cycle, but neither triggers nor causes spatial synchrony of outbreaks. In contrast, a theoretical argument was proposed that migration could be a strong source of population synchrony in spruce budworm outbreaks [17].

In this paper, we re-examine the measurement and interpretation of apparent fecundity in the study of spruce budworm population dynamics. We first describe the reproductive and dispersal ecology of the spruce budworm and how population estimates are made. We then provide a generalized theoretical framework to deduce the relationship between the recruitment rate and population density and identify how environmental factors obfuscate interpretation. We challenge our theoretical constructs with 14 previously unpublished, independent datasets obtained from literature or collected ourselves during outbreaks of spruce budworm. Finally, we analyze patterns of recruitment in different times and places and test the hypothesis that egg recruitment at the scale of outbreaks is a density-dependent process.

2. Materials and Methods

2.1. The Budworm System

Conifer-feeding budworms in the genus *Choristoneura* share common life-history adaptations and ecological relationships resulting in a periodic eruption of population densities over great expanses of their preferred host trees [20]. Adult moths are strong fliers and mated, gravid females can migrate actively several km and passively even further when transported on convective storm systems [1,14,21]. Dispersal may occur at any time in the reproductive cycle following mating and given suitable atmospheric conditions. However, fully-fecund, well-nourished adult females appear less capable of sustained flight, presumably as a result of wing loading, so some portion of their fecundity is realized at the natal location before migration [22,23]. On the other hand, food limitation resulting from severe defoliation reduces the size and fecundity of budworm moths with a commensurate positive effect of their propensity to disperse [24,25].

Dispersal is also a source of mortality, although this has never been measured in budworms. Redistribution within the normal flight range of individuals presents the least risk as spruce budworms are associated closely with very large areas of contiguous host plants and active flight allows expression of adaptive orientation behavior determining when and where to take off, land, and oviposit. Long-distance dispersal, however, is more hazardous as moths are passively transported on weather systems which ultimately may deposit them helplessly in hostile environments where they are lost from the system unless able to re-emigrate and take their chances again.

The data included in this analysis came from studies that measured egg and adult densities, allowing estimation of the per capita rate of egg recruitment or apparent fecundity. The conventional sample unit for spruce budworm life stages is a 45-cm branch tip removed from the mid-crown of host trees [18]. Various expressions of density (per unit area, weight, shoot) can be used and because apparent fecundity is a ratio of densities in successive life stages, the density expression used is of no particular importance, provided both measurements are expressed in the same units.

Egg density is measured routinely by sampling branches after oviposition is complete but before significant loss of vacated egg masses from the foliage. Ideally, all eggs in the egg-mass sample should be counted because egg-mass size varies. More problematic is obtaining sufficient sample sizes at very low densities. If the sampled foliage does not harbor at least one egg-mass, apparent fecundity becomes null. Several sampling schemes were proposed to help determine the minimum sample sizes required [26,27]. This is a significant practical limitation because it is low-density populations where detection of immigration is most important.

Adult density must express the density of moths produced locally because it is the basis from which to assess deviations from the expected fecundity and net migration. The estimate of their density is based on counting pupal exuviae from which local adults have successfully emerged on the branch sample unit. Some have sampled throughout the adult emergence period to construct a daily, cumulative total of pupal exuviae from which they estimated successful eclosion of resident moths to counter the possibility that pupal cases disappeared quickly following adult emergence [11]. However, we did not observe a significant loss of pupal cases between peak adult emergence and the end of the egg hatch period two weeks later [28]. In our own work, we calculated adult density from late-season samples of larvae and pupae taken after 50% adult emergence and reduced those density estimates by recording mortality of remaining larvae and pupae in rearing [28,29].

2.2. Models

We assume seasonal moth migration is a random process such that some proportion of the moths emerging from populations distributed over a large, heterogeneous landscape fly into the surface layer of air above the landscape, mix thoroughly, and land at random anywhere in the landscape. Thus, migrating moths can move short or long distances from source populations and the number of eggs carried by migrating moths and laid in a given location is independent of the density of the receiving population. Moreover, the number of migrants in the airspace in any one year is a regional constant resulting from the average density of populations in the region and weather conditions affecting flight behavior of moths in that year. The immigration rate, *I*, in eggs per unit foliage, can be represented by the function:

$$I = \alpha\kappa F\beta\overline{M} \tag{1}$$

where $\overline{M}$ is the average density of adults over the entire region, F is potential or expected lifetime fecundity of a female moth, and κ is the proportion of females (0.5 in spruce budworm) multiplied by survival from factors other than losses resulting directly from migration (e.g., predation) so that $0 \leq \kappa \leq 1$. Realized fecundity is the product κF. The proportion of fecundity carried away by migrants is $0 \leq \alpha \leq 1$, whether it results from the proportion of moths that emigrate, the proportion of their potential fecundity that moths carry with them when they emigrate, or some combination of the two. The proportion of migrating moths that survive during migration and lay eggs elsewhere is $0 \leq \beta \leq 1$.

The number of eggs E laid in a given location or population is the sum of eggs deposited by immigrants I (Equation (1)) and those laid by resident moths either prior to take-off or by resident moths that do not emigrate at all:

$$E = I + (1 - \alpha)\kappa FM \tag{2}$$

where M is the density of locally-emerged moths.

The relationship between E and M is linear, with a non-zero intercept representing immigration to an empty population (when $M \rightarrow 0$) and a slope equal to the average number of eggs laid by the resident moths prior to take-off: $(1 - \alpha)\kappa F$. In general, this slope is less than realized fecundity κF because $0 \leq \alpha \leq 1$. Apparent fecundity is given by:

$$E/M = \frac{I}{M} + (1 - \alpha)\kappa F \tag{3}$$

Over a region where, by definition $\overline{M}$ is constant, the relationship between E/M and M decreases monotonically from an infinite intercept,

$$\lim_{M \to 0} \left[\frac{I}{M} \right] \to \infty \tag{4}$$

to an asymptotic minimum,

$$\lim_{M \to \infty} \left[\frac{I}{M} + (1 - \alpha)\kappa F \right] = (1 - \alpha)\kappa F. \tag{5}$$

The solid black lines in Figure 1 illustrate Equations (2) and (3). In this example, we set $\kappa F = 0.5 \times$ 218 eggs/adult (assuming a 50% female sex ratio, no local mortality and a fecundity of 218 eggs/female), $\alpha = 0.5$ (50% of moths take off, or moths lay 50% of their eggs before all leaving), and $\beta = 0.5$ (50% of migrating moths are lost during migration). The calculation of average moth density $\overline{M}$ assumes an even distribution across the density range in Figure 1.

In the extreme case of no migration so that all eggs are laid by surviving resident moths (i.e., $\alpha = 0$ and therefore $I = 0$), the relationship between E and M (Equation (2)) becomes $E = \kappa FM$, with intercept 0 and slope κF (dashed line in Figure 1a). Apparent fecundity (Equation (3)) is then a constant with respect to density and equal to realized fecundity, κF. This results in a slope of 0 in the relationship between Log(E/M) and Log(M) (dashed line in Figure 1b). We define the other extreme as "panmixis", when all moths emigrate prior to laying any eggs ($\alpha = 1$). In this case, all eggs observed everywhere are laid by immigrants, and egg density becomes independent of local adult density (i.e., $E = I$, Equation (2)). Under these circumstances, apparent fecundity is inversely proportional to local adult density (($E = I/M$), Equation (3), with 0 as its lower limit when $M \rightarrow \infty$. This extreme case gives a slope of -1 to the relationship between Log(E/M) and Log(M), and is illustrated by the dotted lines in Figure 1a,b.

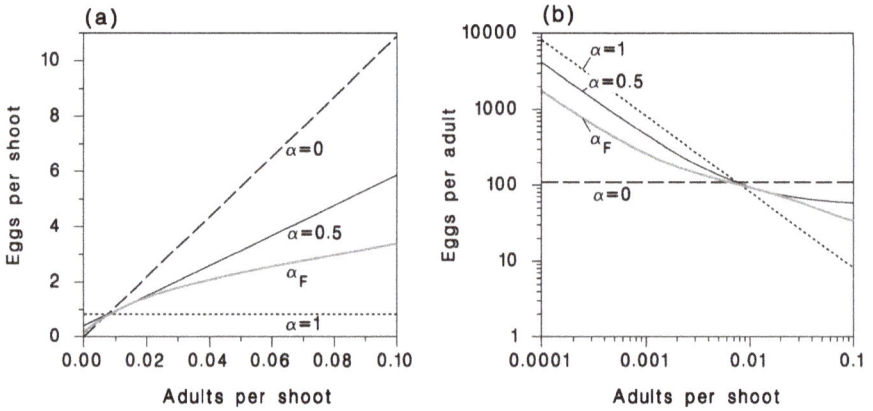

Figure 1. (**a**) Relationship between egg and adult density, illustrating Equations (2) (black lines) and (10) (grey line). (**b**) Relationship between apparent fecundity (eggs/adult) and adult density, illustrating Equations (3) (black lines) and (11) (grey line). Solid line: General case (where $\alpha = 0.5$). Dashed line: No migration ($\alpha = 0$). Dotted line: Panmixis where all moth emigrate ($\alpha = 1$). Grey line: Density-dependent emigration (with $\alpha = 0.5$). In all cases $\beta = 0.5$.

Density-Dependent Emigration

Potential fecundity, F, in spruce budworms is reduced by food limitation associated with defoliation, D [30]. From [31], we have:

$$F_D = 218 - 1.17D \tag{6}$$

We obtained data from [28] to fit an empirical relationship between defoliation D (%) and emerging adult density M (Figure 2):

$$D = \frac{100}{1 + \exp\left(2.622 - 16.659\ \sqrt{M}\right)}. \tag{7}$$

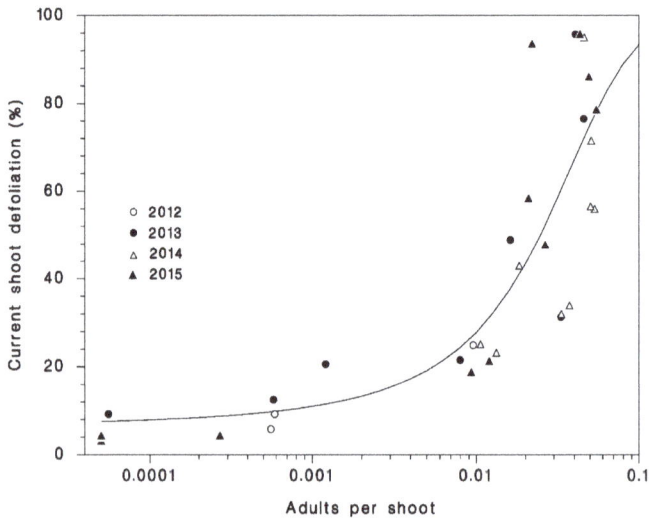

Figure 2. Relationship between adult density and defoliation of current year shoots on balsam fir foliage collected in the Lower St-Lawrence between 2012 and 2015. The line is Equation (7).

Many authors contend budworm moths that are small as a result of food limitation have a greater propensity to fly and may even emigrate without first laying eggs [13,24,25,32,33]. On the other hand, large, fully-fed females may not be able to fly upon emergence and so do not migrate until they have laid at least a portion of their eggs at the natal site [23]. This suggests α, the proportion of moths emigrating, could be a function of F_D:

$$\alpha_F = \alpha\left(\frac{218 - F_D}{218 - 101}\right) \tag{8}$$

that increases linearly from 0 when fecundity if 218 eggs/female (0% defoliation), to α when fecundity drops to 101 eggs/female as occurs with 100% defoliation (Equation (8)).

Because the level of defoliation, and hence female weight, is a function of the density of feeding larvae, the emigration rate is now density-dependent, and so the immigration term (Equation (1)) must also be changed. Assuming a discrete number n of populations in the area concerned, we can define the average density of eggs carried by emigrating moths as

$$\bar{E}_m = \frac{1}{n}\sum_{i=1}^{n}(\alpha_F \kappa F_D M)_i \tag{9}$$

and Equations (2) and (3) become

$$E = \beta\bar{E}_m + (1 - \alpha_F)\kappa F_D M \tag{10}$$

and

$$E/M = \frac{\beta\bar{E}_m}{M} + (1 - \alpha_F)\kappa F_D \tag{11}$$

This is illustrated by the grey lines in Figure 1 (for $\alpha = 0.5$, $\beta = 0.5$ and $\kappa = 0.5$), assuming an even distribution of densities over the range in the figure. Density-dependent emigration makes the relationship between egg and local moth densities curvilinear (Figure 1a) and the resultant slope of the Log(E/M) ratio vs. Log(M) considerable less steep (Figure 1b).

2.3. Datasets

We obtained 13 datasets from spruce budworm and one from western spruce budworm (*C. occidentalis* Free.) populations collected between 1945 and 2015 in Canada. The first four datasets listed below were extracted from published literature and the remainder are presented here for the first time. Each study either provided estimates of apparent fecundity directly or included estimates of adult and egg densities from which apparent fecundity was calculated. Some data were collected during pesticide application trials, but we found no significant effect of treatments on apparent fecundity. All trials involved larvicides, and so affected the current density of the immature population only with no additional effect on adult behavior.

1. Green River 1945–1959. Four multiyear plots (G4, G5, K1, and K2) from the Green River Project, New Brunswick, Canada [34], digitized from (Figures 4 to 7 in [16]).
2. Five New Brunswick locations over four years (Table 2 in [35]). We transformed female density to total adults by assuming a 1:1 sex ratio and used an egg mass size of 18 eggs/mass [31].
3. Multiple-year population data from one location (Black Sturgeon Lake, Ontario) between 1983 and 1998 [36]. Apparent fecundity calculated from egg and adult density in the original dataset.
4. Twenty-two, single-year (1975) plots from an experiment to estimate immigration (Table 1 in [37]). In Area 1 (10 plots), insecticides were applied repeatedly during the season to reduce larval populations as low as possible. Area 2 (nine plots) was located near Area 1 and treated as per operational spruce budworm control. Area 3 (three plots) was not treated.

5. Applications of insecticides in the Gatineau River Valley, Quebec on 12 plots (six controls, six treated) in a single year [38].

6. Applications of insecticides in north-central Ontario on nine treated and three control plots in a single year [39].

7. Applications of insecticides in the Gatineau River Valley, Quebec in 2000 [40].

8. Applications of pesticides in 12 plots plus four controls near Baie-Comeau, Quebec in 2008 [41].

9. Applications of pesticides in three treated and three control plots near Baie Comeau, Quebec in 2010 [42].

10. Lower St-Lawrence 2012–2015. Between six and 13 plots in four years (2012 to 2015) from the Lower St-Lawrence region, Quebec [28].

11. Western spruce budworm in British Columbia 1998–2015. Multiple years (18) and locations (three to 13) on untreated, outbreak populations of the western spruce budworm from Douglas-fir forests of in British Columbia covering an outbreak period from rise to decline [29].

2.4. Data Analysis

We compared two regression analysis methods to estimate the relationship between apparent fecundity and resident moth density. The first is a simple linear regression of observed egg density E on the density of local moths M, Equation (2), and yields an intercept representing the immigration rate (I of Equation (1)) and a slope representing the realized fecundity of moths prior to take-off ($S = (1 - \alpha)\kappa F$). Because the variance of spruce budworm population density estimates increases systematically with the mean [26], a log transformation and non-linear regression approach is used to estimate I and S:

$$\text{Log}(E) = \text{Log}(I + SM) \tag{12}$$

or alternatively:

$$\text{Log}(E/M) = \text{Log}(I/M + S). \tag{13}$$

The second method is to regress Log(E/M) on Log(M):

$$\text{Log}(E/M) = a + b\text{Log}(M). \tag{14}$$

There is no simple mathematical relationship between a and b parameters in Equation (14) and the parameters I and S in Equation (13). While the latter parameters are readily interpretable biologically, the former, in particular, the slope b of Equation (14), has an intuitively simple statistical interpretation that we discuss below. All 13 of the spruce budworm datasets were submitted to regression analysis using both models (12) and (14). For dataset 1 from the Green River Project [34], we compared apparent fecundities in G and K plots by oneway ANOVA on Log(E/M).

The western spruce budworm dataset was used for two purposes. First, we obtained estimates of α and β under both density-independent (Equation (3)) and density-dependent migration (Equation (11)) scenarios. We varied the two parameters systematically between 0 and 1 in steps of 0.01 and selected the pair of values that yielded the highest coefficient of determination (R^2) between the observed apparent fecundity and the values calculated. Regional mean adult density $\overline{M}$ was calculated from all sites available in each year. Second, we investigated the relationship between both local and regional population densities with the value of apparent fecundity in the density-independent (Equation (3)) and density-dependent (Equation (11)) cases, using the parameter values obtained with the procedure above. The range of M was varied systematically between (-4.5, -2.5) and (-2.5, -0.5) in steps of 0.05 (on $\log_{10}$ scale). The expected slope of Log(E/M) vs. Log(M) regressions (b in Equation (14)), was calculated for each density range. As a test of density dependence of the western spruce budworm's emigration rate, we compared the relationship between expected slope and average regional adult density with the annual estimates of slopes of the Log(E/M) vs. Log(M) regression obtained from the western spruce budworm dataset.

3. Results

3.1. Multi-Year Datasets

We regressed egg density with respect to adult density (Equation (12)), and compared apparent fecundity calculated from Equations (13) and (14) with the observations from multiple-year datasets 1, 2, and 3 (Figure 3). In Green River (dataset 1), apparent fecundity was very different in G plots and K plot (Figure 3b). High densities in the K plots were accompanied by low apparent fecundity (39.5 ± 10.9 and 26.4 ± 7.6 eggs/adult in K1 and K2, respectively), low densities in the G plots with very high apparent fecundity (212.2 ± 60.8 and 229.7 ± 73.1 eggs/adult in G4 and G5, respectively). This order-of-magnitude difference in apparent fecundity was highly significant (ANOVA on Log(E/M): $F = 6.8$, $df = 3.36$, $p = 0.001$). This is strong evidence the relatively low-density G plots were migration sinks, receiving more eggs than they exported during the course of the outbreak, while K plots were sources, exporting gravid moths from their outbreak populations. The high variability of this dataset relative to those that follow is due to its multiple-year nature and consequently the effects of annual variation in weather conditions affecting moth dispersal. The slope of the Log(E/M) − Log(M) regression (Equation (14)) was $b = -0.41 \pm 0.07$, and the intercept was $a = 1.93 \pm 0.07$. The corresponding immigration rate (from Equation (13)) was $I = 17.54 \pm 0.01$ eggs/m^2, which translates to about 0.022 eggs/shoot (assuming 800 shoots/m^2). Realized local fecundity was $S = 32.2 \pm 0.1$ eggs/adult.

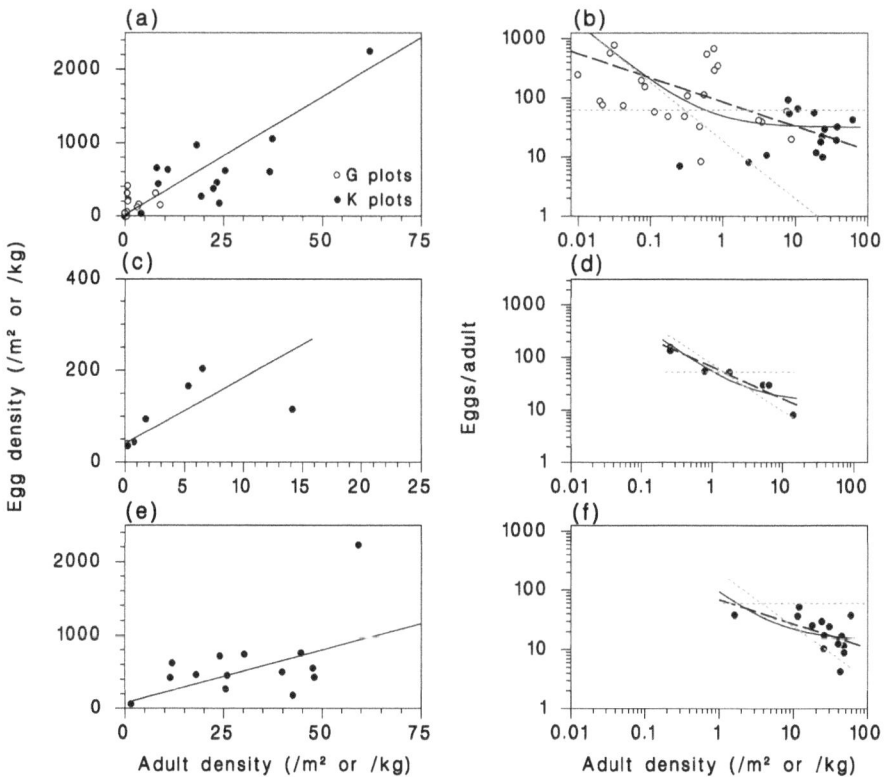

Figure 3. Three multiple-year datasets. Left column: Relationship between egg and adult density (lines are Equation (12)). Right column: Apparent fecundity (solid lines: Equation (13), dashed lines: Equation (14), horizontal dotted line: Expected realized fecundity (60s egg/moth), diagonal dotted lines: −1 slope, under panmixis ($\alpha = 1$). (**a**,**b**) Dataset 1. (**c**,**d**): dataset 2. (**e**,**f**): dataset 3.

Budworm densities recorded in dataset 2 were low and did not span as wide a range as those recorded in the other multiple-year datasets (1 and 3). Nonetheless, as in the other datasets, apparent fecundity was considerably higher in low-density plots than in plots with greater densities (Figure 3c,d). The slope of the Log(*E/M*) − *Log*(*M*) regression (Equation (14)) was *b* = −0.60 ± 0.09, and the intercept was *a* = 1.82 ± 0.06. The corresponding immigration rate (from Equation (13)) was *I* = 40.23 ± 0.01 eggs/m², which translates to about 0.05 eggs/shoot (assuming 800 shoots/m²). Realized local fecundity was *S* = 14.4 ± 0.1 eggs/adult. These parameter values suggest that migration rates to the low-density populations in dataset 2 were even greater than those observed in Green River's dataset 1 (Figure 3b).

Dataset 3 was collected over 13 years from a single site (Black Sturgeon Lake, Ontario) so annual variation in migration rates probably explains the wide scatter of apparent fecundity around the expected lines. Nonetheless, years with lower densities had consistently higher apparent fecundity and vice versa (Figure 3e,f). The slope of the Log(*E/M*) − Log(*M*) regression (Equation (14)) was *b* = −0.40 ± 0.18, and the intercept was *a* = 1.84 ± 0.26. The corresponding immigration rate (from Equation (13)) was *I* = 78.9 ± 0.1 eggs/kg, which translates to about 0.08 eggs/shoot (assuming 1000 shoots/kg). Realized local fecundity was *S* = 14.4 ± 0.1 eggs/adult, identical to the realized fecundity in dataset 2.

3.2. Single-Year Datasets

As with the multiple-year datasets, it is not possible to distinguish statistically between the fits of Equations (13) and (14) to each of the 10 single-year datasets (Figure 4). However, there is still a close relationship between the parameters of the two equations. The slope *S* of Equation (13) is almost perfectly defined by the values of parameters *a* and *b* of Equation (14):

$$\ln(S) = 2.603a - 2.53b \left(R^2 = 0.946, \text{ Figure 5a} \right), \tag{15}$$

while its intercept *I* is obtained, somewhat less precisely, from the slope *b* and the regional-mean density $\overline{M}$ using:

$$\ln(I) = -5.717 - 4.132b + 40.1\overline{M} \left(R^2 = 0.854, \text{ Figure 5b} \right) \tag{16}$$

Therefore, the analysis of apparent fecundity relationships with local density can be performed equally well with both regression approaches. However, the use of Equation (14) (Log(*E/M*) − Log(*M*) regression) is simpler.

The immigration rate *I* (in eggs per unit foliage), can be calculated from the regression parameters *a* and *b* of Equation (14) using Equation (16) (Figure 5a). The realized pre-emigration fecundity *S* can be calculated from the slope *b* of Equation (14) and the regional-mean adult density $\overline{M}$ with Equation (15). Using the Log(*E/M*) − *Log*(*M*) regression (Equation (14)) also has the advantage of giving a direct comparison with the two extreme cases (no migration and panmixis). The closer to −1 the slope is, the more migration has occurred (value of *α* closer to 1).

On average over our 10 single-year datasets, *a* = 0.369, *b* = −0.158 and $\overline{M}$ = 0.027 adults/shoot. These correspond to *I* = 0.126 eggs/shoot and *S* = 14.25 eggs/adult. Two of these datasets have slopes *b* very near −1 (Figure 4a,h) suggesting extensive emigration rates near the extreme case of panmixis when all moths redistribute among all populations (*α* = 1). Three others are close to the opposite case with the apparent absence of emigration (*α* = 0), with slope *b* near 0 (Figure 4d,e,g). The rest (Figure 4b–d,f) are in the intermediate range where *b* ≈ −0.5.

In general, one can expect apparent fecundity to be high in regions undergoing an outbreak (high $\overline{M}$) in years when the emigration rate is high (*α*→1). Under such circumstances, the apparent fecundity in sink populations, where local densities are lower than average, become much higher than expected fecundity. Conversely, apparent fecundity in sources, where densities are above the regional mean, are lower than expected fecundity. The result is homogenization of populations in a highly density-dependent manner with lower-density populations in any one year experiencing greater per capita recruitment rates as a result of net immigration of gravid moths. These moths are leaving high-density populations which then have low per capita egg recruitment rates.

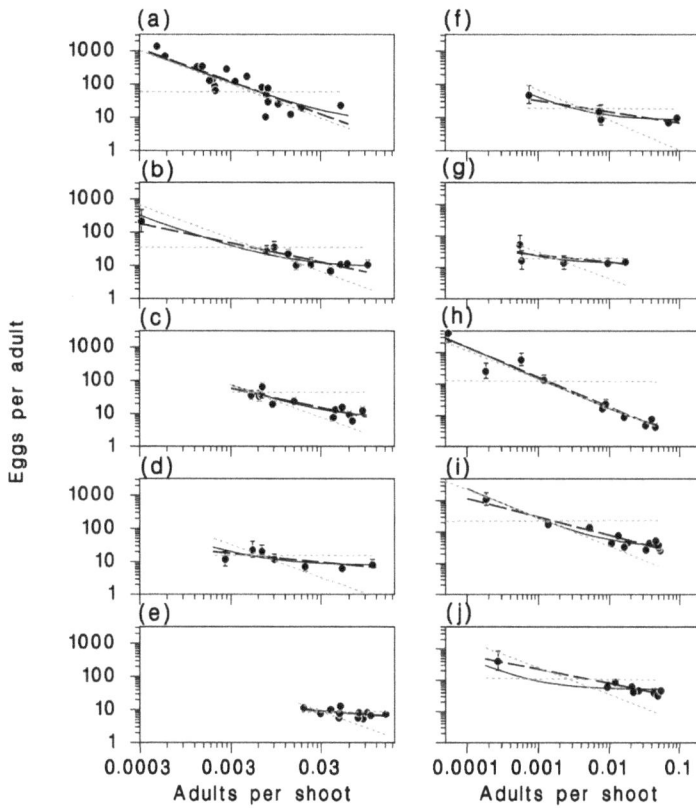

Figure 4. Apparent fecundity in relation to local adult density in ten single-year, multiple-site datasets. Solid lines: Equation (13), dashed lines: Equation (14), horizontal dotted line: Average apparent fecundity, diagonal dotted lines: −1 slope under panmixis $\alpha = 1$). (**a**) Dataset 4, (**b**) dataset 5, (**c**) dataset 6, (**d**) dataset 7, (**e**) dataset 8, (**f**) dataset 9, (**g**–**j**) Lower St-Lawrence datasets 10 to 13.

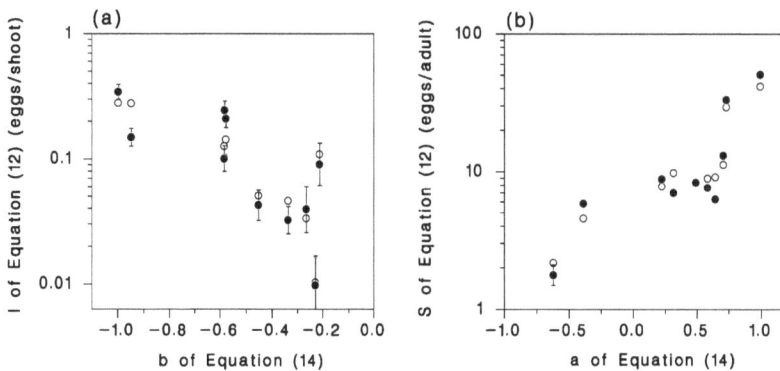

Figure 5. Relationships between parameters of Equations (13) and (14). (**a**) *I*, intercept of Equation (13) (immigration rate in eggs/shoot) is explained mostly by *b*, slope of Equation (14), the rest by $\overline{M}$, the regional-mean density, predicted values calculated with Equation (16). (**b**) *S*, slope of Equation (13) (realized fecundity in eggs/adult) is explained mostly by *a*, intercept of Equation (14), the rest by *b*, its slope, predicted values calculated with Equation (15). Vertical bars are SE of the parameter.

3.3. Western Spruce Budworm

The best fit in this long-term, multi-location dataset was obtained with Equation (3) (density-independent fecundity and emigration a random process) using parameter values α = 0.59, β = 0.27, and constant fecundity F = 218 eggs/female. The resulting R^2 between observed and expected apparent fecundity (on regular scale) is 0.32. Using Equation (11) (density-dependent fecundity and emigration rate) the best fit was obtained with α = 1, β = 0.26 (on regular scale, R^2 = 0.40), a significant improvement in goodness of fit (Mallows C_P = 1.0 with Equation (11) compared to C_P = 18.2 with Equation (3)).

When the range in adult density is varied systematically from low to high using the parameter estimates obtained above, the difference between density-independent and density-dependent migration becomes evident. If fecundity and emigration rates are independent of density (Equation (3)), the slope of the Log(E/M) – Log(M) regression does not vary with average regional density but remains constant at b = −0.45. However, when F and α are density-dependent (Equation (11)), the slope of the density relationship decreases from b = −0.18 when regional mean density is low, to b = −1 when regional density is highest. The western spruce budworm dataset confirms this predicted pattern (Figure 6). Observed annual fluctuations in the slope of the Log(E/M) – Log(M) regression are negatively correlated with fluctuations of regional mean density (Figure 6a). This negative association reflects the prediction (solid grey line in Figure 6b). This constitutes strong evidence that emigration is density-dependent in spruce budworms.

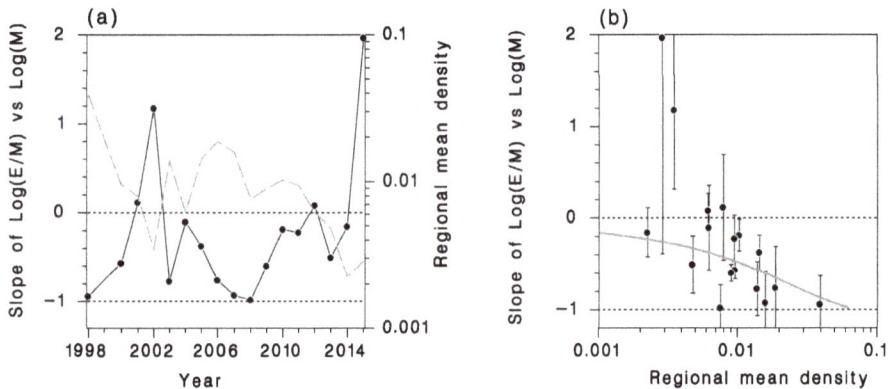

Figure 6. (a) Annual variation of the slope of the Log(E/M) – Log(M) regression (Equation (14), solid line and closed symbols) and of regional-mean adult density (dashed line) in southcentral British Columbia from 1998 to 2015. **(b)** Relationship between the slope of the Log(E/M) – Log(M) relationship and regional mean adult density. Vertical bars are SE of the parameter, solid grey line it the relationship expected from density-dependent emigration described by Equation (11) with α = 1, β = 0.26 and κ = 0.5, horizontal lines are two extreme cases: No migration (α = 0) and panmixis (α = 1).

4. Discussion

Dispersal of winged, adult insects is one of the most important yet enigmatic aspects of insect population ecology. Their capacity to disperse and the fact that they do is obvious to the most casual observer. Massive swarms of forest insects, including spruce budworms, near light sources are impressive, newsworthy events. Anecdotal accounts of outbreaks following such in-flights of moths [43] have been confirmed by direct observation [1]. However, these could be exceptional circumstances, significant at the local scale but transient and inconsequential at the regional scale. For example, observed flights of moths from New Brunswick to Newfoundland in Atlantic Canada in the 1980s resulted in damaging outbreaks which lasted only a few years while the source outbreak on the mainland continued for several more years [21]. Our understanding of the ecological conditions determining if, when, and where spruce budworms fly, are relatively well known (e.g., [44–46]) but the application of this knowledge to the analysis of population patterns is challenging and has only been accomplished recently [47]. As extensive, annual records of defoliation of outbreaks became available, the research approach on dispersal shifted to empirical, geo-statistical techniques to infer patterns of direction and rate of 'spread' in defoliation that might reveal the ecological structure of moth dispersal [4,10,48]. Skeptical of these empirical estimates, theoretical models simplified the difficulty by assigning a random effect with respect to population rates of change [17,49,50].

None of this led to a consensus on the critical question regarding dispersal in spruce budworms, namely, is the large-scale, eruptive dynamics of spruce budworm populations the result of net migration of gravid moths from sources to sinks or are these migrants simply supplementing a general, regional increase population density driven by other large-scale factors such as weather or the availability of susceptible forests? This question has a direct bearing on management strategies. If outbreaks spread from epicenters by moth dispersal, then aggressive suppression of early-rising populations to reduce migrants, irrespective of their location, should be effective [51]. On the other hand, if populations are on the rise everywhere and apparent epicenters are just the first to be detected [17], pest managers have little choice other than to wait until high-valued stands are damaged unacceptably before intervening.

The field data analyzed here provide testable evidence that dispersal of gravid spruce budworm moths is fundamental to population dynamics as it can play a significant role in the initiation of outbreaks and influence the characteristics of outbreak patterns over the entire range of regional outbreaks. We found a consistent, inverse relationship between the rate of egg recruitment and the density of the moth population in independent data from several areas in different time periods and a wide range of local densities. This constitutes statistical density dependence. Spruce budworm populations are a patchwork of sources and sinks which tend to homogenize as mean regional densities increase. The process is not stochastic, high-density populations are sources that export gravid moths to low-density sinks. Losses during migration tend to reduce overall realized fecundity but this mortality may be insignificant at the regional level or more related to weather conditions during flight than population densities. Consequently, population growth rates in low-density populations tend to increase and the lower, realized fecundity in high-density populations may be compensated by reduced mortality among progeny [11,16,29,36]. Further, we demonstrate that biologically density-dependent emigration is a plausible explanation of why the slope of the $Log(E/M) - Log(M)$ relationship decreases (becomes closer to −1) as regional mean density increases. This systematic change can only be observed in the same regional context over several years while population density changes. It cannot be adequately tested from samples taken in different regions although the scale of moth movement that defines the ecological region of interest is very large (100 to 1000 s of km^2).

5. Conclusions

We conclude that migration is central in spruce budworm population dynamics as it is density-dependent in both the statistical and biological sense and not simply a source of vertical perturbation in a random process. It plays a role in synchrony and spread of outbreaks because homogenization is inevitable. Density-dependent dispersal is a sufficiently powerful process to offset local declines in generation survival providing there is a source of emigrants [29]. This phenomenon has been evident to forest pest managers disappointed by re-invasion of forests where pesticide treatments have reduced budworm population densities [37] or populations have recovered immediately following a collapse [31]. It also explains the collapse of outbreaks over a broad region once most populations are experiencing declines in survival such that potential sources are less frequent or more remote [29].

The ubiquitous statistical density dependence of spruce budworm egg recruitment at the regional scale implies that moth dispersal tends to homogenize population. Low-density populations tend to have high growth rates, high-density populations low growth rates. In addition, biological density-dependence implies that as a regional outbreak develops and average density increases, the slope of the statistical density dependence of egg recruitment (and hence population growth rates) becomes steeper (closer to −1). Fluctuations of apparent fecundity have been viewed as a source of stochastic (random) perturbation on an autoregressive, non-linear stochastic process through the Moran effect [52,53]. Many authors have discussed the difficulty of synchronizing metapopulations fluctuations by a Moran effect when inherent cycling frequencies are spatially heterogeneous [54–56], which should be the case in insects like spruce budworms that are distributed over very large and diverse landmasses. However, statistically or biologically density-dependent dispersal introduces additional complexity to its effect on metapopulation synchrony and stability [57,58] that are beyond the aim of this paper and merit thorough investigation.

However, density-dependent apparent fecundity and migration rate may be the strongest arguments in support of an early intervention strategy against spruce budworm outbreaks. Such a strategy, applied in an area-wide fashion, involves detecting and suppressing any high-density populations to prevent their becoming sources. When this is done early, while the regional mean budworm population density is low and survival rates are also low [28], redistribution of eggs through moth dispersal is minimized and it is possible that an outbreak can be prevented, or at least delayed.

Author Contributions: Conceptualization, formal analysis, writing and reviewing, J.R.; data contribution, writing, reviewing, V.G.N.

Funding: The original data and research was funded by members of SERG-international, including SOPFIM, Forest Protection Limited, the governments of Newfoundland, New Brunswick, Quebec, Ontario, Manitoba and Saskatchewan, as well as the USDA Forest Service.

Acknowledgments: We acknowledge Alain Dupont (SOPFIM), Ed Kettela and Kees van Frankenhuyzen for allowing the use of their data. We thank Ariane Béchard for her help in the conduct of the Lower St-Lawrence study and Rod Turnquist for assistance with western spruce budworm.

Conflicts of Interest: The authors declare conflict of interest.

References

1. Greenbank, D.O.; Schaefer, G.W.; Rainey, R.C. Spruce budworm (Lepidoptera: Tortricidae) moth flight and dispersal: New understanding from canopy observations, radar and aircraft. *Mem. Entomol. Soc. Can.* **1980**, *110*. [CrossRef]
2. Hanski, I.; Woiwod, I.P. Spatial synchrony in the dynamics of moth and aphid populations. *J. Anim. Ecol.* **1993**, *62*, 656–668. [CrossRef]
3. Myers, J.H. Synchrony in outbreaks of forest lepidoptera: A possible example of the Moran effect. *Ecology* **1998**, *79*, 1111–1117. [CrossRef]
4. Williams, D.W.; Liebhold, A.M. Spatial synchrony of spruce budworm outbreaks in eastern North America. *Ecology* **2000**, *81*, 2753–2766. [CrossRef]
5. Peltonen, M.; Liebhold, A.M.; Bjornstad, O.N.; Williams, D.A. Spatial synchrony in forest insect outbreaks: Roles of regional stochasticity and dispersal. *Ecology* **2002**, *83*, 3120–3129. [CrossRef]
6. Jardon, Y.; Morin, H.; Dutilleul, P. Périodicité et synchronisme des épidémies de la tordeuse des bourgeons de l'épinette au Québec. *Can. J. For. Res.* **2003**, *33*, 1947–1961. [CrossRef]
7. Safranyik, L.; Carroll, A.L.; Régnière, J.; Langor, D.W.; Riel, W.G.; Shore, T.L.; Peter, B.; Cooke, B.J.; Nealis, V.G.; Taylor, S.W. Potential for range expansion of mountain pine beetle into the boreal forest of North America. *Can. Entomol.* **2010**, *142*, 415–442. [CrossRef]
8. Simard, I.; Morin, H.; Lavoie, C.A. Millenial-scale reconstruction of spruce budworm abundance in Saguenay, Québec, Canada. *Holocene* **2006**, *16*, 31–37. [CrossRef]
9. Hardy, Y.J.; Mainville, M.; Schmidt, D.M. *Spruce Budworms Handbook: An Atlas of Spruce Budworm Defoliation in Eastern North America, 1938–1980*; USDA Forest Service: Beltsville, MD, USA, 1986; p. 52.
10. Bouchard, M.; Auger, I. Influence of environmental factors and spatio-temporal covariates during the initial development of a spruce budworm outbreak. *Lands. Ecol.* **2014**, *29*, 111–126. [CrossRef]
11. Royama, T.; Eveleigh, E.S.; Miron, J.R.B.; Pollock, S.J.; McCarthy, P.C.; McDougall, G.A.; Lucarotti, C.J. Mechanisms underlying spruce budworm outbreak processes as elucidated by a 14-year study in New Brunswick, Canada. *Ecol. Monogr.* **2017**, *87*, 600–631. [CrossRef]

12. Bouchard, M.; Régnière, J.; Therrien, P. Bottom-up factors contribute to large-scale synchrony in spruce budworm populations. *Can. J. For. Res.* **2018**, *48*, 277–284. [CrossRef]

13. Greenbank, D.O. The role of climate and dispersal in the initiation of outbreaks of the spruce budworm in New Brunswick: II. The role of dispersal. *Can. J. Zool.* **1957**, *35*, 385–403. [CrossRef]

14. Sturtevant, B.R.; Achtemeier, G.L.; Charney, J.J.; Anderson, D.P.; Cooke, B.J.; Townsend, P.A. Long-distance dispersal of spruce budworm (*Choristoneura fumiferana* Clemens) in Minnesota (USA) and Ontario (Canada) via the atmospheric pathway. *Agric. For. Meteorol.* **2013**, *168*, 186–200. [CrossRef]

15. Royama, T. Effect of adult dispersal on the dynamics of local populations of an insect species: a theoretical investigation. In *Dispersal of Forest Insects: Evaluation, Theory and Management Implications*; Berryman, A.A., Safranyik, L., Eds.; Wash. State Univ.: Pullman, WA, USA, 1979; pp. 79–93.

16. Royama, T. Population dynamics of the spruce budworm *Choristoneura fumiferana*. *Ecol. Monogr.* **1984**, *54*, 429–462. [CrossRef]

17. Régnière, J.; Lysyk, T.J. Population dynamics of the spruce budworm, *Choristoneura fumiferana*. In *Forest Insects Pests in Canada*; Armstrong, J.A., Ives, W.G.H., Eds.; Natural Resources Canada, Canadian Forest Service: Ottawa, ON, Canada, 1995; pp. 95–105.

18. Morris, R.F.; Miller, C.A. The development of life tables for the spruce budworm. *Can. J. Zool.* **1954**, *32*, 283–301. [CrossRef]

19. Greenbank, D.O. The analysis of moth survival and dispersal in the unsprayed area. In *the Dynamics of Epidemic Spruce Budworm Population*; Morris, R.F., Ed.; The Memoirs of the Entomological Society of Canada: Ottawa, ON, Canada, 1963; pp. 87–99.

20. Nealis, V.G. Comparative ecology of conifer-feeding spruce budworms (Lepidoptera: Tortricidae). *Can. Entomol.* **2016**, *148* (Suppl. 1), S33–S57. [CrossRef]

21. Dobesberger, E.J.; Lim, K.P.; Raske, A.G. Spruce budworm (Lepidoptera: Tortricidae) moth flight from New Brunswick to Newfoundland. *Can. Entomol.* **1983**, *115*, 1641–1645. [CrossRef]

22. Wellington, W.G.; Henson, W.R. Note on the effects of physical factors on the spruce budworm, *Choristoneura fumiferana* (Clem.). *Can. Entomol.* **1947**, *79*, 168–170. [CrossRef]

23. Rhainds, M.; Kettela, E.G. Oviposition threshold for flight in an inter-reproductive migrant moth. *J. Ins. Behav.* **2013**, *26*, 850–859. [CrossRef]

24. Blais, J.R. Effects of the destruction of the current year's foliage of balsam fir on the fecundity and habits of flight of the spruce budworm. *Can. Entomol.* **1953**, *85*, 446–448. [CrossRef]

25. Van Hezewijk, B.; Wertman, D.; Stewart, D.; Beliveau, C.; Cusson, M. Environmental and genetic influences on the dispersal propensity of spruce budworm (*Choristoneura fumiferana*). *Agric. For. Entomol.* **2018**, *20*, 433–441. [CrossRef]

26. Régnière, J.; Sanders, C.J. Optimal sample size for the estimation of spruce budworm (Lepidoptera: Tortricidae) populations on balsam fir and white spruce. *Can. Entomol.* **1983**, *115*, 1621–1626. [CrossRef]

27. Lysyk, T.J.; Sanders, C.J. A method for sampling endemic populations of spruce budworm (Lepidoptera: Tortricidae) based on proportion of empty sample units. *Can. For. Serv. Info. Rep.* **1987**, *O-X-382*, 17.

28. Régnière, J.; Cooke, B.J.; Béchard, A.; Dupont, A.; Therrien, P. Dynamics and management of rising outbreak spruce budworm populations. *Forests* **2019**. under review.

29. Nealis, V.G.; Régnière, J. Ecology of outbreak populations of the western spruce budworm. *Ecol. Monogr.* **2019**. under review.

30. Miller, C.A. The analysis of fecundity proportion in the unsprayed area. In *the Dynamics of Epidemic Spruce Budworm Population*; Morris, R.F., Ed.; The Memoirs of the Entomological Society of Canada: Ottawa, ON, Canada, 1963; pp. 75–87.

31. Nealis, V.G.; Régnière, J. Fecundity and recruitment of eggs during outbreaks of the spruce budworm. *Can. Entomol.* **2004**, *136*, 591–604. [CrossRef]

32. Wellington, W.G. The light reactions of the spruce budworm, *Choristoneura fumiferana* Clemens (Lepidoptera: Tortricidae). *Can. Entomol.* **1948**, *80*, 56–82. [CrossRef]

33. Henson, W.R. The Means of Dispersal of the Spruce Budworm. Ph.D. Thesis, Yale University, New Haven, CT, USA, 1950.

34. Morris, R.F. The dynamics of epidemic spruce budworm population. *Mem. Entomol. Soc. Can.* **1963**, *31*, 30–32. [CrossRef]

35. Miller, C.A.; Greenbank, D.O.; Kettela, E.G. Possible use of canopy light traps in predicting spruce budworm egg-mass counts. *Can. For. Serv. Bi-Mon. Res. Notes* **1979**, *35*, 29–30.

36. Régnière, J.; Nealis, V. Ecological mechanisms of population change during outbreaks of the spruce budworm. *Ecol. Entomol.* **2007**, *32*, 461–477. [CrossRef]

37. Miller, C.A.; Greenbank, D.O.; Kettela, E.G. Estimated egg deposition by invading spruce budworm moths (Lepidoptera: Tortricidae). *Can. Entomol.* **1978**, *110*, 609–615. [CrossRef]

38. Régnière, J. Results of the 1999 Early Intervention Project. SERG-International Annual Workshop Proceedings. Available online: www.serginternational.org (accessed on 16 August 2019).

39. Régnière, J. Early intervention against SBW: Results of the 1999–2001 preliminary trials on Mimic and Bt. SERG-International Annual Workshop Proceedings 2001. Available online: www.serginternational.org (accessed on 16 August 2019).

40. Régnière, J.; Kettela, E.G.; van Frankenhuyzen, K. SERG Project 1999/05: Early Intervention against Spruce Budworm: High dosage Foray 96B in Ontario in 2000. SERG-International Annual Workshop Proceedings 2001. Available online: www.serginternational.org (accessed on 16 August 2019).

41. Trudel, R.; Dupont, A.; Bélanger, A. Experimental Pheromone Applications Using Disrupt Micro-Flakes SBW® for the Control of the Spruce Budworm Populations: Quebec Mating Disruption Trials 2008. SERG-International Annual Workshop Proceedings 2011. Available online: www.serginternational.org (accessed on 16 August 2019).

42. Régnière, J.; Delisle, J.; Pureswaran, D.; Trudel, R. Mate-finding Allee effect in spruce budworm population dynamics. *Entomol. Exp. Appl.* **2012**, *146*, 112–122. [CrossRef]

43. Craighead, F.C. Studies on the spruce budworm [Cacoecia fumiferana Clem.]. Part II. General bionomics and possibilities for prevention and control. *Can. Dep. Agric. Tech. Bull.* **1924**, *37*, 28–57.

44. Boulanger, Y.; Fabry, F.; Kilambi, A.; Pureswaran, D.; Sturtevant, B.R.; Saint-Amant, R. The use of weather surveillance radar and high-resolution three dimensional weather data to monitor a spruce budworm mass exodus flight. *Agric. For. Meteorol.* **2017**, *234-235*, 127–135. [CrossRef]

45. Régnière, J.; Delisle, J.; Sturtevant, B.R.; Garcia, M.; Saint-Amant, R. Modeling migratory flight in the spruce budworm: temperature contraints. *Forest* **2019**. under review.

46. Régnière, J.; Garcia, M.; Saint-Amant, R. Modeling migratory flight in the spruce budworm: circadian rhythm. *Forests* **2019**. under review.

47. Robert, L.E.; Sturtevant, B.R.; Cooke, B.J.; James, P.M.A.; Fortin, M.J.; Townsend, P.A.; Wolker, P.T.; Kneeshaw, D. Landscape host abundance and configuration regulate periodic outbreak behavior in spruce budworm *Choristoneura fumiferana*. *Ecography* **2018**, *40*, 1–16. [CrossRef]

48. Gray, D.R. The relationship between climate and outbreak characteristics of the spruce budworm in eastern Canada. *Clim. Chang.* **2008**, *87*, 361–383, Erratum in **2008**, *89*, 447–449. [CrossRef]

49. Royama, T. *Analytical Population Dynamics*; Chapman Hall. N.Y.: New York, NY, USA, 1992; p. 371.

50. Fleming, R.A.; Barclay, H.J.; Candau, J.N. Scaling-up an autoregressive time-series model (of spruce budworm population dynamics) changes its qualitative behavior. 2002. *Ecol. Model.* **2002**, *149*, 127–142. [CrossRef]

51. Blais, J.R. Trends in the frequency, extent, and severity of spruce budworm outbreaks in eastern Canada. *Can. J. For. Res.* **1983**, *13*, 539–547. [CrossRef]

52. Royama, T. Moran effect on non-linear population processes. *Ecol. Monogr.* **2005**, *75*, 227–293. [CrossRef]

53. Abbott, K.C. Does the pattern of population synchrony through space reveal if the Moran effect is acting? *Oikos* **2007**, *116*, 903–912. [CrossRef]

54. Hugueny, B. Spatial synchrony in population fluctuations: extending the Moran theorem to cope with spatially heterogeneous dynamics. *Oikos* **2006**, *115*, 3–14. [CrossRef]

55. Liebhold, A.M.; Johnson, D.M.; Bjornstad, O.N. Geographic variation in density-dependent dynamics impacts the synchronizing effect of dispersal and regional stochasticity. *Popul. Ecol.* **2006**, *48*, 131–138. [CrossRef]

56. Vasseur, D.A. Environmental colour intensifies the Moran effect when population dynamics as spatially heterogeneous. *Oikos* **2007**, *116*, 1726–1736. [CrossRef]

57. Denno, R.F.; Peterson, M.A. Density-dependent dispersal and its consequences for population dynamics. In *Population Dyanmics: New Approaches and Synthesis*; Cappucino, N., Price, P.W., Eds.; Academic Press: New York, NY, USA, 1995; pp. 113–130. ISBN 0-12-159270-7.
58. Silva, J.A.L.; Giordani, F.T. Density-dependent migration and synchronism in metapopulations. *Bull. Math. Biol.* **2006**, *68*, 451–466. [CrossRef] [PubMed]

Article

Effects of Hardwood Content on Balsam Fir Defoliation during the Building Phase of a Spruce Budworm Outbreak

Bo Zhang [1,*], David A. MacLean [1], Rob C. Johns [2] and Eldon S. Eveleigh [2]

[1]   Faculty of Forestry and Environmental Management, University of New Brunswick, P.O. Box 4400, Fredericton, NB E3B 5A3, Canada; macleand@unb.ca
[2]   Natural Resources Canada, Canadian Forest Service-Atlantic Forestry Centre, P.O. Box 4000, Fredericton, NB E3B 5P7, Canada; rob.johns@canada.ca (R.C.J.); eldon.eveleigh@canada.ca (E.S.E.)
*   Correspondence: bo.zhang@unb.ca

Received: 25 July 2018; Accepted: 30 August 2018; Published: 31 August 2018

Abstract: Defoliation by spruce budworm (*Choristoneura fumiferana* Clem.) on balsam fir (*Abies balsamea* (L.) Mill.) is more severe in fir than in mixed fir-hardwood stands. Previous studies assumed that defoliation in fir-hardwood stands was reduced in proportion to percent hardwood regardless of outbreak severity. We tested the influence of stand composition on defoliation during the first 5 years of a spruce budworm outbreak near Amqui, Quebec, by sampling 27 fir-hardwood plots selected to represent three percent hardwood basal area classes (0%–25%, 40%–65%, and 75%–95%). Balsam fir defoliation was significantly lower ($p < 0.001$) as hardwood content increased, but the relationship varied with overall defoliation severity each year. Annual plot defoliation in fir-hardwood plots, estimated using: (1) defoliation in pure fir plots and the assumption that defoliation in fir-hardwood plots was reduced in proportion to percent hardwood; (2) a generalized linear mixed-effects model with defoliation in pure fir plots, percent hardwood, and interaction as fixed-effects; and (3) Random Forests prediction incorporating 11 predictor variables, resulted in $r = 0.77, 0.87$, and 0.92 versus measured defoliation, respectively. Average defoliation severity in softwood plots and percent hardwood content were the most important variables in Random Forests analysis. Data on average defoliation level in softwood stands, as an indicator of overall outbreak severity, improves prediction of balsam fir defoliation in mixed stands.

Keywords: *Choristoneura fumiferana*; *Abies balsamea*; hardwood content; defoliation prediction

1. Introduction

Effective forest pest management in heterogeneous landscapes and in mixed-species forest stands requires knowledge about how tree diversity affects insect herbivory. A meta-analysis of a worldwide data set of 119 studies by Jactel and Brockerhoff [1] showed a significant reduction in herbivory with increasing forest diversity for oligophagous insects (i.e., species that exploit one or a few closely related genera of hosts). This seems also to be the case for spruce budworm (*Choristoneura fumiferana* Clem.), which is the major defoliator of balsam fir (*Abies balsamea* (L.) Mill.) and spruce (*Picea* spp.) in boreal and New England-Acadian forests in eastern North America [2–4]. Budworm outbreaks are cyclical and have occurred at 30–40-year intervals in eastern Canada during the past century [5,6]. Outbreaks usually last 5–15 years and severe defoliation causes growth loss and tree mortality over large areas [7,8], peaking at over 52 million ha of defoliation of forests in eastern Canada [9]. Spruce budworm defoliation can be assessed by conducing aerial survey, ground assessment with binoculars, and branch sampling with pole pruners, of which branch sampling with pole pruners and rating defoliation on individual shoots is considered to be the most accurate technique [10–12].

Several studies have reported lower spruce budworm-caused defoliation of balsam fir, and lower resulting growth reduction and mortality, in stands or forest landscapes associated with higher percentage of hardwood tree species [13–16]. Mature stands with a large proportion of balsam fir, especially in contiguous softwood landscapes, have the highest susceptibility and vulnerability to spruce budworm outbreaks [17,18]. Balsam fir defoliation assessed using the branch sampling method was 12%-32% in fir-hardwood stands with >40% hardwood content versus 58%-71% in stands with <40% hardwood content [15]. Tree-ring analysis showed that budworm-caused growth reductions averaged 40% in stands with <50% hardwood content versus 20% in stands with >50% hardwood content [16]. Mortality of balsam fir resulting from budworm-caused defoliation was 14%–30% less in fir-hardwood mixed stands (~30% hardwood content) than in fir-dominated stands [13,14].

Two hypotheses have been proposed to explain less severe insect herbivory associated with higher tree diversity. The "natural enemy" hypothesis [19] argues that more diverse plant communities support more abundant natural enemies of herbivore insects by providing alternative prey, more predation opportunities, or better sheltering conditions [20–22]. Alternatively, the "habitat fragmentation" hypothesis argues that reduced host tree availability increases the degree of habitat fragmentation for the insects and creates barriers for foraging, dispersal, and mating success [23–25].

Past studies of how hardwood content influences spruce budworm defoliation have all focused on severely defoliated stands at the peak and declining phases of outbreaks. Su et al. [15] reported that defoliation of balsam fir decreased as hardwood content increased in the declining phase (last 5 years) of the last outbreak, but also noted that the relationship between defoliation and hardwood content may well vary during different stages of outbreak. Information is lacking on the building phase of an outbreak, as budworm populations increase from low to peak density.

Su et al. [15] also proposed a direct linear relationship for use in predicting defoliation in fir-hardwood stands, using percent hardwood and defoliation in a pure fir stand. The relationship can be expressed succinctly as $y = D_0 \times (1 - x)$ (which we term the simplified linear model), where fir defoliation in a mixedwood stand (y) is a function of percent hardwood (x) and fir defoliation level in a pure fir stand (D_0). The relationship was quantified using defoliation data collected in the declining years (1989–1993) of the last outbreak and were subsequently used in Needham et al. [26] and Sainte-Marie et al. [27]. In this study, we examine whether this relationship holds true in the building phase of an outbreak.

Insect herbivory is influenced by other variables in addition to tree diversity. For example, Douglas-fir tussock moth (*Orgyia pseudotsugata* McDunnough) defoliation varied with slope, stand density, and site index [28]. Scots pine (*Pinus sylvestris* L.) defoliation during common pine sawfly (*Diprion pini* L.) outbreaks was correlated with forest site class [29]. Studies conducted in the declining phase of the last spruce budworm outbreak suggested that hardwood content was significant in predicting defoliation [30,31], but other factors can include outbreak stage and severity, soil drainage, and site conditions [30,32]. In the above mentioned simplified linear model, percent hardwood was used as a variable and defoliation in pure fir stand was used as a constant for predicting defoliation in mixedwood stand in a given year. In this study, we also test whether adding other biotic and abiotic variables improve the accuracy of defoliation prediction.

The objectives of this study were to: (1) determine the relationship between balsam fir defoliation and hardwood content during the initiation and building phases (first 5 years) of a spruce budworm outbreak and (2) compare accuracy of predictions of spruce budworm defoliation in fir-hardwood stands based on three alternative model formulations: the simplified linear model; a generalized linear model with mixed-effects; and a machine learning (Random Forests) formulation. We evaluated two predictions: (1) fir-hardwood stands with higher hardwood content will have less severe annual defoliation of balsam fir and (2) the simplified linear model can be used to estimate budworm defoliation in fir-hardwood mixed stands if other predictor variables are not available, but incorporating more variables will improve the accuracy of predictions.

2. Materials and Methods

2.1. Study Area and Stand Sampling

Study sites were located on the north side of Lake Matapédia in the Gaspé area, Quebec (48°32′ N-48°35′ N, 67°25′ W-67°34′ W) (Figure 1). The area is within forest section L6, Témiscouata-Restigouche, in the Great Lakes-St. Lawrence forest region [33]. Aerial survey of spruce budworm defoliation conducted by the Quebec provincial government since 2010 indicated that defoliation was first detected in the study area in 2012 [34]. Defoliation was classified by aerial observers using a three-level scale: light, moderate, and severe. Most of the study area was classified as light or light-moderate defoliation in 2012 and 2013, and as severe or moderate-severe defoliation in 2014 and 2015. Defoliation from the aerial survey declined to light or light-moderate in 2016 [35].

Given that previous studies suggested that less severe defoliation would be observed in fir-hardwood stands with >40%–50% hardwood during a spruce budworm outbreak [15,16,26], mature (age >40 years) fir-hardwood stands were selected within three percent basal area of hardwood content classes: 0%–25% (termed softwood), 40%–65% (mixedwood), and 75%–95% (hardwood) (Figure 1). Nine 12.6 m radius circular (0.05 ha) sample plots were established, at least 50 m apart and away from stand edges within each of the three classes. Within each plot, all trees with diameter at breast height (DBH, 1.3 m above ground) >4 cm were numbered, and DBH and total height were measured. Elevation, slope, tree density, dominant ground vegetation, and GPS coordinates of plot center were recorded. All sample plots were established in spring 2014 except plots 13a, 14a, and 15a, which were established in 2015 to replace the corresponding original plots that were harvested during winter 2014, in an area with similar stand and defoliation conditions. Therefore, defoliation from 2012 to 2014 was collected in plot 13, 14, and 15 whereas defoliation for 2015 and 2016 was collected in plot 13a, 14a, and 15a.

Balsam fir was the dominant softwood species in all 27 plots, ranging from 64%–95% of the basal area in softwood plots, 20%–57% in mixedwood, and 5%–24% in hardwood (Table 1). White spruce (*Picea glauca* (Moench) Voss) occurred in about one-half of the softwood and mixedwood plots, and in one hardwood plot, but comprised <10% of the basal area in most plots. Percent basal area of hardwoods ranged from 1% to 95%, and averaged 12%, 56%, and 88% in the softwood, mixedwood, and hardwood classes, respectively (Table 1). Sugar maple (*Acer saccharum* Marshall) is a late successional species that was commonly found in both hardwood and mixedwood plots but was absent in softwood plots. The most abundant hardwood species in the mixedwood plots were generally yellow birch (*Betula alleghaniensis* Britton), white birch (*Betula papyrifera* Marshall), and red maple (*Acer rubrum* L.). Stand density ranged from 520 to 2975 stems/ha, with an average of 1346 stems/ha. Density of softwood and mixedwood plots was approximately double that of hardwood plots. Mean DBH and height of softwood, mixedwood, and hardwood plots was 15.7, 16.3, and 19.3 cm, and 14.0, 13.6, and 15.4 m respectively. Average diameter of balsam fir always exceeded the plot mean diameter in the softwood plots, but this relationship was variable in the mixedwood and hardwood plots. All plots except three were on flat ground with <10 degrees of slope. Elevation ranged from 165 to 357 m, with an average of 261 m.

Table 1. Description of the 27 plots sampled near Amqui, Quebec.

Plot No. [b]	Density (stems/ha)	DBH [c] (cm)	Ht [c] (m)	DBHbF [c] (cm)	BA [c] (m²/ha)	Species Composition % Basal Area [d]						Total HW [d] (%BA)
						bF	wS	OS	sM	yB	IH	
Softwood [a]												
1	1020	20.0 ± 1.1	16.3 ± 0.6	20.7 ± 1.1	37	78	4	9	-	2	8	9
2	2800	13.6 ± 0.5	12.8 ± 0.4	14.1 ± 0.6	49	78	3	-	-	4	15	19
3	920	21.5 ± 1.4	15.7 ± 0.8	23.4 ± 1.3	39	64	17	16	-	-	3	3
4	2975	13.1 ± 0.5	12.9 ± 0.3	13.1 ± 0.5	47	84	3	12	-	-	1	1
5	1650	16.1 ± 0.9	14.6 ± 0.7	15.8 ± 1.0	40	81	-	5	-	-	14	14
6	2125	15.6 ± 0.5	15.8 ± 0.3	15.8 ± 0.6	44	95	-	2	-	-	3	3
7	2100	15.6 ± 0.8	13.9 ± 0.4	16.0 ± 0.8	50	69	7	8	-	1	16	16
8	1100	16.5 ± 1.7	12.7 ± 0.8	21.5 ± 2.2	38	81	-	-	-	1	19	19
9	1380	16.4 ± 1.1	14.1 ± 0.7	22.5 ± 1.4	39	79	-	-	-	8	13	21
Mixedwood												
10	2540	11.2 ± 0.7	9.7 ± 0.5	7.3 ± 0.7	36	20	10	7	-	34	30	64
11	760	19.2 ± 2.1	12.5 ± 1.0	20.8 ± 2.6	32	36	2	-	29	19	13	62
12	1620	14.1 ± 1.1	10.7 ± 0.6	14.1 ± 1.3	37	45	11	-	1	23	31	55
13	960	20.1 ± 1.5	18.7 ± 1.0	20.9 ± 1.7	39	33	11	7	-	1	49	49
14	980	21.5 ± 1.9	17.5 ± 1.2	19.0 ± 1.8	49	27	1	17	-	7	55	55
15	1280	17.8 ± 1.0	17.5 ± 0.9	18.1 ± 1.5	39	42	5	2	-	7	55	53
13a	1980	11.0 ± 0.7	9.7 ± 0.5	10.0 ± 1.3	26	40	-	-	-	2	59	60
14a	1620	13.7 ± 0.9	11.0 ± 0.6	13.3 ± 1.3	32	44	-	-	-	-	56	56
15a	1780	15.1 ± 0.8	12.0 ± 0.5	14.4 ± 0.9	40	57	4	-	-	5	35	40
16	1260	17.7 ± 1.1	15.0 ± 0.5	19.0 ± 1.2	38	48	-	-	9	16	28	52
17	1200	16.9 ± 1.1	14.5 ± 0.6	18.9 ± 2.3	33	43	-	-	8	22	28	57
18	1200	17.3 ± 1.4	14.0 ± 0.6	16.7 ± 1.9	40	33	4	-	6	17	40	63
Hardwood												
19	800	18.2 ± 1.8	16.7 ± 0.8	26.9 ± 7.3	28	10	-	-	90	-	-	90
20	1080	16.8 ± 1.4	14.5 ± 0.8	21.3 ± 3.7	33	24	-	-	35	9	31	76
21	520	22.8 ± 2.3	16.5 ± 1.0	20.0 ± 2.3	27	12	-	-	51	8	29	88
22	1000	18.5 ± 1.7	12.0 ± 0.7	11.4 ± 2.0	38	12	3	-	34	24	26	84
23	620	22.6 ± 2.7	17.2 ± 1.4	26.8 ± 4.1	35	14	-	-	86	-	-	86
24	640	18.7 ± 2.4	12.6 ± 1.0	25.7 ± 1.2	27	8	-	-	16	28	49	92
25	1120	13.7 ± 1.0	15.2 ± 0.7	22.9 ± 3.9	22	8	-	-	81	2	10	92
26	800	17.2 ± 1.7	16.5 ± 0.8	19.7 ± 4.4	25	5	-	-	79	14	2	95
27	560	24.8 ± 2.3	17.2 ± 0.6	19.6 ± 2.8	33	10	-	-	44	28	18	90

[a] Softwood, mixedwood, and hardwood stand types were classified by hardwood basal area percentage: softwood (0%–25%), mixedwood (40%–65%), and hardwood (75%–95%). [b] The 'a' suffix denotes three sample plots established 1 year after initial sampling to replace original plots which were harvested. [c] Abbreviations: DBH = mean diameter at breast height; Ht = mean total height; DBHbF = mean DBH of balsam fir; BA = basal area. DBH, Ht, and DBHbF are shown as plot average ± one standard error of the mean. [d] Species abbreviations: bF balsam fir; wS white spruce; OS other softwood = black spruce (Picea mariana (Mill.) BSP), eastern white-cedar (Thuja occidentalis L.), and eastern larch (Larix laricina (Du Roi) K. Koch); sM sugar maple; yB yellow birch; IH intolerant hardwood = red maple, white birch, trembling aspen (Populus tremuloides Michx.), and balsam poplar (Populus balsamifera L.), HW hardwood = sM + yB + IH.

2.2. Defoliation Measurements

Current year's defoliation was estimated on four trees per plot, one branch per tree, and 25 current-year shoots per branch after annual defoliation ceased (in August) each year from 2014 to 2016. Mean plot defoliation was computed as the averages of four trees. In each sample plot, four co-dominant balsam fir trees were randomly selected, and one mid-crown branch [36] was collected using pole pruners from each tree. Cardinal direction was not considered as it does not significantly influence spruce budworm density [37] and therefore not defoliation. Twenty-five current-year shoots per branch were randomly selected beginning at the branch tip and back along the branch length and assessed for annual defoliation using the shoot-count (Fettes) method [10]. Total number of needles of an individual shoot was visually compared to the regular pattern of needle positions (phyllotaxis) and the shoot was assigned to one of the seven defoliation classes (0%, 1%–20%, 21%–40%, 41%–60%, 61%–80%, 81%–99%, and 100%) in the field to prevent physical needle removal during transport. The mid-point of each defoliation class was used to calculate mean defoliation per branch and per plot. Required sample sizes (25 shoots per branch) were determined based on MacLean and MacKinnon [11] to achieve 90% accuracy under most defoliation levels.

In 2014, branch samples were also collected in the spring prior to current year defoliation, and used to measure the annual defoliation that occurred in 2012 and 2013. Needle losses on 2012 and 2013 foliage were deemed to be the defoliation that occurred in the corresponding years, since aerial surveys indicated that defoliation was light to light-moderate in both years and back-feeding (late instar larvae feeding on older, non-current-year age classes of foliage) only occurs when very high insect populations consume all current year foliage [38].

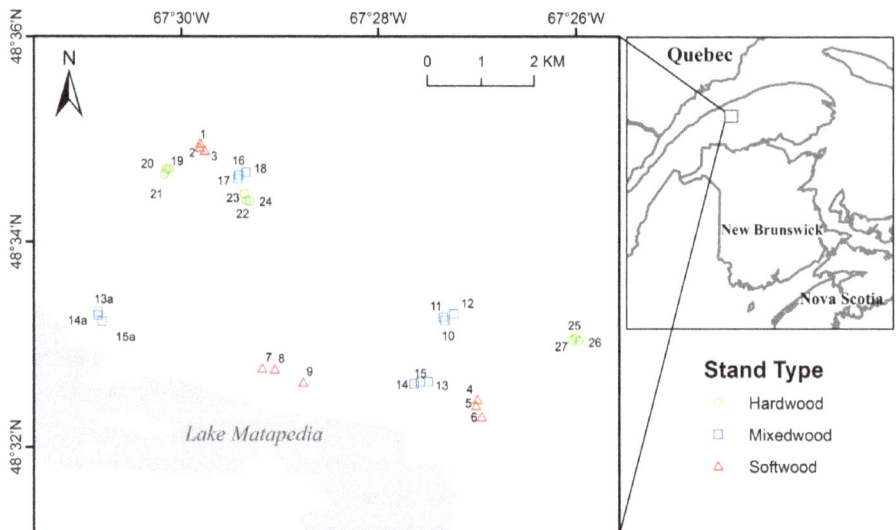

Figure 1. Locations of 27 plots sampled near Amqui, Quebec, by stand type.

2.3. The Simplified Linear Model

As described in the introduction, the simplified linear model was a direct linear relationship proposed by Su et al. [15]:

$$y = D_0 \times (1 - x) \tag{1}$$

where percent hardwood (x) and fir defoliation in a pure fir stand (D_0) were used to predict fir defoliation in fir-hardwood mixed stands (y). We used the average plot defoliation in our softwood plots (i.e., with 0%–25% hardwoods) calculated each year as D_0. In Su et al. [15], intercept from simple linear regression for defoliation as a function of percent hardwood content each year were used as D_0. In our data, preliminary analysis suggested that average plot defoliation in softwood plots was highly correlated ($r = 0.995$) with the intercepts from simple linear regressions. We chose to use the empirical average plot defoliation in softwood plots instead of the theoretical extrapolated "defoliation-with-zero-hardwood" from the statistical models, because it is based on measured data.

2.4. Analyzing the Relationship between Annual Balsam Fir Defoliation and Hardwood Content Using Generalized Linear Mixed-Effects Model

Average balsam fir defoliation at the plot level has two biological limits. At the lower end of the scale, needle loss can occur due to factors other than spruce budworm feeding. This and the lower threshold of measurement (i.e., use of 0% and 1%–20% defoliation classes, the latter with a mid-point of 10% defoliation) prevent defoliation from averaging 0% even in the absence of budworm. At the upper end, competition for food occurs among larvae, and defoliation reaches a near-100% plateau when budworm populations are extreme. Yet, average plot defoliation rarely reaches 100% because usually some needles escape complete defoliation. Hence, while a linear relationship between defoliation and hardwood content could be expected at mid-range defoliation severity levels, non-linear relationships would be expected at very light and at extreme defoliation levels. For these reasons, we analyzed defoliation through a logit link function [39] using a generalized linear mixed-effects model (GLMM):

$$\text{logit}(\text{defoliation}) = \beta_0 + b_0 + \beta_1 x + \beta_2 D_0 + \beta_3 x \times D_0, \tag{2}$$

where logit(defoliation) is the logit link function ($\ln(p/(1 - p))$) for defoliation in a fir-hardwood mixture; x = percent hardwood content; D_0 = average plot defoliation in softwood plots (as in Equation (1)); β_i's are the fixed effects parameter estimates; and b_0 is the random effects. Year as a categorical variable was not included in the model since the annual differences were already incorporated in D_0. Plot was included as random effect and a serial correlation structure CorAr1 was included in the model error term to control for temporal autocorrelation. Averaged annual defoliation for each stand type was calculated and differences among the three stand types were tested using Kruskal-Wallis analysis by ranks.

2.5. The Random Forests Model

Random Forests [40], an ensemble regression tree statistical procedure, was also used to predict spruce budworm defoliation. Random Forests has been used extensively in ecological and forestry studies (e.g., [41–43]) and has shown advantages in dealing with small sample size [44] and complex interactions between factors [45,46]. Random Forests was developed from classification and regression trees [47]. We used the Random Forests routine in R [48] and the default of 500 trees (parameter "ntree") was applied as error rate stabilized at 100–150 trees. A random one-third of predictor variables were used to perform data partitioning at each node (parameter "mtry"). Two-thirds of the overall dataset was randomly selected and used to build the Random Forests model, and the other one-third retained for testing the model. The importance of each predictor variable was measured as the change in prediction accuracy (increase in mean square error, function "importance"), computed by permuting (value randomly shuffled) the variable with out-of-bag data in the Random Forests validation approach [48]. A larger percent increase in mean square error indicates higher importance of a variable in prediction.

Initially, a total of 14 variables were assessed for multicollinearity using a correlation matrix: hardwood percent basal area (HW%), balsam fir percent basal area (bF%), D_0, mean DBH, mean height, basal area (BA), mean DBH of bF, mean height of bF, standard deviation of bF height, mean bF crown base height, standard deviation of bF crown base height, tree density, elevation, and slope. Random Forests does not hold formal distributional assumptions of data and is relatively insensitive to multicollinearity [40], but removing multicollinearity and redundancy to improve predictive power is recommended [49,50]. Correlation analysis indicated that, HW% and bF%, mean DBH of bF and mean height of bF, and mean DBH and density were highly correlated with coefficients (r) of -0.97, 0.87, and -0.84, respectively. HW%, mean DBH of bF, and mean DBH were selected over their counterpart variables because HW% should be included as a predictor in attempts to estimate defoliation reduction caused by hardwood; and mean DBH is readily available in most forest inventories and can be quickly assessed with an efficient sampling design. Therefore, 11 variables were retained and tested using Random Forests: HW%, D_0, mean DBH, mean height, mean DBH of bF, BA, standard deviation of bF height, mean bF crown base height, standard deviation of bF crown base height, elevation, and slope.

2.6. Statistical Analyses

Balsam fir defoliation was estimated using the three alternative models: (1) the simplified linear model; (2) the GLMM model; and (3) the Random Forests model. The annual plot defoliation estimated using each model was plotted against actual defoliation measured in the field. Pearson correlation analysis was used to compare the accuracy of the estimates of the three models to the measured defoliation.

All statistical analyses were performed using R software (R Foundation for Statistical Computing: Vienna, Austria) [51], with $p = 0.05$ used to indicate significance. Kruskal-Wallis test and Pearson correlation analysis were performed using the R "stats" package [51], GLMM model using "nlme" [52], and Random Forests using the "randomForest" [48] packages.

3. Results

3.1. Relationship between Defoliation and Hardwood Content

Both percent hardwood content and D_0 had significant effects on balsam fir defoliation (Table 2), with a significant interaction between percent hardwood and D_0, indicating that the relationship between defoliation and percent hardwood varied significantly with overall defoliation severity each year. Examining the fitted relationships from GLMM, balsam fir defoliation was negatively related to percent hardwood content each year from 2012 to 2016 (Figure 2). The fitted lines indicated that the relationship between defoliation and hardwood amount was weak in 2012 and became stronger in 2013 and 2014, the second and third years of defoliation, then declined 2015 and 2016. Defoliation was highest in softwood plots in 2014 and 2015 (means of 79% and 87%, respectively), and in those years mean defoliation in hardwood plots was 12% and 55%, respectively (Figure 3). In 2012, the first year of defoliation, mean defoliation of balsam fir in softwood, mixedwood, and hardwood plots was 27%, 14%, and 12%, respectively. Average plot defoliation was significantly different among stand types (softwood > mixedwood > hardwood) in 2013 and 2014, when defoliation rapidly increased in softwood and mixedwood plots (Figure 3). Defoliation in softwood was significantly higher than in hardwood plots in all 5 years, but was significantly higher than in mixedwood plots only in 2013 and 2014. Defoliation peaked in 2015 in all three stand types, with means of 87%, 70%, and 55% defoliation in softwood, mixedwood, and hardwood plots, respectively (Figure 3). Defoliation declined in 2016, to 47% and 42% defoliation in softwood and mixedwood and 15% defoliation in hardwood plots, comparable to years prior to 2015. Over the 5 years, defoliation in softwood plots averaged 14% higher than in mixedwood plots, and defoliation in mixedwood plots averaged 20% higher than in hardwood plots. Average fir defoliation in hardwood plots remained below 20% in all years except 2015, the year with the highest defoliation in all stand types (Figure 3).

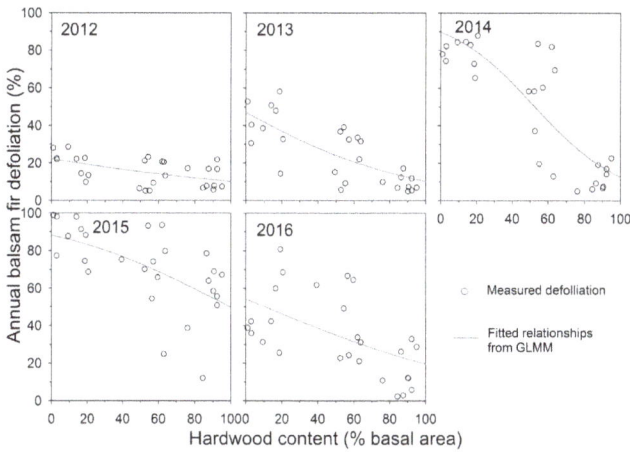

Figure 2. Measured balsam fir defoliation and fitted relationships from generalized linear mixed-effects model to test the effects of percent hardwood content on annual plot defoliation from 2012 to 2016, for 27 plots near Amqui, Quebec.

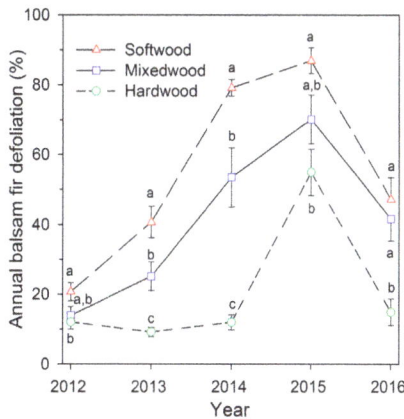

Figure 3. Average annual balsam fir defoliation (± one standard error) from 2012 to 2016, for nine plots in each of three stand types with varied hardwood contents, near Amqui, Quebec. Different letters indicate significant differences among stand types in each year.

Table 2. Results of generalized linear mixed-effects model to test the effects of percent hardwood content (HW%) and average annual defoliation in softwood plots (D_0) on defoliation for 27 plots from 2012 to 2016 in Quebec.

Fixed Effects	Parameter Estimates			Analysis of Deviance	
	Par.	Est.	SE	X^2	p
Intercept	β_0	−2.1223	0.3434		
HW%	β_1	−0.0089	0.0061	46.68	<0.001
D_0	β_2	0.0490	0.0055	175.70	<0.001
HW% × D_0	β_4	−0.0002	0.0001	6.51	0.011

3.2. Defoliation Estimated Using Three Model Formulations

The Random Forests model demonstrated the best performance among all three models, yielding a correlation of 0.92 with defoliation measured in the field (Figure 4C). The GLMM model produced an intermediate correlation ($r = 0.87$) among the three models (Figure 4B), while defoliation estimated using the simplified linear model showed the lowest correlation ($r = 0.77$; Figure 4A). The GLMM described the effects of hardwood content and D_0 on balsam fir defoliation well ($R^2 = 0.85$ considering fixed effects only and 0.94 considering both fixed and random effects [53]).

The simplified linear model that estimated defoliation using Equation (1) largely underestimated defoliation in mixedwood and hardwood plots under moderate and severe defoliation scenarios (Figure 4A). The GLMM performed well but still underestimated defoliation in some hardwood and mixedwood plots under mid-range defoliation level (Figure 4B). Among the three models, the Random Forests model provided the most accurate defoliation estimates, but a slight underestimation in softwood plots when actual defoliation was >80% (Figure 4C). Interestingly, both the Random Forests and GLMM models slightly overestimated defoliation in some hardwood plots under light defoliation levels (<20%) (Figure 4B,C).

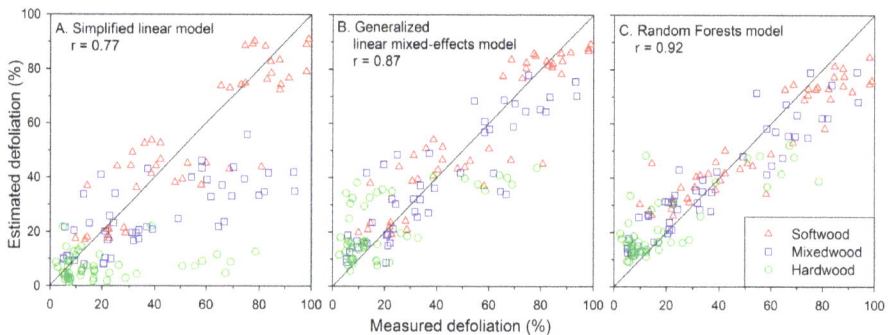

Figure 4. Annual plot defoliation, estimated with: (**A**) the simplified linear model (Equation (1)); (**B**) a generalized linear mixed-effects model; and (**C**) the Random Forests model, plotted against measured defoliation for 27 plots each year from 2012 to 2016.

Amongst the 11 variables used to predict annual defoliation with Random Forests, average annual defoliation in softwood plots (D_0) and percent hardwood content were the most important predictor variables, at increases in mean square error of 43% and 18%, respectively (Figure 5). D_0 was important as an indicator of the overall spruce budworm outbreak severity in a given year, while inclusion of percent hardwood content confirmed its significance in predicting budworm defoliation. Mean DBH, mean height, elevation, and standard deviation of bF height ranked as the third to the sixth most important predictors, at 8%–9% increases in mean square error (Figure 5). Other predictor variables included in the model (BA, slope, mean DBH of bF, standard deviation of bF crown base height, and mean bF crown base height) each had <5% increase in mean square error (Figure 5). We tried running the Random Forests model eliminating some of the variables with little contribution to accuracy (e.g., using only the top several variables in Figure 5), and the resulting correlation between measured defoliation and the model estimate dropped by 2%–3%. Variables in the Random Forests model and the GLMM essentially converged, along with the addition of DBH or Height and Elevation in the Random Forests model.

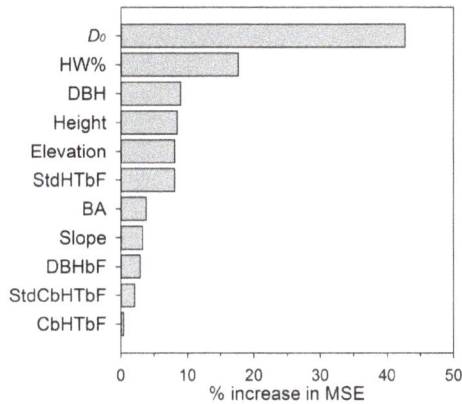

Figure 5. Variable importance (percent increase in mean square error of the Random Forests model when the data for that variable were randomly permuted) of the 11 predictors used to predict spruce budworm defoliation in fir-hardwood mixed stands. High values of percent increase in mean square error indicate more important variables in the Random Forests model. The 11 predictor variables were: average annual defoliation in softwood plots (D_0), percent hardwood basal area (HW%), mean DBH (DBH), mean height (Height), elevation, standard deviation of balsam fir (bF) height (StdHTbF), basal area (BA), slope, mean DBH of bF (DBHbF), standard deviation of bF crown base height (StdCbHTbF), and mean bF crown base height (CbHTbF).

4. Discussion

4.1. Relationships between Defoliation and Hardwood Content during Building Phase of a Spruce Budworm Outbreak

Our results on the relationship between spruce budworm defoliation of balsam fir and hardwood content during the building phase of an outbreak generally conformed to observations in the literature that tree diversity reduces herbivory by oligophagous insects [1]. Defoliation of fir was lower in plots with higher hardwood content in each of the 5 years of our study. Density of softwood and mixedwood plots was approximately twice that of hardwood plots. Under dense host conditions, natural enemies of budworm may have more predation opportunities since budworm as the prey may be more concentrated. Average balsam fir diameter always exceeded the plot mean diameter in the softwood plots, but this relationship was variable in the mixedwood and hardwood plots. Female budworm moths may have less chance landing and laying eggs on balsam fir trees in mixedwood and hardwood plots as balsam fir was less dominant in these plots.

Relationships between defoliation and hardwood content varied significantly with average defoliation in softwood plots each year in the GLMM. The relationship was weak in 2012 and became stronger in 2013 and 2014 before declining again in 2015 and 2016. Similarly, a varying relationship between defoliation and hardwood content was found in the declining phase (1989–1993) of the last outbreak [15]. Relationships were stronger in the first 3 years of the sampling period (1989–1991) and declined gradually in the last 2 years (1992 and 1993). In the building phase of a spruce budworm outbreak, populations increase and eventually reach epidemic level. Likewise, populations gradually decrease to endemic levels in the declining phase. Royama et al. [54] suggested that the budworm outbreak cycle was mainly determined by postdiapause mortality caused by parasitism. The varying relationship between defoliation and hardwood content among years may reflect varying parasitism rates associated with hardwoods. For example, in 2012 when budworm population density was low, assemblage of hyperparasitoids was smaller than in years with high budworm density [55]; parasitism rates may be similar in softwood and hardwood plots. When budworm population density rapidly increased in 2013 and 2014, "switching behavior" of mobile generalist parasitoids was stronger in a heterogeneous environment than in homogeneous forest plots [55,56]; parasitism rates may be higher in plots with more hardwoods than in pure fir plots. Once the budworm population reached its peak in 2015, neither the "birdfeeder effect" [55] nor "switching behavior" could stabilize the food web or regulate the severity of the outbreak under the high prey density; thus, parasitism rates may have been similar again and the defoliation reached its highest level in all three stand types.

Judged based on dispersion of data points around the graphed defoliation versus hardwood content relationships (our Figure 2 versus Figure 2 in Su et al. [15]), our results showed somewhat weaker relationships in the spruce budworm building phase compared with the declining phase. Budworm annual defoliation is significantly affected by budworm outbreak status [32]. It is unclear why a stronger relationship and larger impact between budworm defoliation and hardwood content occurred in the declining phase than it is in the building phase, but there are two possible explanations. First, in the declining phase of a budworm outbreak, sudden declines in the spruce budworm population in a given year can occur (e.g., third to fourth instar larval population decreased about 20 times in 1988 for Plot 1 and Plot 2 in Royama et al. [54]). Given that the density of natural enemies (parasitoids) in year X is determined by their density in the previous year (X − 1) when the budworm population was higher, parasitism and its impact on budworm populations in year X would be disproportionately large [54], resulting in less defoliation and, thus, a strong relationship between defoliation and hardwood content. The second possible explanation is based on the "habitat fragmentation" hypothesis, because tree mortality occurs during the declining phase of a budworm outbreak, resulting in an increasing degree of habitat fragmentation for budworm larvae. After the peak of an outbreak, balsam fir trees and branches as the main food source for budworm would be sparser and patchier than earlier in the outbreak. The worsened host condition in the declining phase could have greater impact on budworm population, and thus defoliation, resulting in a stronger relationship between budworm herbivory and tree diversity. These hypotheses could be tested by further studies on spruce budworm population dynamics.

4.2. Which Model Provides the "Best" Defoliation Estimates?

The Random Forests model yielded the most accurate defoliation estimates among the three tested models. It incorporated 11 predictor variables, with percent hardwood the second most important predictor, following only average annual defoliation in softwood plots (D_0), reflecting average regional severity of defoliation in a particular year. Hardwood content was a significant factor in predicting budworm defoliation, similar to findings of MacKinnon and MacLean [30] and Colford-Gilks et al. [31]. Mean DBH and mean height ranked as the third and fourth important variables, suggesting that average tree size had some importance in predicting budworm defoliation. It has repeatedly been observed that mature and over-mature stands have the highest susceptibility and vulnerability to budworm outbreaks (e.g., [17]). Standard deviation of bF height was the only variable with >5% increase in mean square error among variables related with canopy structure, indicating that canopy structure had little importance in predicting defoliation in these plots, which were mature with a single canopy layer. We conclude that incorporating more independent variables improved the accuracy of defoliation predictions, but at the cost of constructing a larger and more complex model. Such models could be difficult to construct using traditional parametric approaches due to complex interactions among variables and violation of distributional assumptions. Nevertheless, hardwood content and D_0 as the two most important independent variables were not correlated and we suggest that they should be included in any budworm defoliation modeling attempts.

Defoliation estimates from the GLMM were less accurate than the Random Forests model but better than estimates from the simplified linear model. Like the Random Forests model, the GLMM model was constructed using sampled defoliation data from all three stand types, but fewer predictor variables were included. The GLMM modified the form of the relationship and included the different relationships between defoliation and percent hardwood content under varying defoliation severities.

In contrast, the simplified linear model assumed a constant relationship, but did have the benefit of being able to estimate defoliation in fir-hardwood stands without conducting comprehensive sampling, model construction, or parameter estimation. Percent hardwood content is readily available in all forest inventories and D_0 as an indicator of average annual defoliation severity can be estimated relative easily from either regional defoliation surveys carried out by government agencies in softwood stands, or regional spruce budworm population sampling such as second instar larval or moth sampling. With these data and Equation (1), defoliation in fir-hardwood stands could be quickly estimated. However, either traditional parametric (GLMM) model or non-parametric Random Forests model provided more accurate defoliation estimates than the simplified linear model. In similar analyses, we suggest that data about the average defoliation level, or another indicator of annual outbreak severity, should be included in addition to percent hardwood content.

5. Conclusions

Balsam fir defoliation caused by spruce budworm during the initiation and building phase of the outbreak (2012–2016) was significantly lower in plots with higher percent hardwood content. The relationship between defoliation and percent hardwood varied significantly with overall defoliation severity each year. Average defoliation in softwood was significantly higher than in hardwood plots in all years during the studied period, but was significantly higher than in mixedwood plots only in 2013 and 2014 when overall defoliation rapidly increased. Average defoliation severity in softwood plots and percent hardwood content were the two most important variables for predicting balsam fir defoliation caused by spruce budworm. Excluding average defoliation severity in softwood plots as a predictor variable decreased the accuracy of prediction of fir defoliation. A simplified linear model could be used to quickly estimate spruce budworm defoliation in mixedwood stands, but with lower accuracy than parametric or non-parametric modeling approaches that include percent hardwood content and an indicator of average defoliation level or overall outbreak severity. The varying relationship between forest stand composition and defoliation indicated that dynamics and community structure of spruce budworm that was possibly governed by complex ecological processes [57], e.g., "natural enemy" and/or "habitat fragmentation" hypotheses, varied not only in time and with prey density, but also with changing forest condition.

Author Contributions: The authors provided equal contribution towards decisions regarding methodology and study design. B.Z. wrote the manuscript draft; D.A.M., R.C.J., and E.S.E. contributed to manuscript revision and development.

Funding: This research was funded by the Atlantic Innovation Fund project "Early Intervention Strategy to Suppress a Spruce Budworm Outbreak" grant to D.A.M.

Acknowledgments: We thank Wayne E. MacKinnon of the Canadian Forest Service for providing inventory data for stand selection, Chris R. Hennigar for input on statistics, and Annik Proulx of Quebec Ministère des Forêts, de la Faune et des Parcs for excluding our sample plots from harvesting and insecticide treatment. Allison Dykstra, Evan Dracup, Matt Hill, Shawn Donovan, Jessica Cormier, Craig Wall, and Maggie Brewer assisted with field data collection.

Conflicts of Interest: The authors declare no conflicts of interest.

References

1. Jactel, H.; Brockerhoff, E.G. Tree diversity reduces herbivory by forest insects. *Ecol. Lett.* **2007**, *10*, 835–848. [CrossRef] [PubMed]
2. Morris, R.F.; Cheshire, W.F.; Miller, C.A.; Mott, D.G. The numerical response of avian and mammalian predators during a gradation of the spruce budworm. *Ecology* **1958**, *39*, 487–494. [CrossRef]
3. Belle-Isle, J.; Kneeshaw, D. A stand and landscape comparison of the effects of a spruce budworm (*Choristoneura fumiferana* (Clem.)) outbreak to the combined effects of harvesting and thinning on forest structure. *For. Ecol. Manag.* **2007**, *246*, 163–174. [CrossRef]

4. Morin, H.; Jardon, Y.; Gagnon, R. Relationship between spruce budworm outbreaks and forest dynamics in eastern North America. In *Plant Disturbance Ecology: The Process and the Response*; Johnson, E.A., Miyanishi, K., Eds.; Elsevier: New York, NY, USA, 2007; pp. 555–577.

5. Royama, T. Population dynamics of the spruce budworm *Choristoneura fumiferana*. *Ecol. Monogr.* **1984**, *54*, 429–462. [CrossRef]

6. Royama, T.; MacKinnon, W.E.; Kettela, E.G.; Carter, N.E.; Hartling, L.K. Analysis of spruce budworm outbreak cycles in New Brunswick, Canada, since 1952. *Ecology* **2005**, *86*, 1212–1224. [CrossRef]

7. Piene, H.; MacLean, D.A. Spruce budworm defoliation and growth loss in young balsam fir: Patterns of shoot, needle and foliage weight production over a nine-year outbreak cycle. *For. Ecol. Manag.* **1999**, *123*, 115–133. [CrossRef]

8. Hennigar, C.R.; MacLean, D.A.; Quiring, D.T.; Kershaw, J.A. Differences in spruce budworm defoliation among balsam fir and white, red, and black spruce. *For. Sci.* **2008**, *54*, 158–166. [CrossRef]

9. Natural Resources Canada. *Compendium of Canadian Forestry Statistics 1994*; Canadian Council of Forest Ministers: Ottawa, ON, Canada, 1995; pp. 1–217. ISBN 0-662-21710-1.

10. Sanders, C.J. *A Summary of Current Techniques Used for Sampling Spruce Budworm Populations and Estimating Defoliation in Eastern Canada*; Environment Canada, Canadian Forestry Service: Sault Ste. Marie, ON, Canada, 1980; pp. 1–33.

11. MacLean, D.A.; MacKinnon, W.E. Sample sizes required to estimate defoliation of spruce and balsam fir caused by spruce budworm accurately. *North. J. Appl. For.* **1998**, *15*, 135–140. [CrossRef]

12. Donovan, S.D.; MacLean, D.A.; Kershaw, J.A.; Lavigne, M.B. Quantification of forest canopy changes caused by spruce budworm defoliation using digital hemispherical imagery. *Agric. For. Meteorol.* **2018**, *262*, 89–99. [CrossRef]

13. MacLean, D.A. Vulnerability of fir-spruce stands during uncontrolled spruce budworm outbreaks: A review and discussion. *For. Chron.* **1980**, *56*, 213–221. [CrossRef]

14. Bergeron, Y.; Leduc, A.; Joyal, C.; Morin, H. Balsam fir mortality following the last spruce budworm outbreak in northwestern Quebec. *Can. J. For. Res.* **1995**, *25*, 1375–1384. [CrossRef]

15. Su, Q.; Needham, T.D.; MacLean, D.A. The influence of hardwood content on balsam fir defoliation by spruce budworm. *Can. J. For. Res.* **1996**, *26*, 1620–1628. [CrossRef]

16. Campbell, E.M.; MacLean, D.A.; Bergeron, Y. The severity of budworm-caused growth reductions in balsam fir/spruce stands varies with the hardwood content of surrounding forest landscapes. *For. Sci.* **2008**, *54*, 195–205. [CrossRef]

17. MacLean, D.A. Effects of spruce budworm outbreaks on the productivity and stability of balsam fir forests. *For. Chron.* **1984**, *60*, 273–279. [CrossRef]

18. Nealis, V.G.; Régnière, J. Insect host relationships influencing disturbance by the spruce budworm in a boreal mixedwood forest. *Can. J. For. Res.* **2004**, *34*, 1870–1882. [CrossRef]

19. Riihimäki, J.; Kaitaniemi, P.; Koricheva, J.; Vehviläinen, H. Testing the enemies hypothesis in forest stands: The important role of tree species composition. *Oecologia* **2005**, *142*, 90–97. [CrossRef] [PubMed]

20. Siemann, E.; Tilman, D.; Haarstad, J.; Ritchie, M. Experimental tests of the dependence of arthropod diversity on plant diversity. *Am. Nat.* **1998**, *152*, 738–750. [CrossRef] [PubMed]

21. Quayle, D.; Régnière, J.; Cappuccino, N.; Dupont, A. Forest composition, host-population density, and parasitism of spruce budworm *Choristoneura fumiferana* eggs by *Trichogramma minutum*. *Entomol. Exp. Appl.* **2003**, *107*, 215–227. [CrossRef]

22. Cardinale, B.J.; Srivastava, D.S.; Duffy, J.E.; Wright, J.P.; Downing, A.L.; Sankaran, M.; Jouseau, C. Effects of biodiversity on the functioning of trophic groups and ecosystems. *Nature* **2006**, *443*, 989–992. [CrossRef] [PubMed]

23. Kemp, W.P.; Simmons, G.A. Influence of stand factors on survival of early instar spruce budworm. *Environ. Entomol.* **1979**, *8*, 993–996. [CrossRef]

24. Cappuccino, N.; Lavertu, D.; Bergeron, Y.; Régnière, J. Spruce budworm impact, abundance and parasitism rate in a patchy landscape. *Oecologia* **1998**, *114*, 236–242. [CrossRef] [PubMed]

25. Yamamura, K. Biodiversity and stability of herbivore populations: Influences of the spatial sparseness of food plants. *Popul. Ecol.* **2002**, *44*, 33–40. [CrossRef]

26. Needham, T.; Kershaw, J.A.; MacLean, D.A.; Su, Q. Effects of mixed stand management to reduce impacts of spruce budworm defoliation on balsam fir stand-level growth and yield. *North. J. Appl. For.* **1999**, *16*, 19–24. [CrossRef]

27. Sainte-Marie, G.B.; Kneeshaw, D.D.; MacLean, D.A.; Hennigar, C.R. Estimating forest vulnerability to the next spruce budworm outbreak: Will past silvicultural efforts pay dividends? *Can. J. For. Res.* **2014**, *45*, 314–324. [CrossRef]

28. Stoszek, K.J.; Mika, P.G.; Moore, J.A.; Osborne, H.L. Relationships of Douglas-fir tussock moth defoliation to site and stand characteristics in northern Idaho. *For. Sci.* **1981**, *27*, 431–442. [CrossRef]

29. De Somviele, B.; Lyytikäinen-Saarenmaa, P.; Niemelä, P. Sawfly (Hym., Diprionidae) outbreaks on Scots pine: Effect of stand structure, site quality and relative tree position on defoliation intensity. *For. Ecol. Manag.* **2004**, *194*, 305–317. [CrossRef]

30. MacKinnon, W.E.; MacLean, D.A. The influence of forest and stand conditions on spruce budworm defoliation in New Brunswick, Canada. *For. Sci.* **2003**, *49*, 657–667. [CrossRef]

31. Colford-Gilks, A.K.; MacLean, D.A.; Kershaw, J.A.; Béland, M. Growth and mortality of balsam fir-and spruce-tolerant hardwood stands as influenced by stand characteristics and spruce budworm defoliation. *For. Ecol. Manag.* **2012**, *280*, 82–92. [CrossRef]

32. MacLean, D.A.; MacKinnon, W.E. Effects of stand and site characteristics on susceptibility and vulnerability of balsam fir and spruce to spruce budworm in New Brunswick. *Can. J. For. Res.* **1997**, *27*, 1859–1871. [CrossRef]

33. Rowe, J.S. *Forest Regions of Canada*; Environment Canada, Canadian Forestry Service: Ottawa, ON, Canada, 1972; pp. 1–172.

34. Ministère des Forêts de la Faune et des Parcs. *Aires Infestées Par la Tordeuse des Bourgeons de L'épinette au Québec en 2012-Version 1.1*; Gouvernement du Québec, Direction de la Protection des Forêts: Québec, QC, Canada, 2012; pp. 1–19.

35. Ministère des Forêts de la Faune et des Parcs. *Aires Infestées Par la Tordeuse des Bourgeons de L'épinette au Québec en 2016-Version 1.0*; Gouvernement du Québec, Direction de la Protection des Forêts: Québec, QC, Canada, 2016; pp. 1–16. ISBN 978-2-550-7474-6.

36. MacLean, D.A.; Lidstone, R.G. Defoliation by spruce budworm: Estimation by ocular and shoot-count methods and variability among branches, trees, and stands. *Can. J. For. Res.* **1982**, *12*, 582–594. [CrossRef]

37. Morris, R.F. The development of sampling techniques for forest insect defoliators, with particular reference to the spruce budworm. *Can. J. Zool.* **1955**, *33*, 225–294. [CrossRef]

38. Blais, J.R. Effects of the destruction of the current year's foliage of balsam fir on the fecundity and habits of flight of the spruce budworm. *Can. Entomol.* **1953**, *85*, 446–448. [CrossRef]

39. McCullagh, P.; Nelder, J.A. *Generalized Linear Models*, 2nd ed.; Chapman Hall/CRC Press: London, UK, 1989; Volume 37, pp. 476–478. ISBN 0412317605.

40. Breiman, L. Random Forests. *Mach. Learn.* **2001**, *45*, 5–32. [CrossRef]

41. Candau, J.N.; Fleming, R.A. Forecasting the response of spruce budworm defoliation to climate change in Ontario. *Can. J. For. Res.* **2011**, *41*, 1948–1960. [CrossRef]

42. Penner, M.; Pitt, D.G.; Woods, M.E. Parametric vs. nonparametric LiDAR models for operational forest inventory in boreal Ontario. *Can. J. Remote Sens.* **2013**, *39*, 426–443. [CrossRef]

43. Lopatin, J.; Dolos, K.; Hernández, H.J.; Galleguillos, M.; Fassnacht, F.E. Comparing Generalized Linear Models and random forest to model vascular plant species richness using LiDAR data in a natural forest in central Chile. *Remote Sens. Environ.* **2016**, *173*, 200–210. [CrossRef]

44. Chen, X.; Ishwaran, H. Random forests for genomic data analysis. *Genomics* **2012**, *99*, 323–329. [CrossRef] [PubMed]

45. Prasad, A.M.; Iverson, L.R.; Liaw, A. Newer classification and regression tree techniques: Bagging and random forests for ecological prediction. *Ecosystems* **2006**, *9*, 181–199. [CrossRef]

46. Cutler, D.R.; Edwards, T.C.; Beard, K.H.; Cutler, A.; Hess, K.T.; Gibson, J.; Lawler, J.J. Random forests for classification in ecology. *Ecology* **2007**, *88*, 2783–2792. [CrossRef] [PubMed]

47. De'ath, G.; Fabricius, K.E. Classification and regression trees: A powerful yet simple technique for ecological data analysis. *Ecology* **2000**, *81*, 3178–3192. [CrossRef]

48. Liaw, A.; Wiener, M. Classification and regression by randomForest. *R News* **2002**, *2*, 18–22.

49. Murphy, M.A.; Evans, J.S.; Storfer, A. Quantifying *Bufo boreas* connectivity in Yellowstone National Park with landscape genetics. *Ecology* **2010**, *91*, 252–261. [CrossRef] [PubMed]

50. Dormann, C.F.; Elith, J.; Bacher, S.; Buchmann, C.; Carl, G.; Carré, G.; Marquéz, J.R.G.; Gruber, B.; Lafourcade, B.; Leitão, P.J. Collinearity: A review of methods to deal with it and a simulation study evaluating their performance. *Ecography* **2013**, *36*, 27–46. [CrossRef]

51. R Core Team. *R: A Language and Environment for Statistical Computing*; R Foundation for Statistical Computing: Vienna, Austria, 2018.

52. Pinheiro, J.; Bates, D.; DebRoy, S.; Sarkar, D.; R Core Team. *nlme: Linear and Nonlinear Mixed Effects Models*; R package version 3.1-137; 2018. Available online: https://CRAN.R-project.org/package=nlme (accessed on 25 June 2018).

53. Nakagawa, S.; Schielzeth, H. A general and simple method for obtaining R^2 from generalized linear mixed-effects models. *Methods Ecol. Evol.* **2013**, *4*, 133–142. [CrossRef]

54. Royama, T.; Eveleigh, E.S.; Morin, J.R.B.; Pollock, S.J.; McCarthy, P.C.; McDougall, G.A.; Lucarotti, C.J. Mechanisms underlying spruce budworm outbreak processes as elucidated by a 14-year study in New Brunswick, Canada. *Ecol. Monogr.* **2017**, *84*, 600–631. [CrossRef]

55. Eveleigh, E.S.; McCann, K.S.; McCarthy, P.C.; Pollock, S.J.; Lucarotti, C.J.; Morin, B.; McDougall, G.A.; Strongman, D.B.; Huber, J.T.; Umbanhowar, J.; et al. Fluctuations in density of an outbreak species drive diversity cascades in food webs. *Proc. Natl. Acad. Sci. USA* **2007**, *104*, 16976–16981. [CrossRef] [PubMed]

56. Křivan, V.; Schmitz, O.J. Adaptive foraging and flexible food web topology. *Evol. Ecol. Res.* **2003**, *5*, 623–652.

57. Régnière, J.; Nealis, V.G. The fine-scale population dynamics of spruce budworm: Survival of early instars related to forest condition. *Ecol. Entomol.* **2008**, *33*, 362–373. [CrossRef]

forests

MDPI

Article

Spatial-Temporal Patterns of Spruce Budworm Defoliation within Plots in Quebec

Mingke Li *, David A. MacLean, Chris R. Hennigar and Jae Ogilvie

Faculty of Forestry and Environmental Management, University of New Brunswick, P.O. Box 4400, Fredericton, NB E3B 5A3, Canada; macleand@unb.ca (D.A.M.); hennigar@forusresearch.com (C.R.H.); jae.ogilvie@unb.ca (J.O.)
* Correspondence: mingke.li@unb.ca

Received: 24 January 2019; Accepted: 1 March 2019; Published: 6 March 2019

Abstract: We investigated the spatial-temporal patterns of spruce budworm (*Choristoneura fumiferana* (Clem.); SBW) defoliation within 57 plots over 5 years during the current SBW outbreak in Québec. Although spatial-temporal variability of SBW defoliation has been studied at several scales, the spatial dependence between individual defoliated trees within a plot has not been quantified, and effects of defoliation level of neighboring trees have not been addressed. We used spatial autocorrelation analyses to determine patterns of defoliation of trees (clustered, dispersed, or random) for plots and for individual trees. From 28% to 47% of plots had significantly clustered defoliation during the 5 years. Plots with clustered defoliation generally had higher mean defoliation per plot and higher deviation of defoliation. At the individual-tree-level, we determined 'hot spot trees' (highly defoliated trees surrounded by other highly defoliated trees) and 'cold spot trees' (lightly defoliated trees surrounded by other lightly defoliated trees) within each plot using local Getis-Ord Gi^* analysis. Results revealed that 11 to 27 plots had hot spot trees and 27% to 64% of them had mean defoliation <25%, while plots with 75% to 100% defoliation had either cold spot trees or non-significant spots, which suggested that whether defoliation was high or low enough to be a hot or cold spot depended on the defoliation level of the entire plot. We fitted individual-tree balsam fir defoliation regression models as a function of plot and surrounding tree characteristics (using search radii of 3–5 m). The best model contained plot average balsam fir defoliation and subject tree basal area, and these two variables explained 80% of the variance, which was 2% to 5% higher than the variability explained by the neighboring tree defoliation, over the 3–5 m search radii tested. We concluded that plot-level defoliation and basal area were adequate for modeling individual tree defoliation, and although clustering of defoliation was evident, larger plots were needed to determine the optimum neighborhood radius for predicting defoliation on an individual. Spatial autocorrelation analysis can serve as an objective way to quantify such ecological patterns.

Keywords: *Choristoneura fumiferana*; annual defoliation; spatial autocorrelation; spatial-temporal patterns; mixed effect models; intertree variance

1. Introduction

With advances in spatial analysis theory and methodologies, biologists and ecologists have become more interested in analyzing ecological processes from a spatial point of view. Spatial point pattern analyses can indicate how individuals locate with respect to each other, by testing the complete spatial randomness null hypothesis [1]. Spatial autocorrelation analyses and indices can be applied to detect if the observed value of a variable at one site is dependent on values at neighboring sites [2,3]. The first law of geography states "Everything is related to everything else, but near things are more related than those distant ones" [4]. Spatial autocorrelation analyses are used for ecological processes that are distance-related, including speciation, extinction, dispersal, and synchronous population fluctuations [5,6]. Quantitative indices of spatial autocorrelation identify and characterize distribution patterns of objects of interest on the ground, and help to bridge the gap between mechanisms and the behavior of the investigated phenomenon [7,8]. Therefore, spatial autocorrelation analyses are an appropriate method to detect spatial patterns of insects' behavior or of forest dynamics related to insect infestations (e.g., [7,8]).

Spruce budworm (*Choristoneura fumiferana* (Clem.); SBW) outbreaks are a dominant, periodic natural disturbance in eastern Canada [9]. SBW larvae consume the new foliage of balsam fir (*Abies balsamea* (L.) Mill.) and spruce (*Picea* spp. A. Dietr.) trees [10], often for 10 or more years. This results in tree mortality typically ranging from about 40% to 85%, depending upon defoliation severity, tree species, and tree age [11,12]. Baskerville and MacLean [13] observed that SBW-caused tree mortality in immature balsam fir tended to have a strongly contagious spatial distribution. Mortality ranged from 18% to 80% of the total volume per hectare, and tended to occur in distinct patches spreading over time, with almost all trees in these patches killed [13]. Variation in mortality from plot to plot was therefore related to the extent to which such patches occurred in each plot. Spatial variability of mortality has important implications for study of stand vulnerability to SBW attack. If within-stand variability is high and apparently not related to stand characteristics [13], then conventional large scale, between-stand, sampling to establish characteristics of vulnerability could lead to statistically significant relationships between mortality and average stand conditions, even when mortality is not functionally related to average conditions [13].

SBW-caused tree mortality is strongly related to cumulative defoliation [14–16], so one might assume that a clustered spatial distribution of dead trees reflects higher annual defoliation in those trees than in surrounding trees. No studies have determined spatial patterns of SBW defoliation among trees, so in this study, we used spatial autocorrelation analyses to determine patterns of tree locations (clustered, dispersed, or random) and of annual defoliation of trees for 5 years in 57 plots in Gaspé, Québec, Canada. In addition to implications on stand vulnerability to insect-caused mortality, spatial pattern of defoliation is important for projecting effects on stand growth reduction and mortality in distance-dependent or tree-list growth models (e.g., [17]). Relationships between tree mortality and defoliation are non-linear, so use of average defoliation when some trees actually sustain higher or lower levels of defoliation will underestimate or overestimate impacts. For managing defoliated forests, knowledge of tree-to-tree variability of SBW defoliation can also benefit selection of the scale and methods of biological insecticide spray treatment decisions to target moderate or high infestation.

Although the spatial interrelationships of SBW defoliation levels of trees within a plot is not well understood, intertree variance in defoliation has been observed to be greater than intratree variance or variance between plots in a stand [18]. Defoliation levels are clearly a function of SBW population factors including oviposition site selection, larval and moth dispersal, and larval survival, but also may be influenced by tree, stand, site, and topographical factors. At the landscape scale, studies determining "epicenters" of historical defoliation based on aerial survey data have concluded that host species or volume had only minor effects on defoliation [19–21]. Defoliation among plots within stands was typically similar, with low variance (<5%), especially when SBW populations were high, for both eastern [22] and western SBW (*Choristoneura occidentalis* Freeman; [23]). At finer scales, variation of defoliation of branches within a stand was less than that of shoots on a branch [18], because branches from one site tended to have similar defoliation, and distribution of shoot defoliation tended to follow skewed Poisson distributions at both light and severe defoliation levels. Different host species suffer different degrees of defoliation: white spruce (*Picea glauca* (Moench) Voss), red spruce (*Picea rubens* Sarg.), and black spruce (*Picea mariana* (Mill.) B.S.P.) had approximately 72%, 41%, and 28% as much defoliation as balsam fir [10]. Non-host species also influence defoliation: balsam fir defoliation was lower when hardwood content increased, thought to be because of greater diversity and populations of SBW parasitoids [24,25]. In mixed fir-spruce stands, the density of balsam fir increased SBW defoliation severity in fir-dominated and fir-spruce mixed stands, while density of black spruce decreased the infestation in such stands [26].

There are several possible SBW population-related causes contributing to variable defoliation patterns. The frequency distribution of SBW larvae per tree is positively skewed, with a small proportion of trees hosting populations higher than the average [27]. This could result from unequal attraction of female moths to host trees, perhaps based on strength of their stimulus to the olfactory responses of the insect [27], or different exposure to light [28]. In stands without severe defoliation, SBW egg populations were usually higher on taller and dominant trees that were well exposed to light, while in severely defoliated stands, greater defoliation of exposed trees makes them less attractive to female moths, and correlation between the egg population and tree height became negative [27,28]. Larval SBW populations were greater in upper crowns because female moths tended to deposit eggs on the peripheral shoots of the crown [28,29]. Early instar larval dispersal may cause redistribution of larval population within a stand, but mass dispersal movements only occur in severely infested stands where food supply becomes exhausted [27]; such larval dispersal is risky because of high resulting mortality [30]. Although no direct data exist on between-tree spatial distributions of SBW egg masses or larval populations, spatial clustering of other insects has been observed: both adult and larval cereal leaf beetle (*Oulema melanopus* (L.)) were in a clustered distribution within 0.4 hectare areas in wheat fields [31]; and emerald ash borer (*Agrilus planipennis*) larval populations declined with distance, with about 90% of larvae within 100 m of adults' emergence sites [32].

In this study, we evaluated spatial patterns of 5 years of individual-tree SBW defoliation within 57 plots in Québec. Objectives were to: (1) determine whether spatial locations of tree stems within plots were clustered, dispersed, or random, as a baseline for comparing defoliation patterns; (2) quantify spatial aggregation of defoliation within plots, and detect if there was spatial dependence of individual tree defoliation; and (3) test whether the defoliation level of an individual balsam fir tree can be predicted using plot, tree, and neighboring tree defoliation and other characteristics. We evaluated two predictions: (1) defoliation of individual trees within a plot may be spatially clustered in some cases, i.e., highly defoliated trees will be surrounded by highly defoliated trees, and vice versa for lightly defoliated, as would result from preferential selection of oviposition sites by female moths; and (2) defoliation of an individual balsam fir tree within a plot can be predicted using defoliation level and species composition of surrounding neighborhood trees, combined with plot-level defoliation and species composition.

2. Materials and Methods

2.1. Study Area and Data Collection

The study area was located in the central Gaspé Peninsula region of Québec, a balsam fir-white birch (*Betula papyrifera* Marsh.) mixed eco-district [33]. According to aerial forest surveys, light SBW defoliation began in 2012 or 2013 and increased to moderate or severe levels since 2014 [34]. The 57 circular (400 m^2) study plots were established in 2014, with 19 plots about 15 km northwest of Amqui and 38 plots about 40 km southwest of Causapscal. These are a subset of plots studied by Donovan et al. [35] and Zhang et al. [25], and additional description of the sites is available there. Plot locations are shown in Figure 1 of [35]. However, we have omitted 15 plots that were harvested from 2016 to 2018, and three plots missing some defoliation data in 2015. A portion of our studied plots were protected by aerial spraying of *Bacillus thuringiensis* biological insecticide to control defoliation each year: 28% of plots in 2014, 42% in 2015, 30% in 2016, 26% in 2017, and 38% in 2018. The plots included a total of 3693 trees, of which 3200 were balsam fir or spruce.

Plot center coordinates were recorded using survey-grade GPS. Attribute data collected for each tree with diameter at breast height (DBH) ≥10 cm included species, tree azimuth, distance from plot center, DBH, total tree height, height of the live crown base, crown widths in north, east, south, and west directions, and annual ocular estimates of current defoliation using binoculars. Foliage defoliation was classified into seven classes: 0%–10%, 11%–20%, 21%–40%, 41%–60%, 61%–80%, 81%–99%, and 100%, and the midpoint of each class was used for calculations. To calibrate observers and check accuracy, one mid-crown branch was sampled from each of 8 to 15 trees per host species per plot (sample sizes based on [36]), and defoliation of each of 25 current-year shoots was assessed using the same classes as for ocular defoliation. Defoliation data were collected after SBW feeding ceased each year from 2014 to 2018. Criteria for plot selection and methods of data collection were described in more detail in [35].

2.2. Point Pattern Analyses

We first evaluated whether tree stems within plots were clustered, dispersed, or randomly distributed using a point pattern analysis of average nearest neighborhoods [37]. Liu [38] reviewed five methods for spatial pattern analysis, and indicated that when the purpose is to analyze the nearest neighbor, Pielou's statistics [39] and Clark and Evans' statistics (used in this study; [40]) perform the best in avoiding Type II errors, although there was evidence showing that Pielou's statistics [39] have more power for detecting regular and Poisson cluster patterns. The Clark and Evans' statistics calculate the ratio of mean observed distance between all trees and their nearest neighbors, and expected mean nearest neighbor distance for a random arrangement [40]:

$$R = \frac{D_O}{D_E} \tag{1}$$

where $D_O = \frac{\sum_{i=1}^{n} d_i}{n}$ is the observed mean distance between each tree and its nearest neighbor, in which d_i equals the distance between tree i and its nearest neighbor, and n is the number of trees in the plot; $D_E = \frac{0.5}{\sqrt{\frac{n}{A}}}$ is the expected mean distance between each tree and its nearest neighbor given in an ideal random pattern, in which A is the area of minimum enclosing rectangle around all features (set as plot area, 400 m^2, in this study). Generally, if the average distance is less than the average for a hypothetical random distribution ($R < 1$), the distribution is considered to be clustered; if $R > 1$, the distribution tends to be dispersed, and if $R = 1$ the distribution follows a random arrangement. The standard deviation, z-score was calculated as [40]:

$$z = \frac{D_0 - D_E}{SE} \tag{2}$$

where $SE = \frac{0.26136}{\sqrt{\frac{n^2}{A}}}$ is the standard error of the mean, in which the constant is derived from the radius of a circle of unit area and the Poisson probability model. Average nearest neighborhood analyses were done separately on each of the 57 plots, on all host trees, by species (balsam fir, black spruce, and white spruce), and on hardwood trees. Analyses were only applied when number of trees analyzed was ≥5 per plot.

2.3. Spatial Autocorrelation Analyses

Two quantitative indices were used to detect and characterize spatial patterns of defoliation at plot and tree levels. First, global Moran's *I* coefficient [41] was computed, to produce an overall measure of similarities or dissimilarities between neighboring trees within a plot. It helps to estimate the intensity of spatial dependence for the entire plot and summarizes it with a single value [1,42]. Global Moran's *I* was calculated as:

$$I = \frac{n}{S_0} \cdot \frac{\sum \sum W_{ij} Z_i Z_j}{\sum Z_i^2} \tag{3}$$

where W_{ij} is the spatial weight between tree i and tree j using an inverse distance weighting method to conceptualize the spatial relationships between trees, such that numerical weights quantified the proximity of pair-wise observations, with closer trees having higher spatial weights [2]. Z_i or Z_j is the deviation of defoliation for tree i or tree j from the mean of all trees in the plot, calculated as $x_i - \bar{x}$ or

$x_j - \bar{x}$; $S_0 = \sum_{i=1}^{n} \sum_{j=1}^{n} W_{ij}$ is the aggregate of all spatial weights and n is the number of trees in the plot. Global Moran's I ranges from -1 to 1; $I > 0$ or $I < 0$ corresponds to a positive or negative spatial correlation, representing spatial clustering or spatial dispersion, respectively, and $I = 0$ means a random distribution [3]. However, global Moran's I is difficult to interpret unless combined with statistical significance tests, and can only be interpreted within the context of the complete spatial randomness hypothesis. Z-score for global Moran's I was computed as [41]:

$$Z = \frac{I - E(I)}{\sqrt{V(I)}} \tag{4}$$

where $E(I) = \frac{-1}{n-1}$ is the expectation of global Moran's I under the complete spatial randomness hypothesis; when the sample size tends to be infinite, $E(I)$ is zero. $V(I)$ is the variance, computed as $V(I) = E(I^2) - (E(I))^2$. The analyses omitted cases when all trees in the plot had the same level of defoliation.

Following the global Moran's I analysis, Getis-Ord Gi^* analysis, which is a local spatial autocorrelation statistic, was used to determine defoliation patterns of individual trees within each plot. Getis-Ord Gi^* analysis tests the spatial clustering of high or low values of the measured variable around location i, characterizing the internal clustering within each plot by determining the extent to which defoliation of a given tree is surrounded by a cluster of trees with either high or low values [43,44]. Gi^* statistics proportionally compare the local sum for tree i and its neighbors to the sum of all trees, which is computed as [43]:

$$G_i^* = \frac{\sum_j W_{ij} x_j - \bar{X} \sum_j W_{ij}}{S \sqrt{\frac{n \sum_j W_{ij}^2 - (\sum_j W_{ij})^2}{n-1}}} \tag{5}$$

where i is the subject tree, x_j is the defoliation value for one of the neighboring trees j, W_{ij} is the spatial weight between subject tree i and neighboring tree j determined by the fixed distance band method (i.e., neighboring trees within 5 m were set with the same weight), $\bar{X}$ is the average defoliation value of the entire plot, n is the total number of trees in the plot, and S is the standard deviation of the entire plot, given by $S = \sqrt{\frac{\sum_{j=1}^{n} x_j^2}{n} - (\bar{X})^2}$. Gi^* is calculated as sum of the differences between individual values and the mean of all individuals. Therefore, Gi^* is a standard normal distribution z-score. A statistically significant Gi^* results when the difference between calculated local sum and expected local sum is too large to be a random chance. High and positive values of Gi^* indicate 'hot spot trees', severely defoliated trees that are surrounded by severely defoliated trees, whereas low and negative values indicate 'cold spot trees', lightly defoliated trees surrounded by lightly defoliated trees.

Both size and shape of the study plot affect the ability of the Gi^* statistics to accurately estimate the type and significance of a spatial pattern, because objects near the edge would have fewer neighbors than those in the middle of the plot. This is known as edge effect [45]. If the plot does not contain the entire spatial process under study, as was the case in this research, it is important to consider edge effects to increase stability and power of statistics resulting from sampling [46]. In this study, to avoid complicated edge correcting procedures [47], we used an inner buffer zone in each plot to compensate for edge effects. A search radius of 5 m was used, and a corresponding central circular subplot of 6 m was selected inside each plot. Trees in the border area were excluded from the Getis-Ord Gi^* analysis. All spatial autocorrelation analyses were done using ArcMap 10.4 (ESRI, Redlands, CA, USA).

Both global Moran's I and local Getis-Ord Gi^* spatial autocorrelation analyses were conducted separately on each of the 57 sample plots, and results of these and point-pattern tree location analyses were presented as the number and percentage of plots that had significantly clustered results. With such a wide range of defoliation conditions purposefully sampled across plots and years (ranging from <5% to >90% current annual defoliation), it was highly unlikely that there would be a single clustered or not clustered answer for all plots.

2.4. Tree Defoliation Regression Model

To determine whether defoliation levels of individual balsam fir can be predicted by plot or surrounding-tree characteristics, nonlinear mixed effect regression models were fitted, with plots nested within years as random effects. Current-year defoliation of balsam fir was the response variable. Plot-level predictors included average defoliation and % basal area of each host species and hardwoods per plot. Tree-level predictors included basal area of the subject balsam fir, defoliation of the subject balsam fir in the previous year, average defoliation of surrounding (3, 4, and 5 m search radii) trees (calculated for balsam fir, black spruce, white spruce, and for all host trees), total basal area of surrounding trees (balsam fir, hardwood, spruce trees, and for all host trees), total basal area of surrounding trees with basal area higher than the subject balsam fir, representing relative social status of the subject tree (calculated for balsam fir, spruce, and for all host trees). Defoliation of the subject balsam fir in the previous year acted as a temporal variable, and those neighborhood-related predictors at the tree level acted as spatial variables because they represented defoliation levels and species composition around the subject tree. We also considered the possible impact of insecticide spraying by including a dummy variable representing the subject balsam fir being sprayed or not in the corresponding year. All subject trees were balsam fir located in the central 6 m radius subplots, and effects were tested using surrounding 3, 4, or 5 m search radii with at least one neighbor in the neighborhood.

Gradient boosting is a machine learning technique, which identifies the performance of decision trees by using gradients in the loss function (i.e., a measurement of the goodness of coefficients in the model) in a sequential fashion [48]. Gradient Boosting Machine analysis [49] was used to determine the most important variables using the "caret" package [50] within R [51]. One of the important features of Gradient Boosting Machine analysis is variable importance, which was summarized as the ranked variables based on their relative influence (importance) in training the model. Relative influence was computed based on how often a variable was selected for splitting, weighted by the squared improvement to the model in each split [52]. Basal area of the subject tree was included in each tested model because it is related to foliage amount and to avoid having zero variance of fitted individual balsam fir defoliation values within a plot when it was predicted solely using plot-level variables. Except for the subject tree basal area, relatively important predictors were added into candidate models in sequence according to their ranks by Gradient Boosting Machine analysis. The contribution of newly-added predictors was determined by likelihood ratio tests, and based on this, whether the models should include the added predictors was evaluated. Highly correlated variables ($r \geq 0.7$) were avoided in the candidate models. Beta regression was used because the response variable, defoliation of the subject balsam fir, was constrained between 0 and 1 (0% to 100%). A logit-link function was used to set up linear relationships between the response and predictor variables. Likelihood ratio tests were also used to determine whether the random effects from plots or years were significant, by comparing a mixed effect model and a fixed effect model as a null model. The significance of individual predictors in the models was assessed by *t* tests. The "nlme" and "gnls" functions in the "nlme" package [53] were used in fitting mixed effect models and fixed effect models, respectively.

Assumptions of normality and homoscedasticity of residuals were evaluated using residual plots. Goodness of fit of candidate models was assessed and compared by root mean square error (RMSE), adjusted r^2, and mean bias (predicted-observed).

3. Results

3.1. Stand and Plot Characteristics

Characteristics of the 57 plots in the sampled stands ranged from 13.5 to 18.0 m height, 14.9 to 22.2 cm DBH, 1075 to 1919 stem ha^{-1} density, and 34.9 to 47.7 m^2 ha^{-1} basal area (Table 1). Basal area of each host species in the stands was 6% to 98% balsam fir, 0% to 70% black spruce, and 0% to 65% white spruce (Table 1). Plot species composition averaged 64% balsam fir, 19% black spruce, 10%

white spruce, 2% other softwood species, and 5% hardwood species (Table 1). Annual defoliation in the 5 years from 2014 to 2018 averaged 34%, 51%, 28%, 38%, and 32% by ocular estimation, and 45%, 52%, 34%, 54%, and 38% by shoot defoliation estimation on sampled branches. Plot locations were specifically selected in 2014 to represent the full range of defoliation from <10% to 90%–100%, and insecticide sprayed plots were the only way to obtain low defoliation levels.

Table 1. Summary of mean ($\overline{X}$) and standard deviation (σ) characteristics per stand of 57 plots located in 19 stands near Amqui and Causapscal, Québec, Canada.

Stand No.	No. Plots [1]	Density (stem ha^{-1})		DBH [2] (cm)		Height (m)		Basal Area (m^2 ha^{-1})		Species Composition [3] (% Basal Area)				
		$\overline{X}$	σ	$\overline{X}$	σ	$\overline{X}$	σ	$\overline{X}$	σ	BF	BS	WS	HW	OSW
1	4	1919	289	15.0	1.1	13.7	0.9	38.8	3.8	54	30	4	6	7
2	5	1775	417	17.0	1.4	16.5	1.0	42.8	4.3	51	42	3	3	
3	2	1913	636	16.0	1.4	16.0	0.9	41.8	6.7	83	1	11	5	
4	1	1200		20.1		18.0		39.8		6	27	65	2	
5	2	1825	159	15.8	0.1	16.1	0.2	38.1	1.7	98	1		1	
6	5	1600	329	16.5	2.2	14.8	2.0	38.7	4.0	77	7	4	7	7
7	1	1950		14.9		14.8		38.3		78			20	2
8	2	1800	159	15.8	0.4	15.5	0.6	38.1	1.7	70	20	5	6	
9	3	1808	426	16.7	2.6	16.2	1.3	44.2	4.2	86		2	11	1
10	3	1475	275	17.2	0.9	15.9	0.5	36.3	2.7	22	70	7	1	
11	2	1238	106	18.1	1.0	16.5	0.7	34.9	1.8	54	41	2	2	
12	5	1775	191	17.0	0.8	17.3	1.0	43.5	5.7	50	38	9	2	
13	5	1250	317	20.0	1.7	18.4	1.1	42.2	5.5	57		41	2	
14	4	1844	236	16.5	0.8	16.2	0.6	42.4	4.5	90			10	
15	5	1730	224	16.7	0.4	15.2	0.7	40.3	5.3	69	28	2	1	
16	2	1113	177	22.2	1.1	19.2	0.2	47.7	2.7	55		43	2	
17	2	1075	159	20.3	0.9	16.1	0.2	40.0	2.0	71		11	6	15
18	2	1325	265	17.5	0.3	13.5	0.6	44.0	4.5	80			20	
19	2	1763	636	17.3	2.9	14.2	3.0	51.3	8.5	59		11	14	18

[1] We originally (in 2014) established at least three plots per stand [35], but 15 of the original 75 plots were harvested from 2016 to 2018, and three plots had missing defoliation data. Analyses in this paper are all within plots, and stands were used here only to summarize general characteristics. [2] DBH = diameter at breast height. [3] Species abbreviations: BF = balsam fir; BS = black spruce; WS = white spruce; HW = hardwood species, including balsam poplar (*Populus balsamifera*), American mountain ash (*Sorbus americana*), trembling aspen (*Populus tremuloides*), willow (*Salix* spp.), white birch, yellow birch (*Betula alleghaniensis*), red maple (*Acer rubrum*), mountain maple (*Acer spicatum*), and striped maple (*Acer pensylvanicum*); OSW = non-host softwood species, including eastern white cedar (*Thuja occidentalis*), eastern white pine (*Pinus strobus*), and eastern larch (*Larix laricina*).

3.2. Spatial Patterns of Tree Stems within Plots

Average nearest neighborhood analyses for all host species within each plot showed that 35 plots (61%) had randomly distributed host trees, and the remaining 22 plots (39%) had significantly dispersed host trees ($\alpha = 0.05$; Table 2). No plots had significantly clustered host tree stem locations. The average nearest neighborhood analysis for each host species and hardwoods also showed that balsam fir, black spruce, white spruce, and hardwoods were distributed randomly or dispersed in most plots, and few (0% to 5%) plots had significantly clustered tree stems by species (Table 2). Generally, the null hypothesis that there is no difference between observed and random nearest neighbor values was not rejected for almost all plots, either by species or for all host species combined. The results of these average nearest neighborhood analyses set the baseline tree spatial distributions for the following spatial autocorrelation analyses of defoliation levels.

Table 2. Number and percentage of plots with clustered, dispersed, or random tree stem locations based on average nearest neighbor analyses ($\alpha = 0.05$), for balsam fir, black spruce, white spruce, hardwoods, and all host species per plot.

	Balsam Fir	Black Spruce	White Spruce	Hardwoods	All Host Species
Clustered	0	1 (2%)	0	3 (5%)	0
Dispersed	21 (37%)	3 (5%)	1 (2%)	2 (4%)	22 (39%)
Random	35 (61%)	20 (35%)	12 (21%)	25 (44%)	35 (61%)

3.3. Spatial Patterns of Current Year Defoliation for Plots

Global Moran's *I* analyses results for all host species showed that each year from 2014 to 2018, 47%, 28%, 35%, 30%, and 33% of plots, respectively, had significantly clustered intertree defoliation patterns ($\alpha = 0.05$; Table 3). This indicates that in these plots, defoliated trees had a strong tendency to be located closer to trees with similar defoliation values. These results therefore differed substantially from that for the spatial locations of trees, in which no plots had clustered tree spatial distributions (Table 2). In roughly one-third to one-half of the plot-years, defoliation of trees was clustered. A total of 45 plots (79%) exhibited clustered defoliation in at least one of the five sampled years. For balsam fir, 11% to 33% of plots had clustered defoliation distribution, somewhat fewer than for all host species combined (Table 3). With relatively few spruce present, only 2% to 5% and 0% to 4% of plots had clustered defoliation on black spruce and white spruce (Tables 1 and 3). A dispersed defoliation pattern occurred only for black spruce in one plot, with no dispersed defoliation patterns for all host trees or by species.

Table 3. Number and percentage of plots with significantly clustered patterns of defoliation of trees, based on global Moran's *I* analyses among years, for balsam fir, black spruce, white spruce, and all host species in each plot ($\alpha = 0.05$, search radius = 5 m).

Year	Balsam Fir	Black Spruce	White Spruce	All Host Species
2014	19 (33%)	1 (2%)	2 (4%)	27 (47%)
2015	11 (19%)	3 (5%)	0	16 (28%)
2016	13 (23%)	1 (2%)	0	20 (35%)
2017	6 (11%)	2 (4%)	0	17 (30%)
2018	15 (26%)	2 (4%)	0	19 (33%)

We used box plots to compare the distributions of total basal area, average annual plot defoliation, and standard deviation of individual tree defoliation between plots with clustered and non-clustered defoliation on all host trees (Figure 1). Plots were divided into 25% balsam fir plot basal area classes (x-axis) in Figure 1 because the analyzed plots varied widely in species composition. With low balsam fir content (<25% basal area), 18 plot-years with non-clustered defoliation had significantly higher basal area than 12 plot-years with clustered defoliation (mean 46 versus 41 m^2 ha^{-1}; Figure 1a). Overall, annual plot defoliation levels increased with the proportion of balsam fir in plots (Figure 1b). Plots with clustered defoliation had higher defoliation in all balsam fir % basal area classes than plots with non-clustered defoliation, significantly different for 50% to 75% basal area fir plots (43% versus 33% defoliation in clustered and non-clustered plots; Figure 1b). Plots with clustered defoliation consistently had higher standard deviations of tree defoliation than plots with non-clustered defoliation, with significant differences of 9.1% and 6.4% defoliation for plots with 0% to 25% and 50% to 75% balsam fir (Figure 1c).

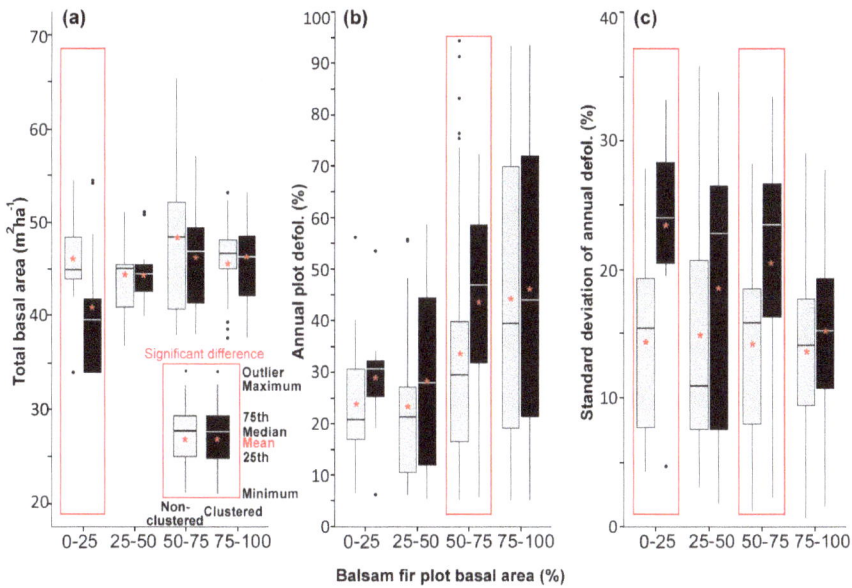

Figure 1. Comparison of plots with and without significant clustering of defoliation (based on results of global Moran's *I* analyses ($\alpha = 0.05$) for all host trees) by 25% balsam fir plot basal area classes for (**a**) total basal area in the plot, (**b**) average current year defoliation, and (**c**) standard deviation of individual tree defoliation within plots.

3.4. Hot Spot and Cold Spot Trees within Plots

In Figure 2, we integrated results of the global Moran's *I* and Getis-Ord *Gi** statistics for all analyzed plots in all years. We ordered the stand_plot number by annual plot defoliation each year, in order to visually convey that although defoliation was clustered in some plots (shown by * above each bar), how and whether hot or cold spots could be detected showed a tendency to be related to defoliation level. Hot spot trees (red bars in Figure 2) tended to be in lightly defoliated plots, and cold spot trees (blue bars) tended to be in highly defoliated plots. Across the five sampled years, minimum 42 and maximum 80 hot spot trees (highly defoliated trees surrounded by other highly defoliated trees), and minimum 44 and maximum 59 cold spot trees (lightly defoliated trees surrounded by other lightly defoliated trees) were detected within the 6 m-radius subplots (Figure 2). Plots with hot or cold spot trees did not necessarily have clustered defoliation (* in Figure 2), although plots with significantly clustered defoliation did tend to have hot or cold spot trees: 25%–45%, 10%–31%, 13%–35%, and 10%–36% of plots with clustered defoliation had hot spot trees, cold spot trees, both hot and cold spot trees, and non-significant spots, respectively, across the five years (Figure 2).

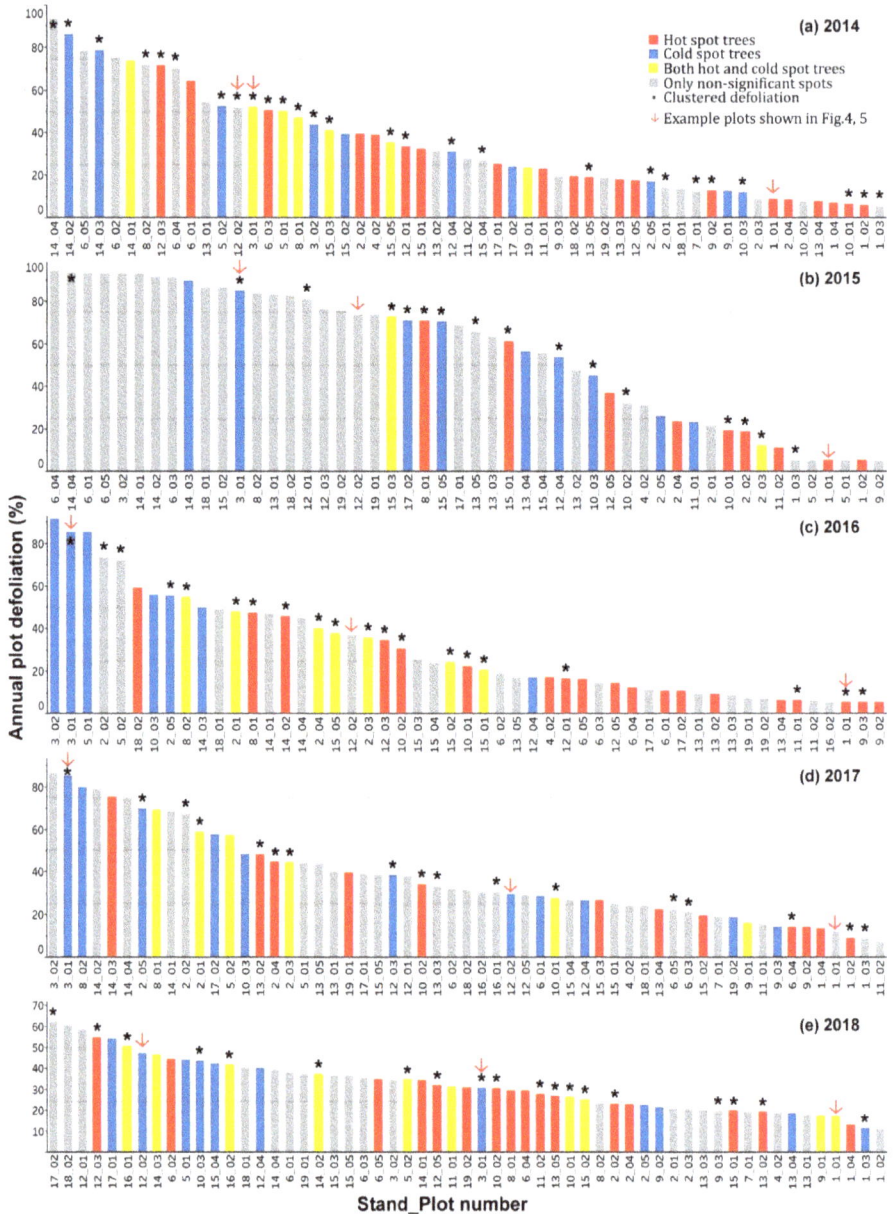

Figure 2. Average current year defoliation of plots, ordered from highest to lowest defoliation each year from 2014 to 2018 (**a–e**), showing plots with significantly clustered (α = 0.05) defoliation (*), with hot spot trees (**red**), cold spot trees (**blue**), both hot and cold spot trees (**yellow**), and only non-significant trees (**grey**), based on results of Getis-Ord Gi^* analyses (α = 0.05).

There was a tendency for highly defoliated plots to have more cold spot trees, and lightly defoliated plots to have more hot spot trees (Figure 2). Over the five sampled years, plots with 0%–25% defoliation had 0–4 cold spots (0.3 on average) and 0–8 hot spots (1.0 on average), while plots with

75%–100% defoliation had 0–12 cold spots (1.7 on average) and virtually no hot spots. Over the 5 sampled years, 27, 11, 26, 18, and 26 plots had hot spot trees (including those with both hot and cold spot trees), while 13 (48%), 7 (64%), 16 (62%), 7 (39%), and 7 (27%) of them were 0%–25% defoliated (Figure 3). Plots with 75%–100% defoliation had either cold spot trees or non-significant spots within them, except one plot in 2017 with >75% defoliation that had one hot spot tree (Figure 3d). That plot experienced 75% annual defoliation, with 100% balsam fir composition, and had only one balsam fir detected as a hot spot tree in 2017.

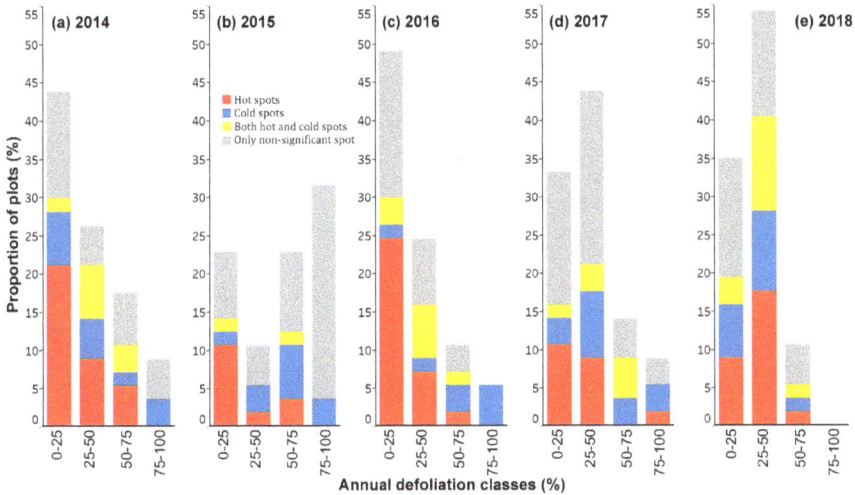

Figure 3. Proportion of plots with hot spot trees, cold spot trees, both hot and cold spot trees, and only non-significant trees (based on results of Getis-Ord Gi^* analyses ($\alpha = 0.05$)) by 25% annual defoliation classes from 2014 to 2018 (**a–e**).

Stem maps of three example plots were selected to represent generally low, moderate, and high defoliation that had hot and cold spot trees (Figures 4 and 5). Plot 1_01 had low defoliation (5%–17%) in all years from 2014 to 2018 (Figure 4), and had 1–5 hot spot trees in four years, but none in 2017 (Figure 5). Interestingly, locations of hot spot trees were the same for only three of the 11 trees, in two years (2015–2016); otherwise they differed. Mean defoliation of trees in plot 12_02 ranged from 29% to 74% over the five years, and covered defoliation classes from 0%–20% to 81%–100% (Figure 4). There were no hot spot trees and only six cold spot trees in two years in this plot (Figure 5). Plot 3_01 had very high defoliation (85%) in three years, and moderate (31% and 52%) defoliation in 2014 and 2018 (Figure 4), resulting in many cold spot trees in all years and two hot spot trees in the first year only (Figure 5). It is noteworthy that at moderate defoliation levels, variation among trees was high: with 31% mean defoliation in 2018 in plot 3_01, trees with all five 20% defoliation classes occurred in the plot (Figure 4).

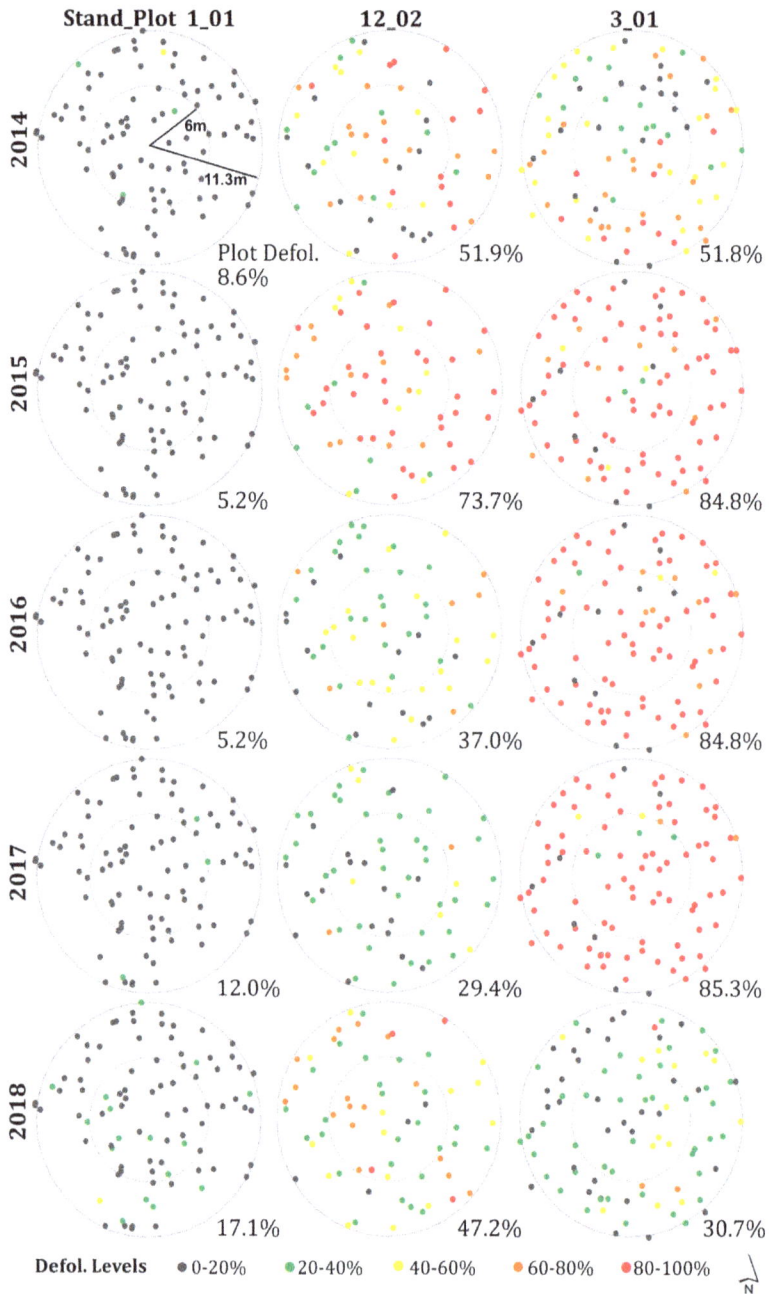

Figure 4. Stem maps of tree locations (diameter at breast height (DBH) ≥ 10 cm) of three example plots for five years, showing spatial distribution of defoliation. The three example plots were selected to represent generally low (1_01), moderate (12_02), and high (3_01) defoliation levels that contained hot spot and cold spot trees (shown in Figure 5).

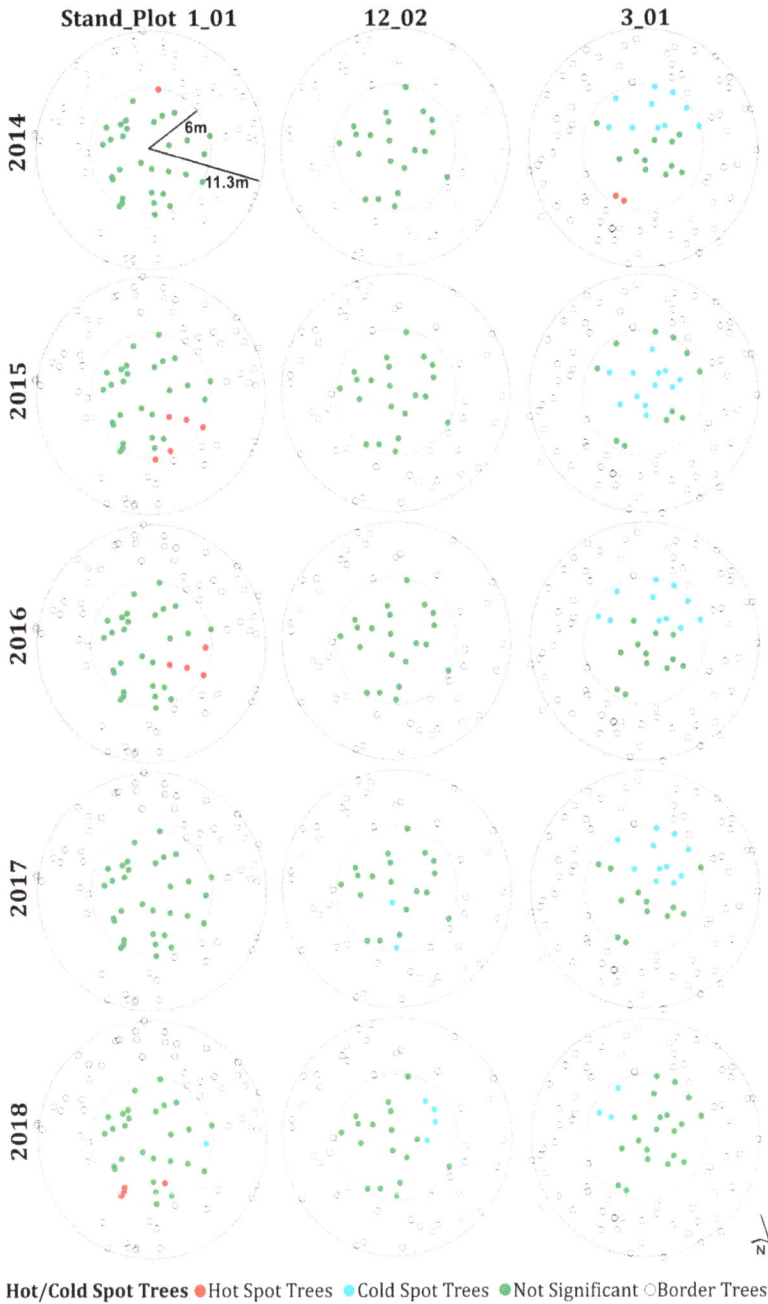

Figure 5. Stem maps of tree locations shown for the inner 6 m center of three example plots (same as in Figure 4) for five years, showing spatial distribution of hot spot trees, cold spot trees, and non-significant trees (based on results of Getis-Ord *Gi** analyses; α = 0.05).

3.5. Prediction of Subject Tree Balsam Fir Defoliation Using Regression Models

Results of Gradient Boosting Machine analysis consistently showed that plot average annual defoliation of balsam fir was the most important predictor for all three neighborhood search radii, with relative influence of 73%, 57%, and 48% for 3, 4, and 5 m search radii, respectively (Figure 6). Given the variability in defoliation among trees within a plot (e.g., Figure 4), it was surprising to us that plot average defoliation was superior to local within-plot defoliation for the three radii circles in predicting defoliation of a single tree. Average annual defoliation of surrounding balsam fir had the highest relative influence of all tree-level predictors, at 15%, 30%, and 41% for 3, 4, and 5 m search radii (Figure 6). Table 4 lists all plot- and tree-level predictors investigated using Gradient Boosting Machine analysis. Correlation analysis of the predictor variables suggested that plot average balsam fir defoliation (i.e., mean of all trees in the plot) was highly correlated ($r > 0.9$) with average defoliation of balsam fir within its neighborhood (i.e., within 3, 4, or 5 m circles). Therefore, average defoliation of balsam fir at the plot and tree level were not both included in the same candidate model. Basal area, representing subject tree size, was included in each model. Total basal area of host trees having higher basal area than the subject tree within 3 and 5 m, and total basal area of host trees within 4 m, which ranked as the sixth most important predictors, were also tested in candidate models. Spraying and defoliation in the previous year variables had little influence (0.3% at most) in the Gradient Boosting Machine analysis. The remaining predictors, including spray and defoliation in the previous year, were dropped in the following process.

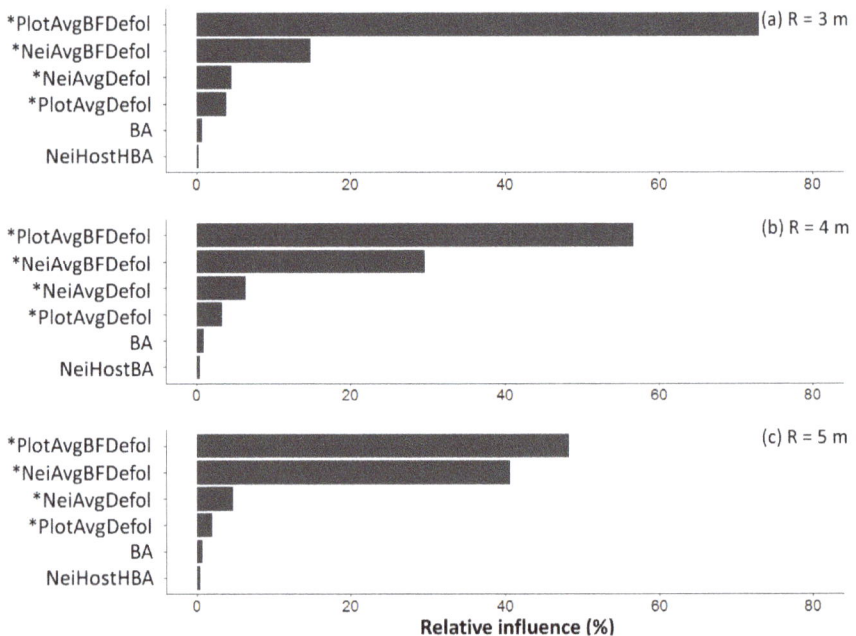

Figure 6. Relative influence (%) of the six most important predictor variables based on Gradient Boosting Machine analysis to predict current year defoliation of individual balsam fir trees (%) with neighborhood tree search radius (R) of (**a**) 3 m, (**b**) 4 m, or (**c**) 5 m. Predictor variable abbreviations are described in Table 4, and predictors marked with * were highly correlated with each other (correlation coefficient $r \geq 0.7$).

Table 4. Abbreviations and description of predictor variables at both plot and tree levels included in Gradient Boosting Machine analysis to determine their relative importance in predicting annual defoliation of a subject balsam fir tree.

Predictor Variables	Description
Plot level	
PlotAvgDefol	Average annual defoliation of all host species per plot (%)
PlotAvgBFDefol	Average annual defoliation of balsam fir per plot (%)
PlotBFBA	% basal area of balsam fir
PlotBSBA	% basal area of black spruce
PlotWSBA	% basal area of white spruce
PlotHWBA	% basal area of hardwoods
Spray	Dummy variable: whether the plot was sprayed by insecticide (1) in corresponding given year or not (0)
Tree level [1]	
BA	Basal area of the subject balsam fir ($m^2\ ha^{-1}$)
PreYearDefol	Annual defoliation of subject balsam fir in previous year (%)
NeiAvgDefol	Average annual defoliation of neighboring[1] host trees (%)
NeiAvgBFDefol	Average annual defoliation of neighboring balsam fir (%)
NeiAvgBSDefol	Average annual defoliation of neighboring black spruce (%)
NeiAvgWSDefol	Average annual defoliation of neighboring white spruce (%)
NeiHostBA	Total basal area of neighboring host trees ($m^2\ ha^{-1}$)
NeiBFBA	Total basal area of neighboring balsam fir ($m^2\ ha^{-1}$)
NeiSPBA	Total basal area of neighboring spruce trees ($m^2\ ha^{-1}$)
NeiHWBA	Total basal area of neighboring hardwoods ($m^2\ ha^{-1}$)
NeiHBA	Total basal area of all trees with basal area greater than the subject balsam fir in the neighborhood ($m^2\ ha^{-1}$)
NeiHostHBA	Total basal area of host trees with basal area greater than the subject balsam fir in the neighborhood ($m^2\ ha^{-1}$)
NeiSPHBA	Total basal area of spruce trees with basal area greater than the subject balsam fir in the neighborhood ($m^2\ ha^{-1}$)
NeiBFHBA	Total basal area of balsam fir with basal area greater than the subject balsam fir in the neighborhood ($m^2\ ha^{-1}$)

[1] Search radii of 3, 4, and 5 m were used for balsam fir trees in a circular subplot of 6 m inside each plot.

Plots nested within years was added as a random effect variable, but likelihood ratio tests suggested that it had little influence, in comparing models with mixed effects versus fixed effects as a null model ($p = 0.9$). This meant that the variance associated with plot-nested-in-year groups could occur by chance. Therefore, the random effects were removed from candidate models, and fit statistics of models with only fixed effects were reported (Table 5).

Results of models to predict subject tree defoliation as a function of plot and neighboring tree variables showed that Model 1 based on plot average balsam fir defoliation and subject tree basal area explained 80% of the total variance in the response variable, with RMSE of 14.1% and bias of 2.8% (Table 5). Variance explained by plot average balsam fir defoliation was slightly higher than that explained by neighboring average balsam fir defoliation within 4 or 5 m (79% and 78%), which were higher than 3 m (75%; Table 5). Coefficients ($p < 0.01$) in these four candidate models (Models 1, 2, 4, and 6) were all positive, indicating that higher subject balsam fir defoliation occurred with higher subject tree basal area, higher plot average defoliation, and higher neighboring balsam fir average defoliation. The other two predictors, total basal area of host trees having higher basal area than the subject tree within 3 m (Model 3) and total basal area of host trees within 4 m (Model 5) were significant ($p < 0.05$), and had slightly better r^2, RMSE, and bias, compared to Model 1. This suggested that including neighboring host tree basal area can slightly improve performance compared to a model that including only plot average balsam fir defoliation.

Table 5. Adjusted r², root mean squared error (RMSE), and mean bias of predictions of individual balsam fir defoliation (%) by candidate models with neighborhood tree search radius equal to 3, 4, and 5 m. Predictor variable abbreviations are described in Table 4.

Candidate Models	Predictors	Fit Statistics [1]		
		Adjusted r²	RMSE	Bias
Model 1	PlotAvgBFDefol + BA	0.8001	0.1411	0.0028
	Search radius = 3 m			
Model 2	NeiAvgBFDefol + BA	0.7539	0.1566	0.0017
Model 3	PlotAvgBFDefol + BA + NeiHostHBA	0.8015	0.1406	0.0025
	Search radius = 4 m			
Model 4	NeiAvgBFDefol + BA	0.7823	0.1473	0.0019
Model 5	PlotAvgBFDefol + BA + NeiHostBA	0.8007	0.1409	0.0027
	Search radius = 5 m			
Model 6	NeiAvgBFDefol + BA	0.7889	0.1450	0.0025

[1] The fit statistics were tested for fixed effect models without random effect terms, which in previous model runs had little contribution to models by likelihood ratio tests ($p = 0.9$).

4. Discussion

4.1. Is Defoliation of Individual Trees Clustered

Generally, defoliation was clustered in some plots and some years, depending on defoliation levels. Plots with clustered defoliation consistently had higher annual defoliation than plots with non-clustered defoliation. Spatial locations of fir and spruce trees within plots were randomly distributed in 61% of plots and dispersed in the remaining 39% of plots; none were clustered. In contrast, an average of 35% of plots had significantly clustered defoliation distributions of balsam fir and spruce, with amount clustered varying from 28% to 47% among years. Plots with clustered defoliation had larger standard deviation of defoliation among trees, reflecting higher variability in defoliation within the plots.

The added value of this study is the first demonstration that there can be spatially clustered defoliation patterns among trees, even though the baseline tree locations are not spatially clustered. Although SBW populations on each tree were not measured, higher defoliation must result from higher SBW populations or higher larval survival. Clustered defoliation within plots probably results from SBW population processes, primarily moth oviposition selection and larval dispersal. SBW egg populations are usually higher on taller and dominant trees that are well exposed to light, but severe defoliation may make such trees less attractive to female moths, and the correlation between egg population and tree height becomes negative [27,28]. Larval dispersal leads to tree-to-tree redistribution of SBW, which can result in reduced tree-to-tree correlation (i.e., more even SBW pressure), because such dispersal occurs rather randomly in direction especially at a small scale [29]. Not much air movement is needed for larval dispersal due to their light body weight [54]. However, a large proportion of the SBW population must be engaged in dispersal processes to contribute to a noticeable change in intertree population distribution, and such mass movements only occur in highly defoliated stands with high competition among feeding larvae [27]. Additionally, SBW larvae tend to avoid risky dispersal behavior because of the high probability of mortality during dispersal [30]. Given that current annual defoliation exceeded 80% in only 8.5% of our 260 plot-years sampled (and most of those were in 2015), larval dispersal probably had minor effects on intertree defoliation distribution in this study.

4.2. Interpretation of Local Hot and Cold Spot Trees

Conceptually, Getis-Ord *Gi** is the ratio of average local defoliation over average global (plot in our case) defoliation. It indicates hot spot (highly defoliated trees surrounded by other highly defoliated) and cold spot (lightly defoliated trees surrounded by other lightly defoliated) trees. Over all plots and years, 28% of plot-years had hot spot trees, 18% had cold spot trees, 12% had both, and 42% had none; in other words, there was significant spatial variability in defoliation in 58% of cases. Whether defoliation is high enough to be a hot spot or low enough to be a cold spot depends on the general defoliation level of the plot. SBW outbreak spatial-temporal dynamics typically begin with patchy low-level but increasing defoliation for several years, observable on individual branches initially and then on trees. It typically takes several years for SBW populations to build sufficiently to result in widespread severe defoliation (>70% of current year foliage on most trees) across stands and regions. Defoliation is most variable among trees and plots at moderate (30%–70% of current year foliage) levels; when low or very high, defoliation tends to be consistent [18]. Plots with severe defoliation tended to have cold spot trees, and plots with light defoliation tended to have hot spot trees. This is because a target highly defoliated tree surrounded by other highly defoliated trees will have a higher value of Gi^* (ratio of average local defoliation over average global defoliation) if it is located in a plot with overall low defoliation, while it is more likely to be non-significant if it is located in a highly defoliated plot. A similar principle applies for cold spot trees.

Highly significant global spatial autocorrelation can lead to overly liberal local spatial autocorrelation, which means significance tests of the local-spatial-autocorrelation coefficients can reject the null hypothesis excessively when a global spatial autocorrelation occurs [42]. Although the local spatial statistic used in this study, Getis-Ord Gi^*, does not need extra significant tests, values are still sensitive to the overall spatial structure of defoliation in plots [1]. Other statistics have been proposed to address such issues [55,56], but a local statistic avoiding influence from or accounting for global spatial autocorrelation is still needed [57]. As suggested by Sokal et al. [42], the major application of local spatial autocorrelation tests should be exploratory instead of significance testing. Our main conclusion here is that in about one-third to one-half of plots over 5 years, clustered defoliation patterns occurred, at either higher (hot spot) or lower (cold spot) levels than the plot average. Therefore the spatially contagious SBW-caused tree mortality observed by Baskerville and MacLean [13] within a uniform immature balsam fir stand, occurring at a scale smaller than the 400 m² plots, may have resulted from higher defoliation in some trees. Over the longer term, mortality creating such 'holes' in stands is also probably exacerbated by windthrow disturbance [58,59].

4.3. Prediction of Subject Balsam Fir Defoliation

Average defoliation for the full plot combined with subject tree basal area explained 80% of the variability of subject balsam fir defoliation, which was 2% to 5% higher than the variability explained by the neighboring tree defoliation, over the 3, 4, and 5 m search radii tested. However, the relative influence of neighboring tree defoliation on subject tree defoliation increased as search radius increased, from 15% to 30% to 41% as search radius increased from 3 m to 4 m to 5 m. At a 5 m search radius, the relative influence of neighboring tree defoliation was nearly as high as plot average defoliation. This suggested that a neighborhood search radius larger than 5 m, and therefore larger overall plot size, was needed, and it might be superior to average plot defoliation for individual-tree-defoliation prediction.

There was wide variation in defoliation among trees in the sampled plots, and although plot mean defoliation was the strongest predictor of subject tree defoliation, inferring similar average defoliation levels for all trees in a plot creates a problem when inputting defoliation values into stand growth models to predict effects of defoliation on growth and mortality. Model driving relationships between growth reduction, mortality, and defoliation are non-linear, such that trees with the highest defoliation will sustain substantially higher rates of growth reduction and mortality than the mean. Recognizing that a portion of trees are in hot or cold spots, with clusters of higher or lower defoliation, implies that model predictions will be more accurate using input distributions of defoliation for tree-list models and spatial defoliation distributions for distance-dependent growth models.

For these primarily balsam fir plots, species composition at either the plot level or tree level was not a meaningful predictor of individual balsam fir defoliation. The stands sampled had 51%–98% fir in all but four plots, which had 77%–92% black and white spruce. The four spruce plots exhibited hot and cold spot trees in 75% of the 20 plot-years sampled. Although hardwood content has been shown to influence SBW defoliation [24,25], hardwoods comprised only an average of 6% of basal area in these plots (range 1%–20%). This gave too few samples to test neighborhood effects of hardwoods on defoliation. Although insecticide spraying was an important predictor when forecasting plot-level annual defoliation [35], it had little impact on tree-level annual defoliation in this study, because the effects of spraying were already reflected by the plot mean defoliation variable. Defoliation of subject trees in the previous year also showed little influence, suggesting that individual-tree defoliation can vary considerably from year to year, as a function of SBW population level each year and insecticide treatments.

5. Conclusions

We evaluated neighborhood effects on individual-tree SBW defoliation, including the detection of defoliation clustering by spatial autocorrelation analysis, and the prediction of subject balsam fir defoliation by mixed effect regression. Key messages from the results include:

1. Including all host species, 47%, 28%, 35%, 30%, and 33% of plots showed significantly clustered defoliation patterns from 2014 to 2018. Plots with clustered defoliation tended to have higher and less uniform defoliation among trees. Results suggested that spatial defoliation patterns resulted from uneven SBW pressure on trees, perhaps from oviposition site selection.
2. Plots with severe defoliation generally tended to exhibit cold spot trees, and plots with light defoliation tended to have hot spot trees, because whether defoliation was high or low enough to be a hot or cold spot depended on the defoliation level of the entire plot.
3. Plot-level average defoliation combined with subject tree basal area explained 80% of the variability of subject balsam fir defoliation, which was 2% to 5% higher than variability explained by the neighboring tree defoliation.
4. Spatial variability of defoliation decreased with larger radius neighborhoods from 3 to 5 m, suggesting that a neighborhood search radius larger than 5 m (and thus plot sizes larger than 400 m^2 (11.3 m radius) to deal with edge effects) may provide better predictions of subject balsam fir defoliation.
5. For these primarily balsam fir plots, species composition at both plot and tree levels were not significant predictors of individual balsam fir defoliation.

Spatial autocorrelation analysis is a useful means to describe and quantify spatial-temporal, ecological patterns of insect defoliation. For managing defoliated forests, knowledge of tree-to-tree variability of SBW defoliation can benefit selection of the scale and methods of biological insecticide spray treatment decisions to target moderate or high infestation. Our results showed that in spite of within-plot variability, there was far more plot-to-plot variability in defoliation. It indicated that a large number of plots is preferred rather than fewer larger plots. On the other hand, understanding spatial variability among defoliation levels of trees can help improve defoliation inputs into distance-dependent or tree-list stand growth models (e.g., [17]). If only the averaged plot defoliation is used as inputs to growth and yield models for all trees in a stand, it will underestimate growth reduction, and especially mortality for those trees suffering from higher defoliation levels (i.e., clusters). Adopting a distribution of defoliation levels of trees as model inputs rather than merely the plot average values would likely result in more accurate forecasts of SBW impacts.

Author Contributions: The authors provided equal contribution towards decisions regarding methodology and study design. M.L. wrote the manuscript draft; D.A.M., C.R.H., and J.O. contributed to manuscript revision and development.

Funding: This research was funded by the Atlantic Innovation Fund project "Spruce Budworm Early Intervention Strategy", grant number 203544 to D.A.M. The Atlantic Innovation Fund was funded by the Atlantic Canada Opportunities Agency.

Forests **2019**, *10*, 232

Acknowledgments: We acknowledge project support from the Healthy Forest Partnership. We appreciate the excellent work of all field assistants involved in the data collection: Shawn Donovan, Sean Lamb, Rebecca Landry, Bo Zhang, Maggie Brewer, Jessica Cormier, Olivia Doran, David Alton, and Kerrstin Trainor.

Conflicts of Interest: The authors declare no conflict of interest.

References

1. Dale, M.R.; Fortin, M.J. *Spatial Analysis: A Guide for Ecologists*; Cambridge University Press: Cambridge, UK, 2014.
2. Sokal, R.R.; Oden, N.L. Spatial autocorrelation in biology 1. Methodology. *Biol. J. Linn. Soc.* **1977**, *10*, 199–228. [CrossRef]
3. Cliff, A.D.; Ord, J.K. *Spatial Processes: Models & Applications*; Pion Ltd.: London, UK, 1981.
4. Tobler, W.R. A computer movie simulating urban growth in the Detroit region. *Geogr. Econ.* **1970**, *46*, 234–240. [CrossRef]
5. Augustin, N.H.; Mugglestone, M.A.; Buckland, S.T. The role of simulation in modelling spatially correlated data. *Environmetrics* **1998**, *9*, 175–796. [CrossRef]
6. Knapp, R.A.; Matthews, K.R.; Preisler, H.K.; Jellison, R. Developing probabilistic models to predict amphibian site occupancy in a patchy landscape. *Ecol. Appl.* **2003**, *13*, 1069–1082. [CrossRef]
7. Ryan, P.A.; Lyons, S.A.; Alsemgeest, D.; Thomas, P.; Kay, B.H. Spatial statistical analysis of adult mosquito (Diptera: Culicidae) counts: An example using light trap data, in Redland Shire, southeastern Queensland, Australia. *J. Med. Entomol.* **2004**, *41*, 1143–1156. [CrossRef] [PubMed]
8. Foster, J.R.; Townsend, P.A.; Mladenoff, D.J. Spatial dynamics of a gypsy moth defoliation outbreak and dependence on habitat characteristics. *Landsc. Ecol.* **2013**, *28*, 1307–1320. [CrossRef]
9. Hardy, Y.; Mainville, M.; Schmitt, D.M. *Spruce Budworms Handbook: An Atlas of Spruce Budworm Defoliation in Eastern North America, 1938–1980*; U.S. Department of Agriculture, Forest Service, Cooperative State Research Service: Washington, DC, USA, 1986.
10. Hennigar, C.R.; MacLean, D.A.; Quiring, D.T.; Kershaw, J.A. Differences in spruce budworm defoliation among balsam fir and white, red, and black spruce. *For. Sci.* **2008**, *54*, 158–166.
11. MacLean, D.A. Vulnerability of fir-spruce stands during uncontrolled spruce budworm outbreaks: A review and discussion. *For. Chron.* **1980**, *56*, 213–221. [CrossRef]
12. MacLean, D.A. Effects of spruce budworm outbreaks on the productivity and stability of balsam fir forests. *For. Chron.* **1984**, *60*, 273–279. [CrossRef]
13. Baskerville, G.L.; MacLean, D.A. *Budworm-Caused Mortality and 20-Year Recovery in Immature Balsam fir Stands*; Inf. Rep. M-X-102; Maritimes Forest Research Centre, Canadian Forestry Service: Fredericton, NB, Canada, 1979; p. 56.
14. MacLean, D.A.; Ostaff, D.P. Patterns of balsam fir mortality caused by an uncontrolled spruce budworm outbreak. *Can. J. For. Res.* **1989**, *19*, 1087–1095. [CrossRef]
15. Erdle, T.A.; MacLean, D.A. Stand growth model calibration for use in forest pest impact assessment. *For. Chron.* **1999**, *75*, 141–152. [CrossRef]
16. Chen, C.; Weiskittel, A.; Bataineh, M.; MacLean, D.A. Even low levels of spruce budworm defoliation affect mortality and ingrowth but net growth is more driven by competition. *Can. J. For. Res.* **2017**, *47*, 1546–1556. [CrossRef]
17. Lamb, S.M.; MacLean, D.A.; Hennigar, C.R.; Pitt, D.G. Forecasting forest inventory using imputed tree lists for LiDAR grid cells and a tree-list growth model. *Forests* **2018**, *9*, 167. [CrossRef]
18. MacLean, D.A.; Lidstone, R.G. Defoliation by spruce budworm: Estimation by ocular and shoot-count methods and variability among branches, trees, and stands. *Can. J. For. Res.* **1982**, *12*, 582–594. [CrossRef]
19. Hardy, Y.J.; Lafond, A.; Hamel, L. The epidemiology of the current spruce budworm outbreak in Quebec. *For. Sci.* **1983**, *29*, 715–725.
20. Gray, D.R.; MacKinnon, W.E. Outbreak patterns of the spruce budworm and their impacts in Canada. *For. Chron.* **2006**, *82*, 550–561. [CrossRef]
21. Zhao, K.; MacLean, D.A.; Hennigar, C.R. Spatial variability of spruce budworm defoliation at different scales. *For. Ecol. Manag.* **2014**, *328*, 10–19. [CrossRef]

22. Ostaff, D.P.; MacLean, D.A. Spruce budworm populations, defoliation, and changes in stand condition during an uncontrolled spruce budworm outbreak on Cape Breton Island, Nova Scotia. *Can. J. For. Res.* **1989**, *19*, 1077–1086. [CrossRef]

23. Alfaro, R.I.; Shore, T.L.; Wegwitz, E. Defoliation and mortality caused by western spruce budworm: Variability in a Douglas-fir stand. *J. Entomol. Soc.* **1984**, *81*, 33–38.

24. Su, Q.; MacLean, D.A.; Needham, T.D. The influence of hardwood content on balsam fir defoliation by spruce budworm. *Can. J. For. Res.* **1996**, *26*, 1620–1628. [CrossRef]

25. Zhang, B.; Maclean, D.A.; Johns, R.C.; Eveleigh, E.S. Effects of hardwood content on balsam fir defoliation during the building phase of a spruce budworm outbreak. *Forests* **2018**, *9*, 530. [CrossRef]

26. Bognounou, F.; Grandpre, L.D.; Pureswaran, D.S.; Kneeshaw, D. Temporal variation in plant neighborhood effects on the defoliation of primary and secondary hosts by an insect pest. *Ecosphere* **2017**, *8*, 1–15. [CrossRef]

27. Morris, R.F.; Mott, D.G. Dispersal and the spruce budworm. In *The Dynamics of Epidemic Spruce Budworm Populations*; Morris, R.F., Ed.; Memoirs of the Entomological Society of Canada: Ottawa, ON, Canada, 1963; Volume 95, pp. 180–189.

28. Morris, R.F. The development of sampling techniques for forest insect defoliators, with particular reference to the spruce budworm. *Can. J. Zool.* **1955**, *33*, 107–223. [CrossRef]

29. Beckwith, R.C.; Burnell, D.G. Spring larval dispersal of the western spruce budworm (Lepidoptera: Tortricidae) in north-central Washington. *Environ. Entomol.* **1982**, *11*, 828–832. [CrossRef]

30. Miller, C.A. The measurement of spruce budworm populations and mortality during the first and second larval instars. *Can. J. Zool.* **1958**, *36*, 409–422. [CrossRef]

31. Reay-Jones, F.P.F. Spatial distribution of the cereal leaf beetle (Coleoptera: Chrysomelidae) in wheat. *Environ. Entomol.* **2010**, *39*, 1943–1952. [CrossRef] [PubMed]

32. Mercader, R.J.; Siegert, N.W.; Liebhold, A.M.; Mccullough, D.G. Dispersal of the emerald ash borer, *Agrilus planipennis*, in newly-colonized sites. *Agric. For. Entomol.* **2009**, *11*, 421–424. [CrossRef]

33. Bélanger, L.; Bergeron, Y.; Camiré, C. Ecological Land Survey in Quebec. *For. Chron.* **1992**, *68*, 42–52. [CrossRef]

34. QMRNF: Québec Ministère des Ressources Naturelles et de la Faune. Aires Infestées par la Tordeuse des Bourgeons de l'épinette au Québec en 2014. Available online: http://www.mffp.gouv.qc.ca/publications/forets/fimaq/insectes/tordeuse/TBE_2014_P.pdf (accessed on 20 October 2017).

35. Donovan, S.D.; MacLean, D.A.; Kershaw, J.A.; Lavigne, M.B. Quantification of forest canopy changes caused by spruce budworm defoliation using digital hemispherical imagery. *Agric. For. Meteorol.* **2018**, *262*, 89–99. [CrossRef]

36. MacLean, D.A.; MacKinnon, W.E. Sample sizes required to estimate defoliation of spruce and balsam fir caused by spruce budworm accurately. *North. J. Appl. For.* **1998**, *15*, 135–140.

37. Ebdon, D. *Statistics in Geography: A Practical Approach*; Basil Blackwell Ltd.: New York, NY, USA, 1985.

38. Liu, C. A comparison of five distance-based methods for spatial pattern analysis. *J. Veg. Sci.* **2001**, *12*, 411–416. [CrossRef]

39. Pielou, E.C. The use of point-to-plant distances in the study of the pattern of plant populations. *J. Ecol.* **1959**, *47*, 607–613. [CrossRef]

40. Clark, P.J.; Evans, F.C. Distance to nearest neighbor as a measure of spatial relationships in populations. *Ecol. Soc. Am.* **1954**, *35*, 445–453. [CrossRef]

41. Moran, P.A.P. Notes on continuous stochastic phenomena. *Biometrika* **1950**, *37*, 17–23. [CrossRef] [PubMed]

42. Sokal, R.R.S.; Oden, N.L.; Thomson, B.A. Local spatial autocorrelation in biological variables. *Biol. J. Linn. Soc.* **1998**, *65*, 41–62. [CrossRef]

43. Getis, A.; Ord, J.K. The analysis of spatial association by use of distance statistics. *Geogr. Anal.* **1992**, *24*, 189–206. [CrossRef]

44. Ord, J.K.; Getis, A. Local spatial autocorrelation statistics: Distributional issues and an application. *Geogr. Anal.* **1995**, *27*, 286–305. [CrossRef]

45. Monserud, R.A.; Ek, A.R. Plot edge bias in forest stand growth simulation models. *Can. J. For. Res.* **1974**, *4*, 419–423. [CrossRef]

46. Sui, D.Z.; Hugill, P.J. A GIS-based spatial analysis on neighborhood effects and voter turn-out: A case study in College Station, Texas. *Polit. Geogr.* **2002**, *21*, 159–173. [CrossRef]

47. Ripley, B.D. Tests of "randomness" for spatial point patterns. *J. R. Stat. Soc.* **1979**, *41*, 368–374. [CrossRef]

48. Friedman, J.H. Greedy function approximation: A gradient boosting machine. *Ann. Stat.* **2001**, *29*, 1189–1232. [CrossRef]

49. Ridgeway, G. Generalized boosted models: A guide to the GBM package. *R Package Vignette.* **2007**.

50. Kuhn, M. Building Predictive Models in R Using the caret Package. *J. Stat. Softw.* **2008**, *28*, 1–26. [CrossRef]

51. R Core Team. *R: A Language and Environment for Statistical Computing*; R Foundation for Statistical Computing: Vienna, Austria, 2018.

52. Elith, J.; Leathwick, J.R.; Hastie, T. A working guide to boosted regression trees. *J. Anim. Ecol.* **2008**, *77*, 802–813. [CrossRef] [PubMed]

53. Lindstrom, M.J.; Bates, D.M. Nonlinear mixed effects models for repeated measures data. *Int. Biom. Soc.* **1990**, *46*, 673–687. [CrossRef]

54. Batzer, H.O. Hibernation site and dispersal of spruce budworm larvae as related to damage of sapling balsam fir. *J. Environ. Entomol.* **1968**, *61*, 216–220. [CrossRef]

55. Kabos, S.; Csillag, F. The analysis of spatial association on a regular lattice by join-count statistics without the assumption of first-order homogeneity. *Comput. Geosci.* **2002**, *28*, 901–910. [CrossRef]

56. Boots, B. Developing local measures of spatial association for categorical data. *J. Geogr. Syst.* **2003**, *5*, 139–160. [CrossRef]

57. Getis, A. A history of the concept of spatial autocorrelation: A geographer's perspective. *Geogr. Anal.* **2008**, *40*, 297–309. [CrossRef]

58. Taylor, S.L.; MacLean, D.A. Spatiotemporal patterns of mortality in declining balsam fir and spruce stands. *For. Ecol. Manag.* **2007**, *253*, 188–201. [CrossRef]

59. Taylor, S.L.; MacLean, D.A. Legacy of insect defoliators: Increased wind-related mortality two decades after a spruce budworm outbreak. *For. Sci.* **2009**, *55*, 256–267.

forests

MDPI

Article

Detection of Annual Spruce Budworm Defoliation and Severity Classification Using Landsat Imagery

Parinaz Rahimzadeh-Bajgiran [1,*], Aaron R. Weiskittel [1], Daniel Kneeshaw [2] and David A. MacLean [3]

[1] School of Forest Resources, University of Maine, 5755 Nutting Hall, Orono, ME 04469, USA; aaron.weiskittel@maine.edu
[2] Department of Biological Sciences, University of Quebec in Montreal, Montreal, QC H3C 3P8, Canada; kneeshaw.daniel@uqam.ca
[3] Faculty of Forestry and Environmental Management, University of New Brunswick, Fredericton, NB E3B 5A3, Canada; macleand@unb.ca
* Correspondence: parinaz.rahimzadeh@maine.edu; Tel.: +1-207-581-2813

Received: 10 May 2018; Accepted: 12 June 2018; Published: 14 June 2018

Abstract: Spruce budworm (SBW) is the most destructive forest pest in eastern forests of North America. Mapping annual current-year SBW defoliation is challenging because of the large landscape scale of infestations, high temporal/spatial variability, and the short period of time when detection is possible. We used Landsat-5 and Landsat-MSS data to develop a method to detect and map SBW defoliation, which can be used as ancillary or alternative information for aerial sketch maps (ASMs). Results indicated that Landsat-5 data were capable of detecting and classifying SBW defoliation into three levels comparable to ASMs. For SBW defoliation classification, a combination of three vegetation indices, including normalized difference moisture index (NDMI), enhanced vegetation index (EVI), and normalized difference vegetation index (NDVI), were found to provide the highest accuracy (non-defoliated: 77%, light defoliation: 60%, moderate defoliation: 52%, and severe defoliation: 77%) compared to using only NDMI (non-defoliated: 76%, light defoliation: 40%, moderate defoliation: 43%, and severe defoliation: 67%). Detection of historical SBW defoliation was possible using Landsat-MSS NDVI data, and the produced maps were used to complement coarse-resolution aerial sketch maps of the past outbreak. The method developed for Landsat-5 data can be used for current SBW outbreak mapping in North America using Landsat-8 and Sentinel-2 imagery. Overall, the work highlights the potential of moderate resolution optical remote sensing data to detect and classify fine-scale patterns in tree defoliation.

Keywords: forest pests; defoliation; spruce budworm; multi-spectral remote sensing; Acadian region; Maine; Quebec

1. Introduction

Northeastern forests of the United States and Canada provide numerous products and services for human livelihood, wildlife, and the environment, including timber, fiber products, firewood, wildlife habitat, watershed protection, carbon storage, and recreation. Northeastern forests have been subjected to several biotic and abiotic stressors. Biotic stressors like the outbreak of pests and pathogens—in particular, spruce budworm (*Choristoneura fumiferana* Clem.; SBW)—have greatly changed forest structure and composition in recent years. SBW is the most damaging forest pest in Northeastern forests. As outbreaks occur on a regional scale, they modify carbon fluxes, as well as the vitality of the forest products industry and regional economics. The last outbreak in the 1970s in Northeastern forests affected 57 million ha of forests [1–3].

Compared to other destructive pests, such as the jack pine budworm (*Choristoneura pinus* Freeman) or gypsy moth (*Lymantria dispar* L.), SBW has longer cyclical and more synchronous outbreaks, occurring at somewhat regular 30–40 year intervals over large areas. SBW outbreak duration is typically 10 years or longer, resulting in widespread tree mortality and a loss of productivity in balsam fir (*Abies balsamea* (L.) Mill.) and spruce (*Picea* spp.) [4–6]. The current SBW outbreak started in 2006 in Quebec, and by the summer of 2017 had defoliated over 7.1 million ha of the province [7]. The first defoliation in northern New Brunswick (NB) was detected in 2015. Currently, the outbreak is expected to affect Maine (ME) in the near future, with about 2.3 million ha of spruce–fir stands being at risk of SBW defoliation [8].

Accurate estimation of annual SBW defoliation extent and severity mapping at the landscape scale is of high interest for SBW risk prediction, better estimation of economic impacts of the outbreak, quantifying wood supply losses and changes in wildlife habitats, and management decisions for forestry practices in order to mitigate the impacts of the infestation. The primary management response is use of biological insecticide *Bacillus thuringiensis* var. *kurstaki* to reduce defoliation and keep trees alive—or, in early intervention attempts, to prevent outbreaks or reduce severity. However, mapping SBW annual defoliation is challenging, due to the short period of optimum visual observation of the damage (red partially-consumed foliage is only visible from the air for about 2–3 weeks, before these reddened needles fall); high year-to-year variation in bud flush timing for balsam fir and spruce, and in turn for defoliation; and the vast geographical extent of defoliation [1]. Current landscape mapping of annual SBW defoliation is mainly based on aerial sketch maps (ASMs), also known as insect and disease surveys (IDS) in the United States. In general, aerial mapping of defoliation is time-consuming, costly, and requires skilled observers; in addition, its overall accuracy varies depending on differences in mapping techniques used in Canada or the United States, which can range from rather fine to coarse spatial resolution [1,9]. The accuracy of ASMs has been quantified in NB, indicating reasonably good accuracy for higher levels of defoliation, but generally lower accuracy in discriminating nil (0–10%) from light (11–30%) defoliation classes [1,10].

SBW defoliation that is of interest to forest management can be evaluated in two different categories: current-year (annual) defoliation and cumulative defoliation. SBW strongly prefers to feed on the new (current-year) foliage age class, but multiple years of feeding results in cumulative removal of all age classes of foliage on balsam fir and spruce trees. Current-year defoliation is quantified as the percentage of the current foliage age class that is removed and is correlated with the reddish discoloration of foliage that provides information on the location and severity of defoliation and distribution of budworm populations in any year. When SBW larvae feed, needles are severed and become entangled with silken threads and frass, later drying out to present a reddish-brown color. Since this phenomenon lasts for only 2–3 weeks, surveys to assess the level of current defoliation must be conducted during this period [1], before the dried foliage is lost. This type of information is normally provided through ASMs [11]. Cumulative defoliation is quantified as the percentage of all foliage on the tree that is removed by successive years of SBW feeding [12]. Cumulative defoliation can be assessed using binocular estimation of individual tree crowns [13], or by summing current annual defoliation over multiple years [14]. Cumulative defoliation information has value as a model input to predict growth reduction and mortality, but current defoliation is what is used in both decision support system (DSS) and protection planning.

Airborne and space-borne optical remote sensing (RS) sensors have been explored in various research in the past for detecting and quantifying insect-induced forest defoliation and mortality, as well as for forecasting future outbreak patterns [11,15,16]. Among them, multi-spectral satellite data, such as Land Remote Sensing Satellite (Landsat), Satellite Pour l'Observation de la Terre (SPOT), Moderate resolution Imaging Spectroradiometer (MODIS), and Sentinel-2 have been used to detect and characterize defoliators like gypsy moth [17–20], forest tent caterpillar (*Malacosoma disstria*) [21], jack pine budworm [22,23], hemlock looper (*Lambdina fiscellaria* (Guenée)) [24], pine-tree lappet (*Dendrolimus pini* L.) [25], and SBW [6,12,26–28]. Compared to other defoliators, remote sensing research on SBW defoliation is less available, and previous attempts to detect and quantify SBW defoliation have been primarily focused on cumulative defoliation [26–31]. Only a few studies have addressed current-year SBW defoliation detection using airborne or satellite imagery like Landsat and SPOT [12,32,33]; however, these did not use the remotely-sensed vegetation indices that are the focus of this work. Other recent remote sensing technologies, such as hyper-spectral sensors with narrow spectral bands and unmanned aerial vehicles, have the potential for forest pest damage and infestation detection and quantification [31–34]. However, methods based on these techniques have not yet been used for SBW current-year defoliation assessment in regional or national forest management and planning. Should remote sensing methodologies prove to be operationally feasible, cost-effective, and accurate, there is interest across several jurisdictions to adopt such methods.

Although multi-spectral satellite imagery has the technical capacity to map both current and cumulative SBW defoliation, detection of current-year defoliation can be problematic, because of the short period of time when detection is possible, generally around July, depending on the SBW and host tree phenology. Cloud contamination and the coarse temporal resolution of some satellites such as Landsat normally limit these type of studies. Satellite-derived data on forest defoliation can potentially have several advantages over ASMs. Errors like missed defoliated regions, which can happen during aerial surveys, may be avoided. Furthermore, the quantification of defoliation can be less subjective compared to ASM-derived data, and the operation can be less expensive and laborious. Long-term archives of Landsat data can also support studies that need accurate information about past forest defoliation, as well as damage extent and dynamics. Accurate historical maps and data on defoliation are needed to (1) better understand the behavior of SBW outbreaks; (2) as input for DSS simulation models, such as the SBW-DSS [35]; and (3) to identify risk factors influencing outbreak severity and future outbreak projections [9,36]. Many historical ASMs, such as historical SBW defoliation maps of Maine produced during the last outbreak in 1970s and 1980s, are coarse-resolution and would benefit from the results of this study.

The objectives of this research were to evaluate remote sensing methods for detection and quantification of current-year (annual) SBW defoliation using Landsat imagery. Two different study areas from Northeastern forests with different disturbance regimes, forest composition, and remote sensing data availability were evaluated. The first study area, in Quebec, was used to evaluate a model for the detection of recent annual SBW defoliation extent and severity using Landsat-5 imagery, with the ultimate goal of generating maps similar to ASMs. The research in the second study area, in Maine, focused on the 1970s–1980s SBW outbreak and evaluated the capability of Landsat-MSS data to detect SBW defoliation.

2. Methods

2.1. Study Area in Quebec, Canada

This study area (~35 × 35 km^2) was located in Quebec's North Shore region, where the current SBW outbreak originated in 2006 (Figure 1). Quebec's North Shore is part of the boreal forest, and the southern portion of the study area is within the balsam fir–white birch (*Betula papyrifera* Marsh.) domain, while the northern area is part of the black spruce (*Picea mariana* (Mill.) B.S.P))–moss domain [37]. Our study area was an ecotone between the two domains, where black spruce and balsam fir are the dominant tree species. White spruce (*Picea glauca* (Moench) Voss), trembling aspen (*Populus tremuloides* Michx.), jack pine (*Pinus banksiana* Lamb.), and white birch are also present across the study area. A large portion of the landscape is free from logging activities (unmanaged). In comparison with continental Canada, this region has a lower fire frequency, and thus it is dominated by old-growth forests, along with irregularly shaped patches of younger forests recovering from fire or other severe disturbances [38,39]. The major prominent natural disturbances in the region are periodic SBW defoliation and fire.

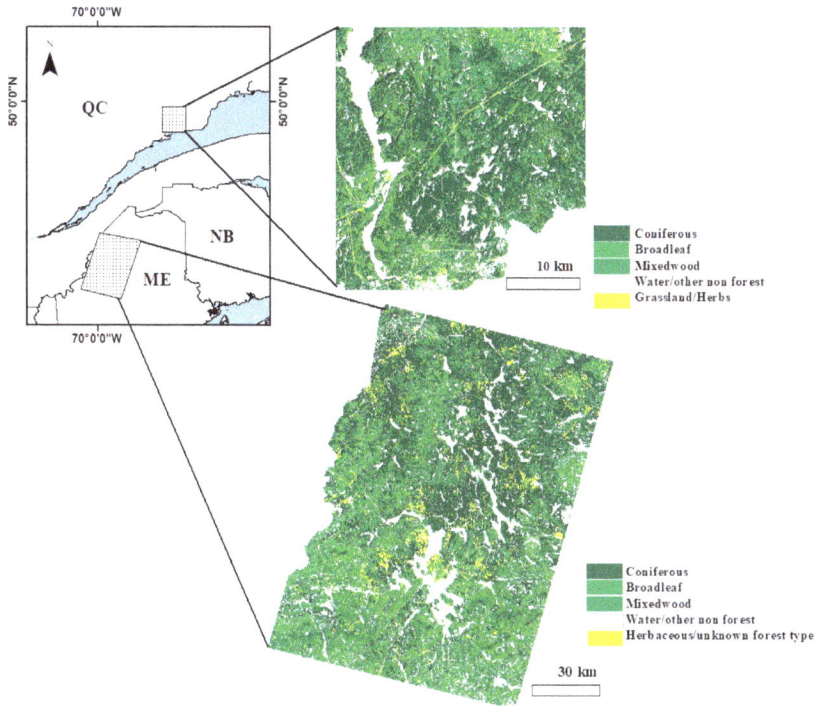

Figure 1. Location of the study areas in Quebec (QC), Canada and Maine (ME), United States. The Quebec study area (~35 × 35 km²) was located in Landsat-5 scene 12/26, and the Maine study area (~100 × 150 km²) was located in Landsat-MSS scene 13/28 [40].

2.2. Study Area in Maine, United States

This study area (~100 × 150 km²) was located in the northern part of Maine (Figure 1). This region is part of New England Acadian forests, which are a transition zone between boreal spruce–fir forests to the north and deciduous forests to the south. It is relatively flat, with low mountains and abundant lakes, ponds, and streams. The forest cover type is composed of coniferous species, in particular balsam fir and red spruce, deciduous species of red maple (*Acer rubrum* L.), sugar maple (*Acer saccharum* Marshall), yellow birch (*Betula alleghaniensis* Britton), white birch, American beech (*Fagus grandifolia* Ehrh.), and mixed stands of coniferous and deciduous trees. Over 90% of the forestlands are privately owned and are of commercial value. Intensive clear-cutting during the SBW outbreak between 1970s and 1980s and SBW-induced defoliation were the major landscape-scale causes of change in the region. Later, in the 1990s, management practices shifted to clear cutting with protection of advance regeneration [40]. The last outbreak in Maine was particularly severe. Forest conditions in Maine have changed considerably as a result of SBW-induced spruce–fir stand mortality, which killed between 72.5 and 90.6 million m³ of fir [41], and intensive salvage logging.

2.3. Satellite Data Acquisition and Pre-Processing

For the study area in Quebec, the Landsat-5 Thematic Mapper (TM) atmospherically corrected images of surface reflectance, and spectral indices products with 30 m spatial resolution were obtained from the United States Geological Survey (USGS) for study years 2004, 2005, 2008, and 2009 (Table 1). Although defoliation had been detected since 2006 in the study area, the full range of all defoliation severity levels were better observed from 2008 on. Landsat-8 images available since 2013 were too cloudy in the study area to be used. Landsat-7 data also suffered from striping problems and could not be used. Image selection was based on using multiple-year data that compared healthy forest conditions with no defoliation as the base years, versus defoliated years. For healthy years (i.e., before the outbreak), two cloud-free images for 2004 and 2005 during the summer were selected. To minimize the effect of phenology and dry/wet years on vegetation vigor, vegetation indices values of 2004 and 2005 were averaged and used as the base-year vegetation index values. Data for a specific period of time in early to late July were needed to map forest foliage discoloration (red-brown) similar to ASMs; this makes obtaining cloud-free imagery challenging in this region.

Timing of current-year defoliation was determined based on vegetation and SBW phenology for 2008 and 2009, as the defoliated years. SBW phenology was simulated using the BioSIM tool. The BioSIM is a climatically-driven model for pest management applications, is based on climatic data and degree–day compilation from a suite of weather stations [42], and has been validated and used to simulate annual pest phenology [15,42]. The BioSIM simulation software (v.10) [43] was used for 2008 and 2009 to estimate SBW larvae phenology to model peak SBW larval feeding time. In both years, peak feeding time was modeled as occurring around day of year (DOY) 170, when red-brownish defoliation was expected to start and last for a 2–3 week window (Figure 2). For vegetation phenology, a time series of Landsat TM data for DOY 160 to DOY 280 in a healthy year (2005) and defoliated year (2009) (five images for each year) were used (row and path 11/26 and 12/26). Landsat-derived vegetation indices were calculated for the 2005 healthy year, as well as for the 2009 defoliated year for the Quebec study area. Defoliated areas were determined based on ASMs. In the defoliated areas from 2009, vegetation indices started to decline around DOY 170 to 190, and leveled off until DOY 205, at which point they increased slightly until around DOY 220. However, in moderate and severely defoliated areas, the max vegetation index values did not reach the peak that was observed in 2005. Based on the BioSIM simulation results and Landsat-derived vegetation phenology data, two images for DOY 189 and DOY 191 for 2008 and 2009, respectively, were selected for the defoliated years. Clouds, cloud shadows, and water bodies were delineated and masked out prior to change detection analysis.

For the study area in Maine, relative radiometric normalized Landsat-MSS imagery for a pre-defoliation years (1972 and 1973), two defoliated years (1975 and 1982), and a Landsat-derived forest cover type map for 1975 with 60 m spatial resolution [40] were acquired (Table 1). BioSIM simulation was not used for image selection in Maine because of limited satellite image availability. For 1975 and 1982, two images of DOY 221 and DOY 211 were available and were used for defoliation detection. Cloud and cloud shadows were removed using automated cloud cover identification [44]. Because the northern part of the study area was found to be moderately defoliated in 1973, based on historical ASMs and SBW egg mass data [35], in order to produce pre-defoliated imagery, an image from early September 1972 for row 12/28 was acquired, radiometrically normalized, and applied to replace spectral band values in the northern part of Landsat-MSS scene 13/28 of 1973.

Table 1. Landsat images and acquisition dates used for annual defoliation detection in Quebec, Canada and Maine, United States.

Study Area	Imagery Date	Landsat Sensor	Path/Row
Quebec	12 July 2004	TM5	12/26
	15 July 2005	TM5	12/26
	8 July 2008	TM5	12/26
	10 July 2009	TM5	12/26
Maine	2 September 1972	MSS1	12/28
	23 July 1973	MSS1	13/28
	9 August 1975	MSS2	13/28
	30 July 1982	MSS2	13/28

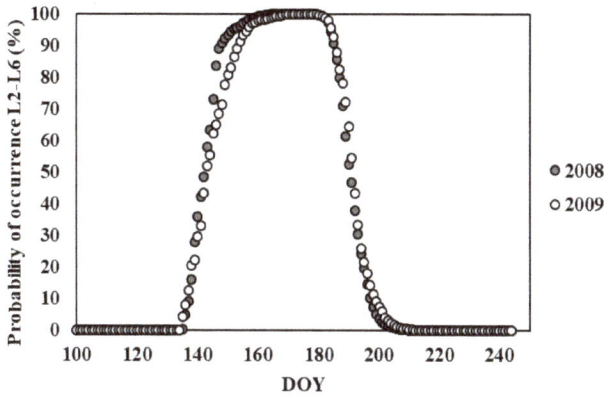

Figure 2. Spruce budworm (SBW) probability of occurrence (L2–L6) in 2008 and 2009 for the Quebec study area, based on the BioSIM simulation: L2 and L6 are the second and sixth instar in the lifecycle of spruce budworm, respectively. DOY 170–190 represent 100% occurrence of SBW larvae and timing of stage 6 larva that do most (87%) of the defoliation, as described by Miller [45].

2.4. Spruce Budworm Defoliation Detection and Severity Level Estimation for the Quebec Study Area Current Spruce Budworm Outbreak

Remote sensing of insect defoliation can be based on single- or multiple-date image analysis [11]. Our method was based on multi-date change detection using vegetation indices (VIs) [6,20,46]. The Landsat-5 TM sensor has six spectral bands (blue, green, red, near infrared (NIR), and two shortwave infrared (SWIR) bands) with a spatial resolution of 30 m; therefore, several common vegetation indices can be estimated. A wide range of vegetation indices, such as visible (VIS)-near infrared (NIR) [19,46–48] or NIR-SWIR indices [17,20,28,49], have been tested in various forest defoliation detection studies. In this study, seven vegetation indices, including normalized difference vegetation index (NDVI) [50], enhanced vegetation index (EVI) [51,52], green chlorophyll index (Chlgreen) [53,54], greenness normalized difference vegetation index (GNDVI) [55], normalized difference moisture index (NDMI) [56], normalized burn ratio1 (NBR1) [57], and normalized burn ratio2 (NBR2) [58] were tested for their capacity to detect and quantify defoliation (Table 2). These indices have information on vegetation pigment content (NDVI, GNDVI, and Chlgreen), water content (NDMI, NBR1 and NBR2), and foliage structure and amount (NDVI and EVI). Defoliation can be detected by studying reflectance changes in defoliated forest stands compared to their healthy condition before the damage occurred. Defoliated forest stands exhibit progressive decreases in near-infrared reflectance but an increase in short-wave infrared and visible reflectance, due to changes in canopy cover pigment content, water content, and foliage amount [12].

Eco-forest maps from the Quebec Ministry of Forests, Wildlife and Parks (3rd Inventory) with 25 m spatial resolution were used to extract information about susceptible forest stands [59]. Five species groups (balsam fir, black spruce, spruce mixed with other conifers, balsam fir mixed with other conifers, and balsam fir mixed with broad-leaved species) were selected. Annual ASMs of SBW defoliation (e.g., [7]) were used as our field data, to train the remote sensing model for defoliation severity classification. ASM maps have been used for the same purpose by others [24,60,61].

The random forest (RF) non-parametric method [62] was employed to evaluate the performance of the VIs for SBW defoliation detection and severity classification. In the RF method, the variables in a dataset can be ranked, and the most influential variables can be selected and used for defoliation detection and classification. The RF training algorithm applies a bagging (bootstrap aggregation) operation, where a number of decision trees are created based on a random subset of samples derived from the training data. The RF algorithm gives an error rate—called the out-of-bag (OOB) error—for each input variable, using the data not used in deriving the decision trees. The "RandomForest" [63] library in R statistical software v.3 [64] was applied to implement the statistical analysis.

Table 2. Landsat remotely-sensed indices evaluated in the study for defoliation detection. Landsat TM vegetation indices (Vis) were used for the Quebec study area, and Landsat MSS normalized difference vegetation index (NDVI) was used for the Maine study area.

Landsat Sensor	Index	Acronym and Formulation	Reference
TM	Enhanced Vegetation Index	EVI = 2.5 * (NIR − Red)/(NIR + 6 * Red − 7.5 * Blue + 1)	[51,52]
	Normalized Difference Vegetation Index	NDVI = (NIR − Red)/(NIR + Red)	[50]
	Green Chlorophyll Index	Chlgreen = (NIR/Green) − 1	[53,54]
	Greenness Normalized Difference Vegetation Index	GNDVI = (NIR − Green)/(NIR + Green)	[55]
	Normalized Difference Moisture Index	NDMI = (NIR − SWIR1)/(NIR + SWIR1)	[56]
	Normalized Burn Ratio1	NBR1 = (NIR − SWIR2)/(NIR + SWIR2)	[57]
	Normalized Burn Ratio 2	NBR2 = (SWIR1 − SWIR2)/(SWIR1 + SWIR2)	[58]
MSS	Normalized Difference Vegetation Index	NDVI = (NIR2 − Red)/(NIR2 + Red)	[50]

Four hundred samples (200 samples for each year) were collected using a stratified random sampling method. Samples were extracted for the five above-mentioned species groups (40 samples per species) from non-defoliated and defoliated areas in three defoliation severity classes (light, moderate, and severe) [7]. From 400 samples, two-thirds were used as training data to create an in-bag partition to construct the decision tree, while one-third were used for validation and OOB estimation. Seven VIs (Table 2) and species types were input variables, and the OOB error was used to assess classification and choose the number of variables that yielded the smallest error rate. The Gini importance measure was used to determine the order of importance of the variables. Finally, the best VIs were selected for defoliation detection and severity classification. Both single and combinations of VIs were tried for defoliation detection and severity classification. Although previous studies have shown that there is no need for cross-validation or separate tests, because this is already achieved by OOB estimates [65], a confusion matrix was also constructed for further validation of our best model, using an additional 100 samples (50 per year in 2008 and 2009) that were not used for the modeling [40,65]. Confidence intervals were calculated using Wald's method [66].

2.5. Spruce Budworm Defoliation Detection for the Maine Study Area Past Spruce Budworm Outbreak

The method for the Maine study area was also based on multi-date change detection using VIs [6,20]. However, Landsat-MSS sensors only had four spectral bands (green, red, and two NIR) with a spatial resolution of 60 m, so that many common vegetation indices could not be estimated; therefore, change detection was based only on NDVI. Among the different spectral bands and VIs that could be used for foliage damage detection using Landsat MSS, the red and NIR2 bands (2 and 4), as well as NDVI are suggested as the best for vegetation change studies [67]. The historical ASMs of Maine were very coarse in spatial resolution, and thus were not suitable as a measure of defoliation data. Expected defoliation levels derived from SBW egg-mass data [35] were used instead for comparison with Landsat-MSS derived defoliation maps. A total of 349 and 247 egg-mass data plots were used for the years 1975 and 1982, respectively. Egg-mass data were converted to defoliation levels, using the method outlined in Simmons (1974) [68] and the equation presented in Hennigar et al. [35]. Ordinal regression [69] was used to evaluate the relationship between expected defoliation levels and NDVI changes in both years. Any reduction in NDVI larger than 0.05 was considered as defoliation, and SBW defoliation maps were produced from NDVI data for the years 1975 and 1982. The percentage of correctly identified defoliated areas was determined by comparing defoliation information derived from egg-mass data and those derived from Landsat-MSS.

2.6. Detecting Other Disturbances in Spruce Budworm-Defoliated Forests

In the Quebec study area, no fire or harvest disturbances were detected. The Maine study area included intensive harvest activities, including clear-cuts in the 1970s that changed fir–spruce forests, while SBW defoliation was developing. Producing SBW defoliation maps thus required differentiating harvest-related changes. To avoid confounding information, past Maine harvest data from [40] were used to remove all areas with harvest activities from the Landsat-MSS imagery. The remaining changes detected over the host forests were assumed to be related to SBW defoliation, which was the dominant natural disturbance in the area.

3. Results

3.1. Spruce Budworm Defoliation Detection and Severity Level Classification in the Current Quebec Outbreak

Figure 3 compares the performance of the seven single VIs and three best combinations thereof, to detect defoliated versus non-defoliated forests. All indices were able to identify defoliated stands with greater than 80% producer's accuracy, but identification of non-defoliated stands was much more variable, as were errors in detection. The best indices to detect and classify SBW defoliation, in descending order, were NDMI, NBR1, EVI, and NDVI. NDMI had 90% accuracy in detecting defoliated pixels and 76% accuracy in detecting non-defoliated pixels. Different combinations of the four best vegetation indices were also evaluated. The results showed that the combination of NDMI, EVI, and NDVI reduced the error rate by 5% for SBW defoliation detection compared to NDMI only.

Based on the results of accuracy estimations in Figure 3, the four best indices and their combinations were selected for severity classification analysis. The combination of NDMI, EVI, NDVI, and NBR1 reduced the OOB error rate by 10%, compared with the use of the single best index NDMI or NBR1 (Figure 4). Addition of a fourth VI only marginally improved estimation error when compared with the three-VI approach. As NBR1 and NDMI contain similar information about canopy water content, we selected the combination of NDMI, EVI, and NDVI for defoliation severity classification. Landsat-derived SBW defoliation severity maps for the years 2008 and 2009 produced using the three-VI classification approach for parts of the Quebec study area, along with ASMs in the same years, are presented in Figure 5.

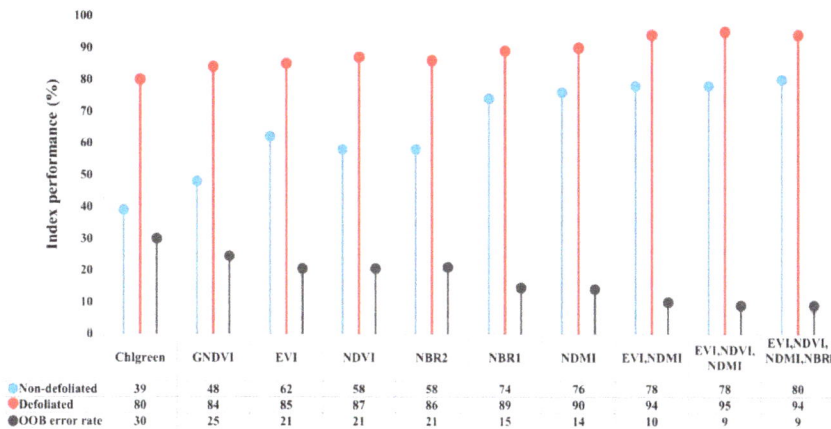

	Chlgreen	GNDVI	EVI	NDVI	NBR2	NBR1	NDMI	EVI,NDMI	EVI,NDVI,NDMI	EVI,NDVI,NDMI,NBRI
Non-defoliated	39	48	62	58	58	74	76	78	78	80
Defoliated	80	84	85	87	86	89	90	94	95	94
OOB error rate	30	25	21	21	21	15	14	10	9	9

Figure 3. Comparison of the performance of seven vegetation indices (VIs) and best combinations thereof to detect defoliated versus non-defoliated forests in the current spruce budworm outbreak in Quebec, using the random forest method. OOB is the out-of-bag error rate.

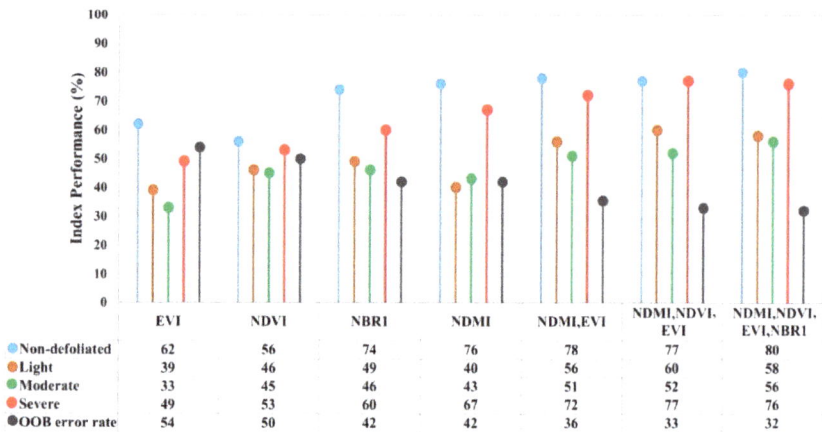

	EVI	NDVI	NBR1	NDMI	NDMI,EVI	NDMI,NDVI, EVI	NDMI,NDVI, EVI,NBR1
Non-defoliated	62	56	74	76	78	77	80
Light	39	46	49	40	56	60	58
Moderate	33	45	46	43	51	52	56
Severe	49	53	60	67	72	77	76
OOB error rate	54	50	42	42	36	33	32

Figure 4. Comparison of the performance of best single VIs and best combinations thereof to classify different severities of defoliation in the current spruce budworm outbreak in Quebec, using the random forest method. OOB is the out-of-bag error rate.

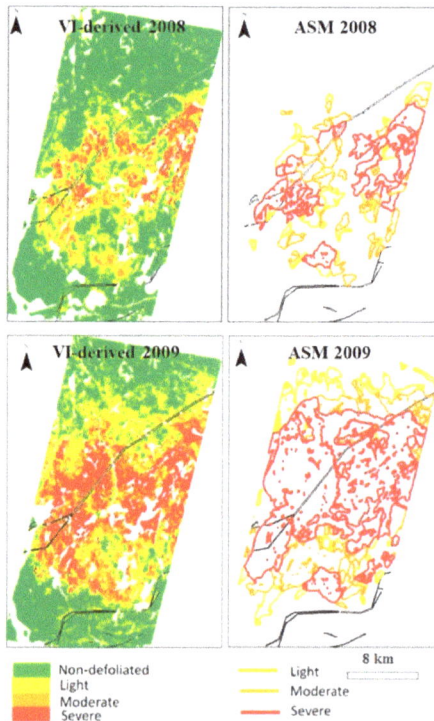

Figure 5. Landsat-derived annual (current-year) defoliation severity maps for 2008 and 2009, at 30 m spatial resolution, for a portion of the Quebec study area, using a combination of NDMI, EVI, and NDVI, compared with aerial sketch maps (ASMs) for the same area. White pixels in VI-derived defoliation maps are water bodies, clouds, or cloud shadows.

Results of defoliation classification accuracy for 2008 and 2009 are presented in Table 3. The overall accuracies and Kappa coefficients for the three VI-derived defoliation detection were, respectively, 93% and 0.84 for 2008 and 90% and 0.65 for 2009, indicating that the suggested method was able to detect defoliated areas. The overall accuracies and Kappa coefficients for defoliation severity classification were, respectively, 72% and 0.63 for 2008 and 64% and 0.50 for 2009. Less than 1% of the severely defoliated forests were classified as lightly or non-defoliated in both years. Less than 1% and 6% of pixels with moderate defoliation were classified as non-defoliated in 2009 and 2008, respectively. When misclassified, light defoliation was either classified as "no defoliation" or "moderate defoliation" in both years. Therefore, it seems that misclassification primarily occurred only between two adjacent categories.

Table 3. Error matrix and accuracy estimate (%) for the Quebec study area for the 2008 and 2009 SBW defoliation maps, derived from VIs versus ASMs (PA: producer's accuracy; UA: user's accuracy). Bold values are % correct classes. Confidence intervals are calculated at a 95% confidence level.

	No Defoliation	Light Defoliation	Moderate Defoliation	Severe Defoliation	PA	UA	PA Conf. Interval	UA Conf. Interval
2008								
No defoliation	**93.4**	15.7	6.2	0.1	93.4	85.9	92–94	84–87
Light defoliation	6.6	**67.3**	35.4	6.0	67.3	55.0	65–70	52–57
Moderate defoliation	0.0	14.9	**53.1**	25.8	53.1	56.3	50–56	54–59
Severe defoliation	0.0	2.0	5.4	**68.1**	68.1	91.0	66–70	89–93
Overall Acc. and Kappa Coeff.: 72.4%, 0.63								
2009								
No defoliation	**62.2**	20.1	0.5	0.0	62.2	81.9	60–64	80–84
Light defoliation	24.2	**49.5**	4.4	0.6	49.5	55.3	47–52	53–58
Moderate defoliation	13.1	29.0	**67.8**	21.5	67.8	48.0	65–70	46–50
Severe defoliation	0.5	1.2	27.3	**78.1**	78.0	65.7	75–80	63–68
Overall Acc. and Kappa Coeff.: 64%, 0.50								

3.2. Spruce Budworm Defoliation Detection in the Maine Past Outbreak

The relationship between defoliation levels estimated from egg mass data and change in mean NDVI values was weak, but statistically significant (Table 4). Not much variation in defoliation levels was explained by NDVI variation, as indicated by low pseudo-R^2 values. On average, 57% and 47% of plots were correctly identified as either defoliated or non-defoliated for 1975 and 1982, respectively. In both years, the identification accuracy was considerably higher at greater defoliation levels. Due to the weak statistical relationship between the expected defoliation data and NDVI in Maine, but better accuracy for defoliation identification (% correctly identified data), only defoliated versus non-defoliated classes were mapped.

Table 4. Results of regression analysis between egg-mass-derived expected defoliation levels and Landsat-MSS-derived NDVI and defoliation occurrence analysis (Tot.: Total; Ave.: Average).

Year	Egg Mass Counts Class/100 ft²	Expected Defoliation Class (%)	Samples per Egg-Mass Class	% of Total Egg-Mass Samples	% Correctly Identified	p-Value	Pseudo R^2 (Nagelkerke)
	0	0	3	1	67		
	1–50	1–12	29	8	41		
1975	51–170	13–42	87	25	45	0.001	0.038
	171–320	43–78	87	25	68		
	321–+400	79–100	143	41	60		
	-	-	Tot. 349	Tot. 100	Ave. 57		
	1–50	1–12	52	21	42		
	51–170	13–42	55	22	36		
1982	171–320	43–78	48	20	56	0.002	0.041
	321–400	79–98	92	37	53		
	-	-	Tot. 247	Tot. 100	Ave. 47		

Figure 6 shows maps of SBW occurrence derived from NDVI for the years 1975 and 1982 in the Maine study area, along with the historical ASMs and expected defoliation calculated from egg-mass data. Although both ASMs and egg-mass-derived defoliation maps presented in Figure 6 contain information about defoliation severity levels, because our Landsat MSS-derived map only presents defoliation occurrence, our results only focus on defoliation occurrence and not severity. In general, the Landsat-MSS-derived map of defoliation showed good spatial agreement with expected defoliation in both years. Looking at the forest cover type map of 1975 in Figure 1, more defoliation would be expected in the northern, western, eastern, and central parts of the study area in Maine, where SBW host species were dominant, whereas less defoliation would be expected in the southern parts, where hardwoods were more dominant. This trend is evident in both Landsat-MSS-derived SBW defoliation maps and expected defoliation data, but not clearly in ASMs. Apparently, forest composition data were not incorporated into the ASM defoliation estimates [35].

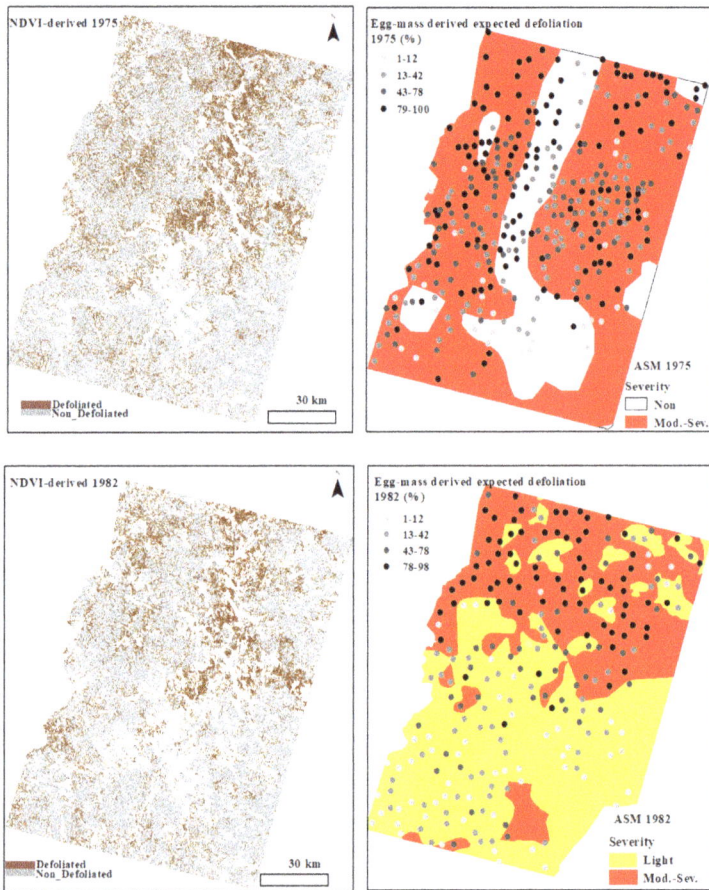

Figure 6. (**Top**) Landsat-MSS SBW defoliation occurrence map for 1975 at 60 m spatial resolution, and expected defoliation (%) derived from egg-mass counts in 1975 overlaid on an ASM SBW defoliation map for 1975. (**Bottom**) Landsat-MSS SBW defoliation occurrence map for 1982 at 60 m spatial resolution, and expected defoliation (%) derived from egg mass counts in 1982 overlaid on an ASM SBW defoliation map for 1982. White pixels in the Landsat-derived maps are water bodies, clouds, cloud shadows, and harvested forests.

In general, defoliation severity was higher in 1975 than in 1982. Looking at the ASM for 1975, the central and the southern parts of the Maine study area were identified as non-defoliated, while the central parts of the region should have been moderately defoliated, based on expected defoliation data retrieved from the egg-mass counts of 1975. Those areas were also mapped as defoliated in the Landsat-MSS derived defoliation map of 1975. In 1985, defoliation was more severe in the northern parts of the study area and lighter in the central and southern parts, based on ASM and expected defoliation data. This can be also observed in the Landsat-MSS defoliation map of 1985. It is evident that Landsat-MSS and expected defoliation data present more details about the status of defoliation over coniferous forests than ASM.

4. Discussion

Using Landsat-5 data, results indicated that NDMI was the best single index to detect defoliation across the studied forests in Quebec. In general, there is little literature on multi-VI defoliation detection performance in coniferous forests. Our findings are in agreement with studies performed on the detection of defoliation in deciduous forests where vegetation water indices like NDMI and NBR1, which are based on NIR/SWIR bands, performed better than NDVI or EVI, which are based on VIS/NIR bands [17,20]. Other studies applied single vegetation water indices for defoliation detection over coniferous forests but did not compare them with other indices [6,24]. Our results showed that the combination of NDVI, EVI, and NDMI can reduce the OOB rate of error for both SBW defoliation detection and classification. NDVI is more sensitive to canopy chlorophyll content, while EVI responds to canopy structural characteristics, such as leaf area index and canopy type. The combined application of these two indices has been suggested as a complementary tool for vegetation change detection [70]. Chlgreen and GNDVI were applied in this study as better indices, with wider dynamic ranges and higher sensitivity than NDVI, to evaluate changes in canopy chlorophyll concentration [53–55]; however, neither Chlgreen nor GNDVI were able to detect defoliated regions better than NDVI, and their performance to differentiate non-defoliated pixels was lower than NDVI.

It was evident that using the combination of Landsat-5-derived NDVI, EVI, and NDMI at 30 m resolution, as suggested in this research, provides a potentially useful tool for SBW defoliation detection, but differentiation of lightly defoliated from non-defoliated areas is still challenging. Using the model for severity classification, misclassification mostly occurred at lower defoliation levels (Table 3), with different levels of accuracy between the two years. This can be related to the broad range of defoliation categories of ASMs and their inherently lower accuracy for light defoliation detection [10]; for example, the lower range of defoliation in the medium-defoliation class can be confounded with the higher range in the light-defoliation class, making these variations inevitable. In addition, possible limitations of Landsat-derived vegetation indices to detect very light defoliation (e.g., <15%) can be attributed to both radiometric and spatial resolutions of the data. Minor disturbances, such as wind and frost-related damage, might also contribute to errors in non-defoliated forest detection. In terms of future work, models derived from satellite sensors, such as Landsat and Sentinel-2, should be trained with more accurate tree- and plot-measured defoliation data, in order to provide better estimates of defoliation severity.

Townsend et al. [20] suggested the viability of NDVI for mapping defoliation where SWIR spectral bands are not available. This was the situation for the past SBW outbreak in Maine using Landsat-MSS data. We found weak but statistically significant relationships between historically expected defoliation classification data and Landsat-MSS-derived NDVI change maps, indicating a potential mismatch between defoliation severity classes and NDVI-derived defoliation severity. The weak relationship can be attributed to differences between the egg-mass plot area size (~10 m^2) that was used to estimate defoliation levels and the Landsat-MSS NDVI data pixel size (3600 m^2). In addition, Landsat-MSS images for early August and late July were used for SBW defoliation detection in 1975 and 1982, respectively, because of the unavailability of data in mid-July, which was not an ideal time for annual defoliation detection. Finally, NDVI is known to saturate over dense forest canopy cover [55], and therefore, a slight decrease in foliage pigment content and structure may not always be detected.

In the Maine study area, on average, 57% and 47% of plots were correctly identified as either defoliated or non-defoliated in 1975 and 1982, respectively, with higher identification accuracy at greater defoliation levels. Therefore, we can conclude that Landsat-MSS NDVI is capable of detecting defoliation from historical Landsat imager—in particular, moderate and severe defoliation. In general, there was good spatial agreement with forest cover maps where SBW hosts were located, as well as expected defoliation levels derived from egg-mass data and Landsat-derived defoliation maps. Landsat-MSS defoliation occurrence and higher defoliation levels were concentrated in the northern and central parts of the study area in Maine for both years where SBW tree host species were dominant, while less defoliation was observed in the south. It is generally assumed that spruce–fir stands receive certain protection from hardwood stands, attributed to SBW dispersal and migration losses [71].

In addition to spectral information and the type of VI, data timing and radiometric consistency within the images were key factors for change detection in this analysis. For defoliation detection for the current outbreak in Quebec study area, Landsat-5 surface reflectance data were used, and BioSIM was applied to simulate SBW phenology for each year. Both DOY 189 and 191 for 2008 and 2009, respectively, appeared to be sound dates to detect SBW-induced foliage discoloration, and were reasonable dates for SBW annual defoliation detection. For defoliation detection during the past outbreak in Maine, relative radiometrically-corrected imagery was used, but because of unavailability of Landsat-MSS imagery for mid-July, available cloud-free images in late July and early August were used. Although these dates were not ideal times for annual defoliation detection, as discolored foliage might have fallen off, they still contain useful information on SBW defoliation extent, and how two different types of disturbances (i.e., SBW defoliation and harvest) were intensively altering Maine's spruce–fir forest landscape. In addition, in multi-date change detection, the correct assignment of a base (healthy) image as reference is important, because it will affect change detection accuracy. Not much has been previously discussed about how effective the base image can be for the evaluation of defoliation severity. For a mature forest, it seems that the average of a few years of VIs in healthy condition in early–mid July would be ideal for current-year defoliation detection. This ensures the inclusion of vegetation phenology and the effect of dry/wet years on vegetation vigor. The fact that bud flush in balsam fir normally occurs, on average, two weeks earlier than in black spruce, and that the entire foliage discoloration process may take about three weeks, should be taken into consideration if only one date per year is used for image analysis and change detection.

5. Conclusions

Landsat data have the potential to be used for fine-scale annual SBW defoliation detection and quantification. In this work, we used historical Landsat-MSS and Landsat-5 data for annual SBW defoliation detection and classification. Landsat-5 imagery can be used to detect and classify current-year SBW defoliation, but more research is needed to improve the accuracy of the developed method, so that light and moderate defoliation can be better differentiated. Landsat-MSS also proved able to provide valuable information about annual SBW defoliation extent in a past outbreak, to complement (and potentially improve) historical coarse resolution ASMs and field data, such as egg-mass survey data.

The unavailability of cloud-free satellite imagery/pixels during the biological window to observe foliage discoloration is one of the main reasons SBW annual defoliation detection using remote sensing techniques has not been well-evaluated. It is expected that multi-spectral sensors will be available more widely in future, since for the first time, three free-of-charge, fine-resolution, multi-spectral satellites with harmonized spectral bands have recently become available. A combination of Landsat-8 with Sentinel-2A and 2B satellite data promises 2–3 day temporal resolution, and should provide sound data collection for defoliation detection and monitoring at a scale comparable with ASMs.

This fine-resolution, satellite-derived, annual SBW defoliation information can be used for sustainable forest management and planning at the regional and local scales.

Author Contributions: P.R., A.W., D.K., and D.M. conceived and designed the study; P.R. developed the methodology; P.R., A.W., and D.K. analyzed the data; D.K. and D.M. contributed research materials and field data; P.R. wrote the paper.

Funding: Funding for this research was provided in part by the Cooperative Forestry Research Unit (CFRU), the USDA Forest Service Northeastern States Research Cooperative (NSRC), and the USDA National Institute of Food and Agriculture, McIntire-Stennis project number #ME041516 through the Maine Agricultural and Forest Experiment Station (Maine Agricultural and Forest Experiment Station Publication Number 3607).

Acknowledgments: The authors would also like to thank Kasey Legaard for providing the Landsat-MSS derived forest cover/harvest data for the past outbreak in Maine.

Conflicts of Interest: The authors declare no conflict of interest.

References

1. MacLean, D.A.; MacKinnon, W.E. Accuracy of aerial sketch-mapping estimates of spruce budworm defoliation in New Brunswick. *Can. J. For. Res.* **1996**, *26*, 2099–2108. [CrossRef]
2. MacLean, D.A.; MacKinnon, W.E. Effects of stand and site characteristics on susceptibility and vulnerability of balsam fir and spruce to spruce budworm in New Brunswick. *Can. J. For. Res.* **1997**, *27*, 1859–1871. [CrossRef]
3. MacLean, D.A. Effects of spruce budworm outbreaks on the productivity and stability of balsam fir forests. *For. Chron.* **1984**, *60*, 273–279. [CrossRef]
4. Gray, D.R. The relationship between climate and outbreak characteristics of the spruce budworm in eastern Canada. *Clim. Chang.* **2008**, *87*, 361–383. [CrossRef]
5. Williams, D.W.; Liebhold, A.M. Spatial synchrony of spruce budworm outbreaks in eastern North America. *Ecology* **2000**, *81*, 2753–2766. [CrossRef]
6. Hall, R.; Filiatrault, M.; Deschamps, A.; Arsenault, E. Mapping eastern spruce budworm cumulative defoliation severity from Landsat and SPOT. In Proceedings of the 30th Canadian Symposium on Remote Sensing, Lethbridge, AB, Canada, 22–25 June 2009; pp. 22–25.
7. Ministère des Forêts de la Faune et des Parcs. *Aires Infestées Par la Tordeuse des Bourgeons de L'épinette au Québec en 2017-Version 1.0*; Gouvernement du Québec, Direction de la Protection des Forêts: Québec, QC, Canada, 2017; p. 16.
8. Wagner, R.G.; Bryant, J.; Burgason, B.; Doty, M.; Roth, B.E.; Strauch, P.; Struble, D.; Denico, D. *Coming Spruce Budworm Outbreak: Initial Risk Assessment and Preparation & Response Recommendations for Maine's Forestry Community*; Cooperative Forestry Research Unit, University of Maine: Orono, ME, USA, 2015; p. 77.
9. Gray, D.R.; Régnière, J.; Boulet, B. Analysis and use of historical patterns of spruce budworm defoliation to forecast outbreak patterns in Quebec. *For. Ecol. Manag.* **2000**, *127*, 217–231. [CrossRef]
10. Taylor, S.L.; MacLean, D.A. Validation of spruce budworm outbreak history developed from aerial sketch mapping of defoliation in New Brunswick. *North. J. Appl. For.* **2008**, *25*, 139–145.
11. Hall, R.; Castilla, G.; White, J.; Cooke, B.; Skakun, R. Remote sensing of forest pest damage: A review and lessons learned from a Canadian perspective. *Can. Entomol.* **2016**, *148*, S296–S356. [CrossRef]
12. Leckie, D.; Teillet, P.; Ostaff, D.; Fedosejevs, G. Sensor band selection for detecting current defoliation caused by the spruce budworm. *Remote Sens. Environ.* **1988**, *26*, 31–36. [CrossRef]
13. MacLean, D.A.; Ostaff, D.P. Patterns of balsam fir mortality caused by an uncontrolled spruce budworm outbreak. *Can. J. For. Res.* **1989**, *19*, 1087–1095. [CrossRef]
14. Ostaff, D.P.; MacLean, D.A. Patterns of balsam fir foliar production and growth in relation to defoliation by spruce budworm. *Can. J. For. Res.* **1995**, *25*, 1128–1136. [CrossRef]
15. Rullan-Silva, C.; Olthoff, A.; de la Mata, J.D.; Pajares-Alonso, J. Remote monitoring of forest insect defoliation—A Review. *For. Syst.* **2013**, *22*, 377–391. [CrossRef]
16. Brovkina, O.; Cienciala, E.; Zemek, F.; Lukeš, P.; Fabianek, T.; Russ, R. Composite indicator for monitoring of Norway spruce stand decline. *Eur. J. Remote Sens.* **2017**, *50*, 550–563. [CrossRef]
17. De Beurs, K.; Townsend, P. Estimating the effect of gypsy moth defoliation using MODIS. *Remote Sens. Environ.* **2008**, *112*, 3983–3990. [CrossRef]

18. Dottavio, C.L.; Williams, D.L. Satellite technology: An improved means for monitoring forest insect defoliation. *J. For.* **1983**, *81*, 30–34.

19. Hurley, A.; Watts, D.; Burke, B.; Richards, C. Identifying gypsy moth defoliation in Ohio using Landsat data. *Environ. Eng. Geosci.* **2004**, *10*, 321–328. [CrossRef]

20. Townsend, P.A.; Singh, A.; Foster, J.R.; Rehberg, N.J.; Kingdon, C.C.; Eshleman, K.N.; Seagle, S.W. A general Landsat model to predict canopy defoliation in broadleaf deciduous forests. *Remote Sens. Environ.* **2012**, *119*, 255–265. [CrossRef]

21. Hall, R.; Crown, P.; Titus, S. Change detection methodology for aspen defoliation with Landsat MSS digital data. *Can. J. Remote Sens.* **1984**, *10*, 135–142. [CrossRef]

22. Leckie, D.G.; Cloney, E.; Joyce, S.P. Automated detection and mapping of crown discolouration caused by jack pine budworm with 2.5 m resolution multispectral imagery. *Int. J. Appl. Earth Obs. Geoinf.* **2005**, *7*, 61–77. [CrossRef]

23. Radeloff, V.C.; Mladenoff, D.J.; Boyce, M.S. Detecting jack pine budworm defoliation using spectral mixture analysis: Separating effects from determinants. *Remote Sens. Environ.* **1999**, *69*, 156–169. [CrossRef]

24. Fraser, R.; Latifovic, R. Mapping insect-induced tree defoliation and mortality using coarse spatial resolution satellite imagery. *Int. J. Remote Sens.* **2005**, *26*, 193–200. [CrossRef]

25. Hawryło, P.; Bednarz, B.; Wężyk, P.; Szostak, M. Estimating defoliation of Scots pine stands using machine-learning methods and vegetation indices of Sentinel-2. *Eur. J. Remote Sens.* **2018**, *51*, 194–204. [CrossRef]

26. Franklin, S.; Fan, H.; Guo, X. Relationship between Landsat TM and SPOT vegetation indices and cumulative spruce budworm defoliation. *Int. J. Remote Sens.* **2008**, *29*, 1215–1220. [CrossRef]

27. Franklin, S.; Waring, R.; McCreight, R.; Cohen, W.; Fiorella, M. Aerial and satellite sensor detection and classification of western spruce budworm defoliation in a subalpine forest. *Can. J. Remote Sens.* **1995**, *21*, 299–308. [CrossRef]

28. Leckie, D.; Teillet, P.; Fedosejevs, G.; Ostaff, D. Reflectance characteristics of cumulative defoliation of balsam fir. *Can. J. For. Res.* **1988**, *18*, 1008–1016. [CrossRef]

29. Hall, R.J.; Skakun, R.S.; Arsenault, E.J. Remotely sensed data in the mapping of insect defoliation. In *Understanding Forest Disturbance and Spatial Pattern: Remote Sensing and Gis Approaches*; Wulder, M.A., Franklin, S.E., Eds.; Taylor and Francis, CRC Press: Boca Raton, FL, USA, 2006; pp. 85–111.

30. Leckie, D.G.; Yuan, X.; Ostaff, D.P.; Piene, H.; MacLean, D. Analysis of high resolution multispectral MEIS imagery for spruce budworm damage assessment on a single tree basis. *Remote Sens. Environ.* **1992**, *40*, 125–136. [CrossRef]

31. Goodbody, T.R.; Coops, N.C.; Hermosilla, T.; Tompalski, P.; McCartney, G.; MacLean, D.A. Digital aerial photogrammetry for assessing cumulative spruce budworm defoliation and enhancing forest inventories at a landscape-level. *ISPRS J. Photogramm. Remote Sens.* **2018**, *142*, 1–11. [CrossRef]

32. Chalifoux, S.; Cavayas, F.; Gray, J.T. Map-guided approach for the automatic detection on Landsat TM images of forest stands damaged by the spruce budworm. *Photogramm. Eng. Remote Sens.* **1998**, *64*, 629–635.

33. Franklin, S.; Raske, A. Satellite remote sensing of spruce budworm forest defoliation in western Newfoundland. *Can. J. Remote Sens.* **1994**, *20*, 37–48.

34. Campbell, P.E.; Rock, B.; Martin, M.; Neefus, C.; Irons, J.; Middleton, E.; Albrechtova, J. Detection of initial damage in Norway spruce canopies using hyperspectral airborne data. *Int. J. Remote Sens.* **2004**, *25*, 5557–5584. [CrossRef]

35. Hennigar, C.R.; MacLean, D.A.; Erdle, T.A. *Potential Spruce Budworm Impacts and Mitigation Opportunities in Maine*; Cooperative Forest Research Unit, University of Maine: Orono, ME, USA, 2013; p. 68.

36. Gray, D.R. The influence of forest composition and climate on outbreak characteristics of the spruce budworm in eastern Canada. *Can. J. For. Res.* **2013**, *43*, 1181–1195. [CrossRef]

37. Saucier, J.; Bergeron, J.; Grondin, P.; Robitaille, A. Les régions écologiques du Québec méridional: Un des éléments du systeme hiérarchique de classification écologique du territoire mis au point par le Ministere des Ressources Naturelles. *L'Aubelle* **1998**.

38. Bélisle, A.C.; Gauthier, S.; Cyr, D.; Bergeron, Y.; Morin, H. Fire regime and old-growth boreal forests in central Quebec, Canada: An ecosystem management perspective. *Silva Fenn.* **2011**, *45*, 889–908. [CrossRef]

39. Boucher, D.; De Grandpré, L.; Kneeshaw, D.; St-Onge, B.; Ruel, J.-C.; Waldron, K.; Lussier, J.-M. Effects of 80 years of forest management on landscape structure and pattern in the eastern Canadian boreal forest. *Landsc. Ecol.* **2015**, *30*, 1913–1929. [CrossRef]

40. Legaard, K.R.; Sader, S.A.; Simons-Legaard, E.M. Evaluating the impact of abrupt changes in forest policy and management practices on landscape dynamics: Analysis of a Landsat image time series in the Atlantic Northern Forest. *PLoS ONE* **2015**, *10*, e0130428. [CrossRef] [PubMed]

41. Maine Forest Service. *Assessment of Maine's Wood Supply*; Maine Forest Service, Department of Conservation: Augusta, ME, USA, 1993; p. 38.

42. Régnière, J. Generalized approach to landscape-wide seasonal forecasting with temperature-driven simulation models. *Environ. Entomol.* **1996**, *25*, 869–881. [CrossRef]

43. Régnière, J.; Cooke, B.; Bergeron, V. *BioSIM: A Computer-Based Decision Support Tool for Seasonal Planning of Pest Management Activities*; User's Manual; Canadian Forest Service Information Report LAU-X-116; Canadian Forest Service Publications: Sainte-Foy, QC, Canada, 1995; 1996p.

44. Braaten, J.D.; Cohen, W.B.; Yang, Z. Automated cloud and cloud shadow identification in Landsat MSS imagery for temperate ecosystems. *Remote Sens. Environ.* **2015**, *169*, 128–138. [CrossRef]

45. Miller, C.A. The feeding impact of spruce budworm on balsam fir. *Can. J. For. Res.* **1977**, *7*, 76–84. [CrossRef]

46. Eklundh, L.; Johansson, T.; Solberg, S. Mapping insect defoliation in Scots pine with MODIS time-series data. *Remote Sens. Environ.* **2009**, *113*, 1566–1573. [CrossRef]

47. Spruce, J.P.; Sader, S.; Ryan, R.E.; Smoot, J.; Kuper, P.; Ross, K.; Prados, D.; Russell, J.; Gasser, G.; McKellip, R. Assessment of MODIS NDVI time series data products for detecting forest defoliation by gypsy moth outbreaks. *Remote Sens. Environ.* **2011**, *115*, 427–437. [CrossRef]

48. Olsson, P.-O.; Lindström, J.; Eklundh, L. Near real-time monitoring of insect induced defoliation in subalpine birch forests with MODIS derived NDVI. *Remote Sens. Environ.* **2016**, *181*, 42–53. [CrossRef]

49. Thomas, S.J.; Deschamps, A.; Landry, R.; van der Sanden, J.J.; Hall, R.J. Mapping insect defoliation using multi-temporal Landsat data. In Proceedings of the CRSS/ASPRS Specialty Conference: Our Common Borders—Safety, Security and the Environment through Remote Sensing, Ottawa, ON, Canada, 28 October–1 November 2007.

50. Rouse, J.W., Jr.; Haas, R.; Schell, J.; Deering, D. *Monitoring Vegetation Systems in the Great Plains with ERTS*; NASA: Washington, DC, USA, 1974; pp. 309–317.

51. Huete, A.; Justice, C.; Leeuwen, W. *MODIS Vegetation Index (MOD13), EOS MODIS Algorithm*; Theoretical Basis Document; NASA Goddard Space Flight Center: Greenbelt, MD, USA, 1996.

52. Huete, A.R.; Liu, H.; van Leeuwen, W.J. The use of vegetation indices in forested regions: issues of linearity and saturation. In Proceedings of the 1997 IEEE International Geoscience and Remote Sensing—A Scientific Vision for Sustainable Development, Singapore, 3–8 August 1997; pp. 1966–1968.

53. Gitelson, A.A.; Vina, A.; Arkebauer, T.J.; Rundquist, D.C.; Keydan, G.; Leavitt, B. Remote estimation of leaf area index and green leaf biomass in maize canopies. *Geophys. Res. Lett.* **2003**, *30*, 1248. [CrossRef]

54. Gitelson, A.A.; Vina, A.; Ciganda, V.; Rundquist, D.C.; Arkebauer, T.J. Remote estimation of canopy chlorophyll content in crops. *Geophys. Res. Lett.* **2005**, *32*, L08403. [CrossRef]

55. Gitelson, A.A.; Kaufman, Y.J.; Merzlyak, M.N. Use of a green channel in remote sensing of global vegetation from EOS-MODIS. *Remote Sens. Environ.* **1996**, *58*, 289–298. [CrossRef]

56. Hunt, E.R., Jr.; Rock, B.N. Detection of changes in leaf water content using near-and middle-infrared reflectances. *Remote Sens. Environ.* **1989**, *30*, 43–54.

57. Hardisky, M.; Klemas, V.; Smart, M. The influence of soil salinity, growth form, and leaf moisture on the spectral radiance of *Spartina alterniflora* canopies. *Photogramm. Eng. Remote Sens.* **1983**, *49*, 77–83.

58. Key, C.; Benson, N. Measuring and remote sensing of burn severity. In Proceedings of the Joint Fire Science Conference and Workshop, Boise, ID, USA, 15–17 June 1999; University of Idaho and International Association: Moscow, ID, USA; p. 284.

59. Ménard, S.; Darveau, M.; Imbeau, L.; Lemelin, L.-V. *Méthode de Classification des Milieux Humides du Québec Boréal à Partir de la Carte Écoforestière du 3e Inventaire Décennal*; Canards Illimités Canada: Quebec City, QB, Canada, 2006.

60. Franklin, S.; Wulder, M.; Skakun, R.; Carroll, A. Mountain pine beetle red-attack forest damage classification using stratified Landsat TM data in British Columbia, Canada. *Photogramm. Eng. Remote Sens.* **2003**, *69*, 283–288. [CrossRef]

61. Vogelmann, J.E.; Tolk, B.; Zhu, Z. Monitoring forest changes in the southwestern United States using multitemporal Landsat data. *Remote Sens. Environ.* **2009**, *113*, 1739–1748. [CrossRef]

62. Breiman, L. Random forests. *Mach. Learn.* **2001**, *45*, 5–32. [CrossRef]

63. Liaw, A.; Wiener, M. Classification and regression by randomForest. *R News* **2002**, *2*, 18–22.

64. R Core Team. *R: A Language and Environment for Statistical Computing*; R Foundation for Statistical Computing: Vienna, Austria, 2013.

65. Adelabu, S.; Mutanga, O.; Adam, E.; Sebego, R. Spectral discrimination of insect defoliation levels in mopane woodland using hyperspectral data. *IEEE J. Sel. Topics Appl. Earth Obs. Remote Sens.* **2014**, *7*, 177–186. [CrossRef]

66. Sauro, J.; Lewis, J.R. Estimating completion rates from small samples using binomial confidence intervals: Comparisons and recommendations. In Proceedings of the Human Factors and Ergonomics Society Annual Meeting, Los Angeles, CA, USA, 26–30 September 2005; pp. 2100–2103.

67. Lyon, J.G.; Yuan, D.; Lunetta, R.S.; Elvidge, C.D. A change detection experiment using vegetation indices. *Photogramm. Eng. Remote Sens.* **1998**, *64*, 143–150.

68. Simmons, G. *Influence of spruce budworm moth dispersal on suppression decisions, In A Symposium on the Spruce Budworm*; United States Department of Agriculture, Forest Service: Alexandria, VI, USA, 1974.

69. Harrell, F.E. Ordinal logistic regression. In *Regression Modeling Strategies*; Springer: New York, NY, USA, 2001; pp. 331–343.

70. Huete, A.; Didan, K.; Miura, T.; Rodriguez, E.P.; Gao, X.; Ferreira, L.G. Overview of the radiometric and biophysical performance of the MODIS vegetation indices. *Remote Sens. Environ.* **2002**, *83*, 195–213. [CrossRef]

71. Chen, C.; Weiskittel, A.; Bataineh, M.; MacLean, D.A. Evaluating the influence of varying levels of spruce budworm defoliation on annualized individual tree growth and mortality in Maine, USA and New Brunswick, Canada. *For. Ecol. Manag.* **2017**, *396*, 184–194. [CrossRef]

Article

Modeling Migratory Flight in the Spruce Budworm: Temperature Constraints

Jacques Régnière [1,*], Johanne Delisle [1], Brian R. Sturtevant [2], Matthew Garcia [3] and Rémi Saint-Amant [1]

[1]   Natural Resources Canada, Canadian Forest Service, Québec, QC G1V 4C7, Canada;
     johanne.delisle@canada.ca (J.D.); remi.st-amant@canada.ca (R.S.-A.)
[2]   USDA-Forest Service, Northern Research Station, Rhinelander, WI 54501, USA; brian.r.sturtevant@usda.gov
[3]   Department of Forest and Wildlife Ecology, University of Wisconsin-Madison, Madison, WI 53706, USA;
     Matt.E.Garcia@gmail.com
*   Correspondence: Jacques.Regniere@Canada.ca; Tel.: +1-418-648-5257

Received: 30 July 2019; Accepted: 9 September 2019; Published: 13 September 2019

Abstract: We describe an individual-based model of spruce budworm moth migration founded on the premise that flight liftoff, altitude, and duration are constrained by the relationships between wing size, body weight, wingbeat frequency, and air temperature. We parameterized this model with observations from moths captured in traps or observed migrating under field conditions. We further documented the effects of prior defoliation on the size and weight (including fecundity) of migrating moths. Our simulations under idealized nocturnal conditions with a stable atmospheric boundary layer suggest that the ability of gravid female moths to migrate is conditional on the progression of egg-laying. The model also predicts that the altitude at which moths migrate varies with the temperature profile in the boundary layer and with time during the evening and night. Model results have implications for the degree to which long-distance dispersal by spruce budworm might influence population dynamics in locations distant from outbreak sources, including how atmospheric phenomena such as wind convergence might influence these processes. To simulate actual migration flights en masse, the proposed model will need to be linked to regional maps of insect populations, a phenology model, and weather model outputs of both large- and small-scale atmospheric conditions.

Keywords: spruce budworm; *Choristoneura fumiferana*; moth; Lepidoptera; forest protection; early intervention strategy; migration; simulation; aerobiology

---

1. Introduction

Long-distance migration and dispersal behaviors are fundamental life history traits across a broad range of insect taxa [1]. Long-distance movements enable insect species to accommodate seasonal phenology of food resources and to escape local predation pressure, inadequate or devastated resources, and other environmental stressors [2], and thus constitute a fundamental mechanism for "spreading the risk" throughout the population [3]. Migration behavior is typified by temporary suspension of other base functions such as foraging, habitat-searching, and mate-finding to allow sustained and directionally consistent movements [4], generally during the winged (adult) development stage [1]. Virtually any insect species entering flight will be subject to a series of physical and physiological constraints affecting migration success, resulting in diverse physiological and behavioral adaptations. Radar studies across several decades have demonstrated the prevalence of long-distance migration aided by high wind speeds near the top of the atmospheric boundary layer, especially for nocturnal insect flights [5,6]. Understanding the main factors that control insect aerial migration could provide a way to predict migratory movements, which is particularly important in the case of economically important outbreak species exhibiting long-distance migration and dispersal behavior such as acridid locusts [7] and other agricultural pests such as the fall armyworm (*Spodoptera frugiperda* Smith) [8].

One of the most intensely studied forest insects in terms of its aerobiology is the spruce budworm (SBW), *Choristoneura fumiferana* (Clem.) [9], the larvae of which periodically defoliate spruce (*Picea* spp.) and balsam fir (*Abies balsamea* L.) across broad regions of the North American boreal forest [10]. Endemic populations of the SBW are subject to mate-finding and demographic Allee effects [11,12], and immigrant males may improve the odds of local females finding a mate. Immigrant gravid females can also help the local population overcome the demographic Allee effect [12]. Both processes could facilitate the spatial spread of outbreaks. Understanding the main factors that control long-distance transport of SBW and other species can thus help interpret the spatiotemporal dynamics of insect outbreaks. SBW moths do not feed, and females emerge with their full egg complement. Females mate prior to migration, and both males and females participate in mass migratory behavior [9]. Observations indicate that fully gravid females are generally too heavy to fly, and thus lay part of their eggs locally prior to undertaking long-distance flight [13,14]. Exceptions can occur in conditions of highly depleted food resources, where starved females carrying smaller egg complements may emigrate without first laying eggs [15,16].

SBW falls into a class of flying insects that initiate exodus flight around sunset and dusk (see the work of [17]) and tend to fly within and near the relatively stable nocturnal boundary layer, often associated with a near-surface temperature inversion [18]. Budworm moths are strong fliers, launching into the wind during a rapid-ascent flight phase before transitioning to a common downwind orientation [9]. As in other nocturnal migrants, migrating SBW moths have been observed to stratify within one or more vertical layers in the atmospheric boundary layer [6,18]. Those altitudes often correspond with higher wind speeds that increase displacement velocity and ultimate dispersal distance (e.g., the work of [19]), generally over tens to hundreds of kilometers in a single night [20,21].

In previous research, Sturtevant et al. [21] synthesized knowledge of SBW aerobiology to develop a Lagrangian agent-based model of long-distance aerial dispersal. That model produced realistic moth flight trajectories and deposition patterns that were consistent with ground-based trapping surveys. However, the authors applied several simple empirical functions to accommodate some highly uncertain processes, including static distributions of both cruising altitudes and durations of moth flight. Improvements on that model would relate the ability of moths to a rise in the air profile according to the underlying moth physiology and atmospheric structure that drive such flight behavior, leading to more dynamic vertical distributions of migrating SBW, as observed in nature [9]. The compositions of migrant layers in terms of sex ratio, size, and egg load carried by females in flight then become emergent properties of the individual-based model, rather than prescribed parameters.

In this paper, we describe a combined mathematical and empirical framework for the simulation of insect migratory flight behaviors including liftoff conditions, flight altitude, and landing based on air temperature and adult SBW physical characteristics including wing size and body weight. SBW is highly suitable for this approach because so much is known about its migratory behavior and there is a renewed appreciation of the importance of migration in its population dynamics. This is the first in a series of papers where we build upon a previous modeling framework [21] to develop a more sophisticated process-based solution with which we can simulate moth migration events. The ultimate objective of this work is to develop a simulation model that will allow the prediction, in near-real-time, of the spread of the insect's populations through moth migration and subsequent oviposition. We aim for a model that takes into account source population distribution and density, seasonal phenology including the progression of reproduction, circadian rhythm, and interactions with meteorological conditions at the surface (i.e., in the host forest stands) and within the lowest 1 km of the atmospheric boundary layer.

2. Model Construction

2.1. Modeling Approach

The physics of insect flight is a complex subject [22]. As a simplifying assumption, to avoid a very large number of details for which little data exist, we consider three main factors that determine the ability of a winged moth to generate the forces required to lift off and move by flight: body weight, wing surface area, and wingbeat frequency. We then relate wingbeat frequency to temperature to obtain a temperature-dependent model of flight capacity. In the SBW, several factors affect body size (wing area) and weight, and thus flight. Males are typically smaller and lighter than females. In females, body weight is also determined by reproductive status. Gravid females deposit their eggs gradually in successive masses of diminishing size [23], such that the weight of a female drops considerably during her lifetime [24]. In addition, fecundity (and thus body weight) depends on the quality and quantity of food the female was able to acquire during larval development [25]. Food availability is strongly affected in turn by defoliation intensity, a function of population density, which influences adult size and weight in both sexes [15]. SBW moths also lose weight as they consume stored energy reserves, a topic of current research not included in this model. Our individual-based model thus requires both empirical relationships and associated distributions among these SBW adult moth morphometrics and their underlying drivers.

2.2. Morphometric Relationships

We obtained data (weight, forewing surface area, fecundity) on individual moths either collected as pupae from host foliage or caught daily in canopy traps suspended well above the top of host trees at Lac des Huit-Milles, Quebec, Canada, in the summers of 1989 and 1990. The canopy traps used were described in detail by Eveleigh et al. [24]. We also collected moths in light traps (Model 2851U, BioQuip Products, Rancho Dominguez, CA, USA) at 2- to 4-day intervals in the lower St. Lawrence region of Quebec, Canada, between 2010 and 2015. Throughout this paper, we use dry weight as a measure of insect mass. All weight measurements were obtained after desiccating the insects for 24 h at 70 °C. The values of all weight-related parameters thus implicitly account for the missing water content of moths. This assumes that the relative water content of moth bodies remains constant. We also use the area of a single forewing as an index of total wing area, realizing that actual wing area in moths is composed of two forewings and two hindwings.

2.2.1. Female Fecundity, Weight, Wing Area, and Influence of Defoliation

We collected SBW pupae in the Lac des Huit-Milles stand on host foliage during 1989–1990 to determine the forewing surface area and dry weight of 122 fully gravid and 132 fully spent females (which were allowed to lay eggs until their death in the laboratory). Host defoliation (current-year foliage) averaged 64% in 1989–1990 in the stand. From the fully spent females in this sample, we established a relationship between potential fecundity (*E*, in eggs per female) and forewing surface area (*A*, in cm^2) as follows:

$$E = 739.2A^{1.758}\varepsilon \left(R^2 = 0.45\right),\tag{1}$$

(Figure 1a) where ε is a lognormal error term (Anderson-Darling test of normality [26] AD = 0.324; $p = 0.52$). Observed fecundity E_d, where the level of defoliation is known, can be corrected to provide an estimate of potential fecundity in the absence of defoliation, E_0, using the relationship between defoliation d (a proportion between 0 and 1) and fecundity reported in the work of [25]:

$$E_0 = F_d/(1 - 117d/216.8),\tag{2}$$

which is represented for $d = 0.64$ by the dotted line in Figure 1a.

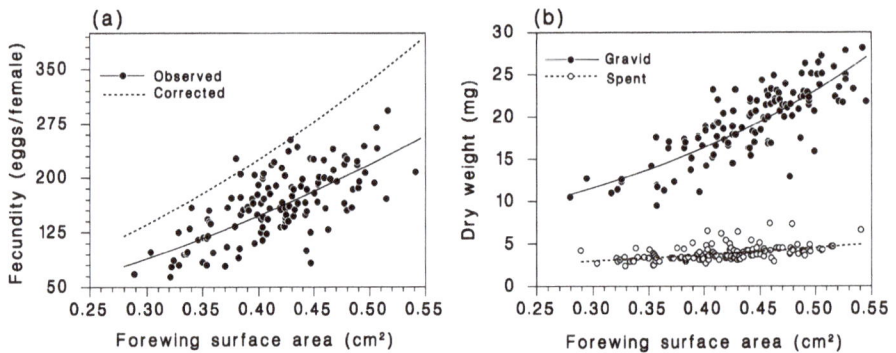

Figure 1. (a) Relationship between forewing surface area and lifetime fecundity among females collected as pupae on balsam fir foliage (64% defoliation) at Lac des Huit-Milles, 1989–1990 (solid line: Equation (1); dotted line: after correction for defoliation using Equation (2)). (b) Relationship between forewing surface area and dry weight of fully gravid and spent females (lines: Equation (3)), pupae collected at Lac des Huit-Milles, 1989–1990.

The dry weights (*M*, in g) of gravid and spent females were vastly different because of the weight of eggs (Figure 1b), with fully-gravid females at 0.0194 ± 0.0042 g (standard deviation) and spent females at 0.0039 ± 0.0009 g. A mixed-effects regression accurately described the relationship between dry weight and wing area among the fully gravid and fully spent females:

$$M = e^{-6.4648+0.9736G+2.14A+1.3049GA} \varepsilon \left(R^2 = 0.95 \right), \tag{3}$$

using gravidity $G = E/E_0$ as a continuous variable in the range 0 (fully spent, $E = 0$) to 1 (containing its full potential fecundity, $E = E_0$), and with ε near-lognormal with mean = 1 and standard deviation 0.16 ($n = 254$; AD = 1.675; $p < 0.005$). In the model, the weight of females is related to their remaining fecundity (unlaid eggs), and because defoliation decreases initial fecundity (by the ratio E_d/E_0), it reduces their weight correspondingly. The weight of males is also reduced by the same ratio ($1-117d/216.8$) in areas of known defoliation [15].

Among female SBW moths caught in light traps in the lower St. Lawrence region of Quebec, Canada, between 2010 and 2015, the year of capture explained 37% of the total variability in forewing surface area (analysis of variance (ANOVA) F = 6.74; df = 5, 47; $p < 0.001$). Although statistically significant, the relationship with stand-level defoliation explained only 8.6% of the total variability (F = 7.84; df = 1, 47; $p = 0.007$). There was no significant interaction between year and defoliation (F = 1.53; df = 1, 42; $p = 0.202$). Because the effect of defoliation on forewing area was small, only 0.02 cm^2 over the range of 0% to 100% defoliation (Figure 2), we chose to ignore this factor in the final flight model. Thus, in our model, defoliation has an influence only on the moth weight, and individuals that were submitted to starvation due to overcrowding have larger wings relative to their body weight than well-fed ones, and are thus more apt to emigrate.

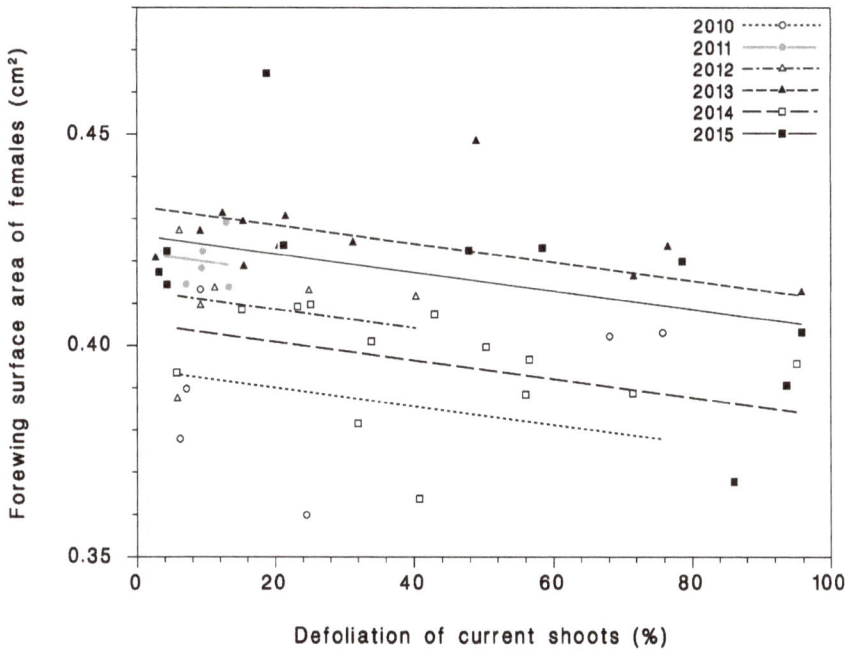

Figure 2. Effect of year and current-shoot defoliation on forewing surface area among female moths caught in light traps in the lower St. Lawrence region, 2010–2015. Lines are linear least-squares regressions fitted to the observations available for each year.

2.2.2. Moths Captured in Canopy Traps

Moths collected in canopy traps placed several meters above the forest canopy are believed to be representative of emigrating SBW adults [24]. The sex, forewing length and area, and weight of SBW moths caught in those traps at Lac des Huit Milles in 1989–1990 were recorded. The forewing surface area averaged 0.361 ± 0.047 cm^2 ($n = 1044$) in males and 0.421 ± 0.063 cm^2 ($n = 1024$) in females, a significant difference (F = 599; df = 1; $p < 0.001$). The forewing surface area was normally distributed in males (AD = 1.039; $p = 0.01$) and near-normally distributed in females (AD = 0.587; $p = 0.126$). For the same moths, dry weight averaged 0.00475 ± 0.00143 g in males and 0.00684 ± 0.00287 g in females. We found a curvilinear relationship between wing surface area and dry weight, with a significant effect of sex on intercept (F = 7.15; df = 1, 2061; $p = 0.008$), but no effect of sex on slope (F = 1.56; df = 1, 206; $p = 0.212$):

$$M = \begin{cases} e^{-6.697+3.626A} \; \varepsilon \text{ for males} \\ e^{-6.582+3.626A} \; \varepsilon \text{ for females} \end{cases} \tag{4}$$

where the distribution of error ε is approximately lognormal with mean of 1 and standard deviation $\sigma_\varepsilon = 0.206$ for males (AD = 7.3; $p < 0.005$) and $\sigma_\varepsilon = 0.289$ for females (AD = 6.4; $p < 0.005$) ($R^2 = 0.95$ for both; Figure 3).

Both the forewing surface area and dry weight of moths caught in canopy traps decreased during the flight season for both sexes (Figure 4a). It has been suggested that this gradual drop in wing surface area during the flight season results from slower development and generally later emergence of smaller individuals as a possible result of parasitism, poor food quality, and/or low food quantity in the larval stages [24]. Among males, dry weight decreased at nearly the same rate as wing area, but among females, the dry weight loss was steeper than the decrease of wing area, which we attribute to the females losing substantial body weight with oviposition over the course of the flight season.

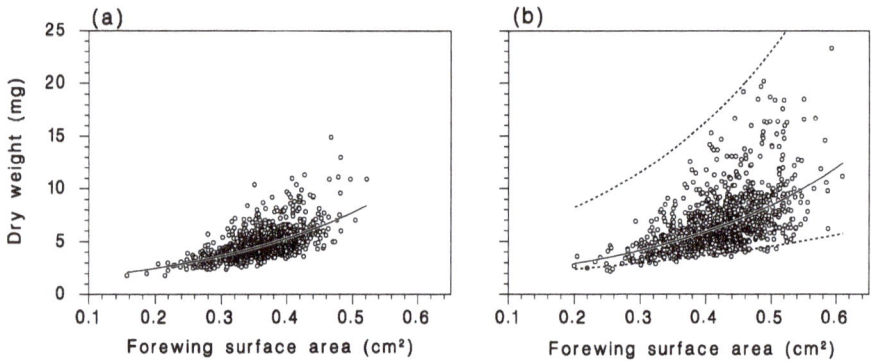

Figure 3. Relationship between dry weight (g) and forewing surface area (cm²) in (**a**) male and (**b**) female spruce budworm (SBW) moths caught in canopy traps over the entire flight season at Lac des Huit-Milles, 1989–1990. Solid lines represent the fitted relationships for (**a**) males and (**b**) females, as in Equation (4). Dotted lines in (**b**) represent theoretical bounds for females based on gravidity using Equation (3).

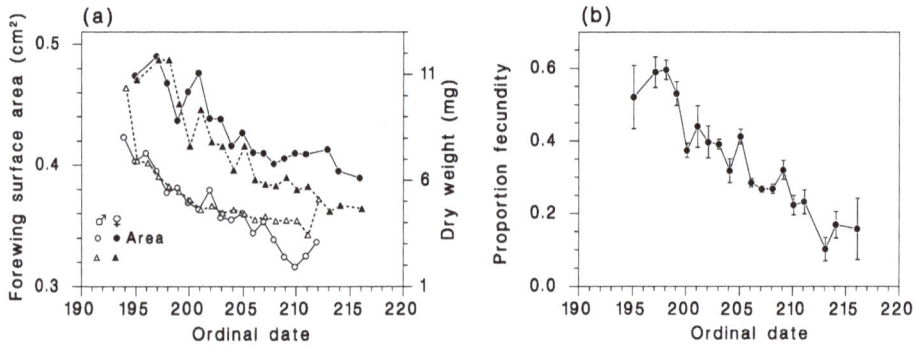

Figure 4. (**a**) Changes in average forewing surface area and dry weight over the flight season in males and females. (**b**) Gravidity of females calculated with Equation (5) (vertical bars represent the standard error of the mean). Moths caught in canopy traps at Lac des Huit-Milles, 1989–1990.

Rearranged, Equation (3) was used to estimate the diminishing gravidity G of females collected over time in the canopy traps at Lac des Huit-Milles during the moth flight season (Figure 4b):

$$G = \frac{\ln(M) + 6.4648 - 2.14A}{0.9736 + 1.3049A} \qquad (5)$$

The observed distribution of gravidity among those females had an overall mean of $G = 0.33 \pm 0.19$ ($n = 1023$), very similar to the distribution reported in the work of [9] (their Table VI) with $G = 0.31 \pm 0.17$ ($n = 854$), assuming an average fecundity of 200 eggs/female.

2.2.3. Defoliation at the Individual Level

Individual SBW larvae cause and experience defoliation at the tree shoot (branchlet) level. To enable the generation of realistic individual-level defoliation exposures from stand-level defoliation estimates, we measured defoliation using Fettes' method [27] in the lower St. Lawrence stands during 2010–2015 from 45 cm branch samples taken at the end of the SBW egg hatch. Defoliation was assessed on the 20 most apical shoots of each branch. These data were used to relate average (stand-level) defoliation to the frequency distribution of defoliation at the shoot level (and, by extension, the individual insect's level) using a Beta (α, β) distribution [28]. The parameters of this distribution are estimated by

$$\alpha = \mu_d\left(\frac{\mu_d(1-\mu_d)}{\sigma_d^2} - 1\right) \text{ and } \beta = (1-\mu_d)\left(\frac{\mu_d(1-\mu_d)}{\sigma_d^2} - 1\right) \tag{6}$$

where μ_d and σ_d are the mean and standard deviation of defoliation d (a proportion between 0 and 1) measured at the shoot level. We also obtained an empirical relationship between the mean and variance of defoliation values from these foliage samples using ordinary least-squares regression:

$$\sigma_d^2 = 0.008101 + 0.5289\mu_d - 0.5228\mu_d^2 \ (R^2 = 0.953), \tag{7}$$

where mean defoliation μ_d is expressed as a proportion rather than a percentage (Figure 5a). The observed and corresponding Beta distributions of shoot-level defoliation match very well (Figure 5b–g). The results indicate that there is large variability in the degree of food competition among larvae when defoliation is in the range $d = 0.2$–0.8. When defoliation is extreme ($d > 0.9$), the vast majority of individuals experience food limitation. In the individual-based model, these defoliation-dependent scaling distributions affect the distributions of moth weight and fecundity across a range of source populations and landscapes.

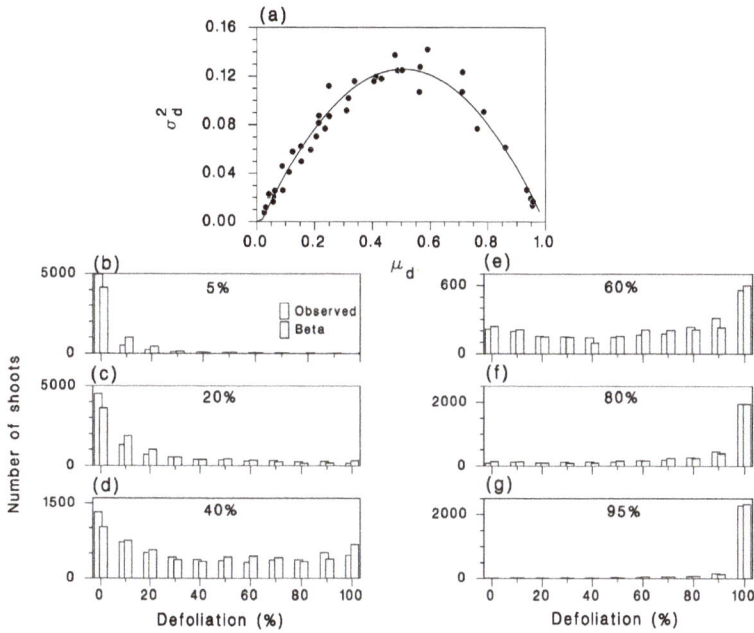

Figure 5. (a) Relationship between stand-level variance σ_d^2 and mean μ_d defoliation expressed as a proportion, with the line from Equation (7). (b–g) Observed and Beta frequencies of shoot-level defoliation for ranges of stand-level defoliation from 5% to 95%. Data from balsam fir in the lower St. Lawrence region, 2012–2015.

2.3. Flight Model

Using a table of in-flight wingbeat frequencies from 160 species of flying insects [29], Deakin [30] developed a simple double-allometric relationship to describe these observations with high accuracy:

$$v_L = K\frac{\sqrt{M}}{A} \tag{8}$$

where K is a proportionality constant (Hz cm^2 g$^{-1/2}$), M is mass (g), A is the surface area of a single forewing (cm^2), and v_L (Hz) is the wingbeat frequency recorded at 20–24 °C. We assume here that v_L is the minimum wingbeat required for the insect to lift off.

The wingbeat frequency of moths observed by the authors of [9] in sustained flight over radar was in the range $v_S = 25$–42 Hz. We used an iterative numerical optimization procedure to maximize the overlap between this range and the wingbeat frequencies calculated with Equation (8). This procedure involved calculating the liftoff wingbeat frequency v_L of all moths in the samples collected from canopy traps in Lac des Huit-Milles in 1989–1990, varying the value of K between 165 and 175 (in steps of 0.1), and selecting the value that provided the maximum overlap with the range $v_S = 25$–42 Hz. With this procedure, we obtained $K = 167.5 \pm 0.05$ Hz cm^2 g$^{-1/2}$, yielding 94.4% of overlap. The distribution of liftoff wingbeat frequencies for male and female moths in our sample from canopy traps is very similar (Figure 6a). Using this estimate of K, the liftoff wingbeat frequencies of females from our sample of pupae illustrate that liftoff is far easier for spent females than for gravid females (Figure 6b); while spent females require typical liftoff wingbeat frequencies $v_L < 30$ Hz, fully gravid females can have liftoff wingbeat frequencies $v_L > 50$ Hz, beyond the range observed by the authors of [9] and supporting the observation that gravid females must typically deposit their first eggs in the natal site before attempting migratory flight.

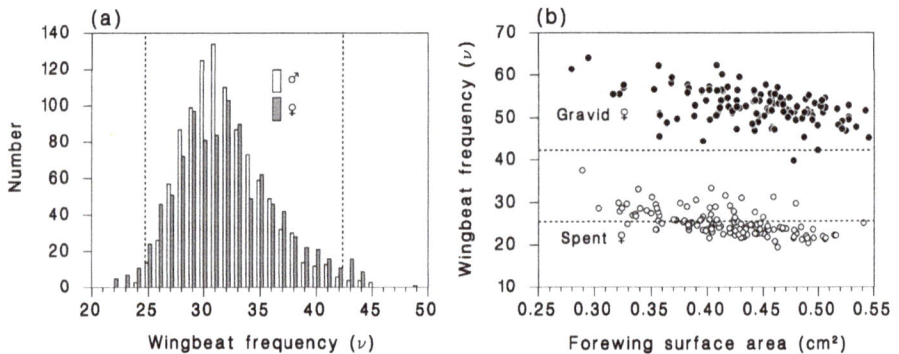

Figure 6. Liftoff wingbeat frequencies calculated from dry weight and forewing surface area with Equation (8) using $K = 167.5$ Hz. (**a**) Distribution of wingbeat frequencies of male and female moths caught in canopy traps. Vertical dotted lines: observed range of wingbeat frequencies of migrating budworm moths ($v = 25$–42 Hz according to the authors of [9]). (**b**) Liftoff wingbeat frequencies of fully gravid (●) and spent (○) females from observations at Lac des Huit-Milles, 1989–1990.

As with many insect physiological processes, the rate at which an insect can beat its wings is a function of ambient temperature [31–37]. There is some evidence that wingbeat frequency may actually decrease at high temperatures in some moths [37], but this has not been seen in other insects. Because thermal responses in insects are non-linear [38], we use a sigmoid-shaped logistic curve where wingbeat increases exponentially at low temperatures, then asymptotically approaches a maximum (v_{max}) at higher temperatures:

$$v(T) = \frac{v_{max}}{1 + \exp^{-b(T-a)}} \tag{9}$$

where T is ambient temperature (°C), v_{max} is a species-specific maximum wingbeat frequency (Hz), a is the midpoint temperature of the response (°C), and b is the spread of that response with respect to temperature (°C^{-1}). In small moths, it seems that wing fanning does not generate enough heat to raise the insect's body temperature significantly above that of the air around it, so thermoregulation can be ignored [32]. In our model, we ignore SBW thermoregulation.

Here, we assume that the wingbeat of an airborne individual determines whether it ascends (when $v(T) \geq v_L$) or descends (when $v(T) < v_L$) in the atmospheric boundary layer with a given temperature profile. In this model, moths cannot lift off at a temperature $T < T_L$ because they cannot beat their wings fast enough. This leads to limitations on the time of SBW liftoff as the near-surface temperature decreases from sunset through the evening [17].

After liftoff, an airborne moth climbs through the air column until it reaches the altitude at which its wingbeat frequency matches v_L, which occurs at the temperature T_L:

$$T_L = a - b \, \ln\left(\frac{A \, v_{max}}{K \, \sqrt{M}} - 1\right) \tag{10}$$

Once the moth has reached that altitude, we introduce an energy conservation factor Δ_v that allows the insect to settle into sustained flight at an altitude where the temperature T_S is somewhat higher than T_L, but without increasing its wingbeat frequency as in Equation (9). We define $0 < \Delta_v \leq 1$ as a proportional reduction from physiological maximum wingbeat frequency v_{max}, such that

$$v_L = v(T_S) \, \Delta_v \tag{11}$$

Thus, the altitude at which sustained flight occurs is that where

$$T_S = a - b \, \ln\left(\frac{\Delta_v}{K} \frac{A \, v_{max}}{\sqrt{M}} - 1\right) \tag{12}$$

If the air temperature changes while the insect is in flight, the individual must settle at a new altitude that satisfies Equation (12). For a typical temperature profile above the nocturnal inversion, this higher temperature occurs at a slightly lower flight altitude. However, as the ambient temperature drops below the moth's required T_S at all reachable altitudes, the airborne insect will continue to descend in search of T_S until it lands. Equation (12) tells us that the temperature at which an insect can fly depends on its weight and wing surface area, and that, other than K in Equation (8), there are four parameters that need to be estimated to complete this description: v_{max}, a, and b in Equations (9) and (10) and Δ_v in Equations (11) and (12).

We are aware of three sets of data with flight activity observations recorded over a wide range of temperature for the SBW. Flights by mated, egg-laying females have been observed in the laboratory [39], where the lower temperature threshold for flight was 15 °C. However, these were not observations of migratory flight. The late C.J. Sanders observed from tall scaffolds the number of males "buzzing" around the upper crowns of selected balsam fir trees in a mixed stand near Black Sturgeon Lake, Ontario, in 1987 (unpublished data). Observations of buzzing males and the ambient temperature were noted during five-minute periods, replicated several times per hour, during the peak evening period of moth activity on successive nights. From those observations, we calculated the average number of moths buzzing for temperature classes of 2 °C width in the range 12–30 °C, with very few males observed buzzing below 14 °C. However, these observations do not provide information on the value of v_{max}, and it is not yet entirely clear how the wingbeat frequency of male buzzing at upper tree crowns is related to migration flight.

The third dataset provides more useful information. Through a complex procedure of foliage sampling and radar observations, the proportion of available egg-laying females that emigrated on several evenings during peak flight activity in 1973–1974 was estimated for two locations in central New Brunswick [40]. The authors defined as "available" those females that were at least two days old and had begun oviposition. They also recorded the top-of-canopy temperature at 20:40 (sunset) every evening (n = 24). While these observations do not provide wingbeat frequencies, the observers were specifically concerned with females attempting migratory flight. Using our morphometric measurements (M and A) from females caught in canopy traps at Lac des Huit-Milles in 1989–1990 that were also (presumably) attempting emigration, we calculated the proportion of females that could lift off as a function of temperature and compared those calculations with the work of [40].

Finally, we used a grid-search optimization method to estimate values for the three unknown parameters of Equation (9): ν_{max}, a, and b at resolutions of 0.5, 0.1, and 0.005, respectively. We selected the values that minimized the residual sum of squares between the observed proportions [40] and those calculated from the liftoff wingbeat frequencies provided by Equation (8) for females caught in canopy traps at Lac des Huit-Milles in 1989–1990. We used the value of K obtained from Equation (8) above in our parameter estimation for Equation (9). Our best parameter estimates were ν_{max} = 72.5 ± 0.5 Hz, a = 23.0 ± 0.1 °C, and b = 0.115 ± 0.005 °C^{-1}, yielding the lowest residual sum of squares (R^2 = 0.561; line in Figure 7a). There was a high correlation between calculated wingbeat frequency and the observations of male SBWs buzzing around the crowns of host trees near Black Sturgeon Lake in 1987 (r = 0.89; Figure 7b).

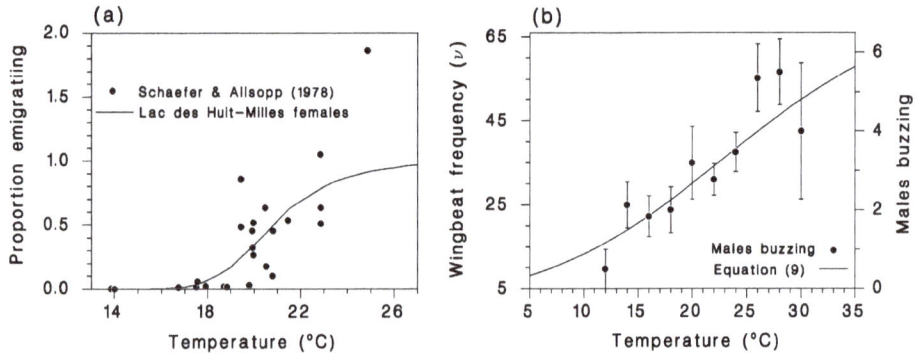

Figure 7. (**a**) Influence of temperature on proportion of females emigrating (circles: observations [39], line: prediction for females caught at Lac des Huit-Milles, 1989–1990); (**b**) observed temperatures and numbers of "buzzing" males (circles, from unpublished data of C.J. Sanders) and corresponding wingbeat frequency calculated with Equation (9) (solid line).

2.4. Flight Model Simulations

We used the above relations to simulate flight for 10,000 individual SBW moths (50% female) in idealized atmospheric boundary layer conditions. Each individual was assigned a random value of forewing surface area according to sex from normal distributions based on the observations for male and female moths collected in canopy traps at Lac des Huit-Milles in 1989–1990 (males 0.361 ± 0.047 cm^2; females 0.421 ± 0.063 cm^2). From those assigned forewing surface areas, we calculated the dry weight for each individual with Equation (4) using a lognormally-distributed ε error term with mean of 1 and standard deviations of 0.206 for males and 0.289 for females. We calculated female fecundity using Equations (1) and (2), and gravidity using Equation (5). We then calculated the liftoff wingbeat frequency ν_L for each moth using Equation (8), the liftoff temperature T_L with Equation (10), and the sustained flight temperature T_S with Equation (12) for two values of Δ_ν (1.0 and 0.85). When Δ_ν = 1, the liftoff temperature T_L and sustained flight temperature T_S are identical. As Δ_ν decreases, $T_S > T_L$ and sustained flight requires increasingly warmer temperatures relative to liftoff conditions.

To simulate the progression of flight under realistic temperature conditions, we generated an hourly time series of idealized air temperature profiles for the lower 1500 m of the atmospheric boundary layer from 20:00 (sunset) through 05:00 the next morning (Figure 8). This profile time series imitates the evolution of a nocturnal temperature inversion (i.e., where temperature increases with altitude) owing to surface radiative cooling that is typical of a calm, clear summer night in temperate North America. This inversion appears near the surface at sunset, then increases in depth and dissipates gradually overnight (Figure 8; cf. observed profiles in the work of [41]).

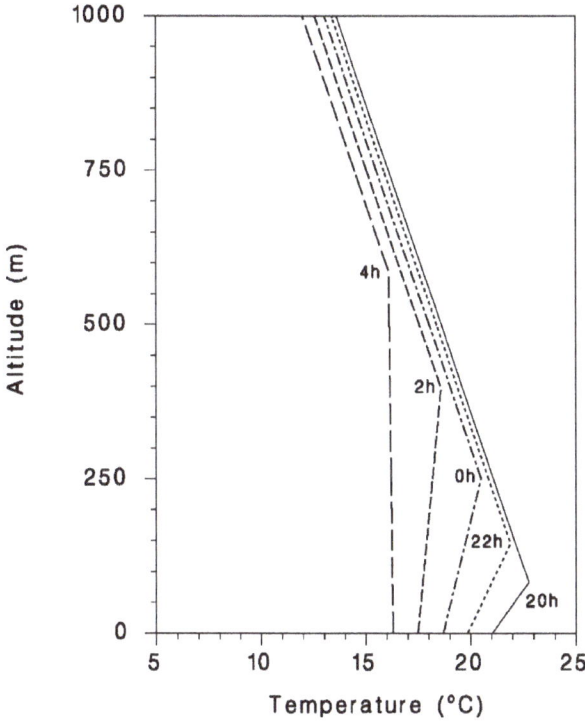

Figure 8. Idealized (archetypal) evening transition and nocturnal boundary layer temperature profile for a clear, calm summer night in northern temperate latitudes. The inversion develops from surface radiative cooling that begins near sunset, increases in depth, and dissipates gradually by early morning.

The evening boundary layer transition and evolution of the nocturnal inversion, as well as their potential roles in SBW migratory flight around sunset, are discussed further in a companion paper [17]. Here, we have specified no wind profile for our flight simulations, choosing instead to focus solely on temperature-related influences on SBW flight. We specify that all moths attempt to lift off at sunset. Those moths that can lift off, where $T > T_L$ at the surface, either reach the altitude where temperature allows their sustained flight or, not finding such a level, return to the ground immediately. At each hour of the simulation, we recorded the vertical distributions of moth density, sex ratio, and gravidity (proportion of initial fecundity) carried by flying females. In the simulations presented here, we allow the moths to reach their cruising altitude immediately upon liftoff, as the time step is 1 h, long enough for them to reach it. When a temperature inversion exists, the simulated moths settle into sustained flight above the inversion. In a more realistic simulation context, with actual temperature data, we use a 5–10 min time step. In that case, the ascent rate is proportional to the difference between the wingbeat frequency for the air temperature at the current location $v(T)$ and the sustained flight wingbeat frequency v_S as $Vz = \alpha[v(T) - v_S]$ where $\alpha = 0.11$ m s^{-1} Hz^{-1}, yielding a range of about 0–2 m/s, commensurate with observed ascent rates [9].

3. Simulation Results

The vertical distribution of flying moths varied over time and with the value of Δ_ν (Figure 9a,d). Overall, females tended toward sustained flight at higher altitudes than males (Figure 9b,e), although the heavier females (i.e., those with greater gravidity) generally remained at lower flight altitudes (Figure 9c,f) because of their greater weight relative to partially or totally spent females. In the $\Delta_\nu = 1$ simulation, nearly 60% of moths achieved sustained flight, compared with only 40% when $\Delta_\nu = 0.85$. The number of flying moths decreased gradually overnight as the air column cooled, and the mode (peak concentration) of flying SBW ascended with time. With $\Delta_\nu = 1$, the mode of the vertical distribution of moths in flight remained ~200 m above the top of the surface inversion layer, while it descended to the top of the inversion with $\Delta_\nu = 0.85$. With $\Delta_\nu = 0.85$, the flight cloud lost males more quickly and became proportionally more populated by low-gravidity females over the evening.

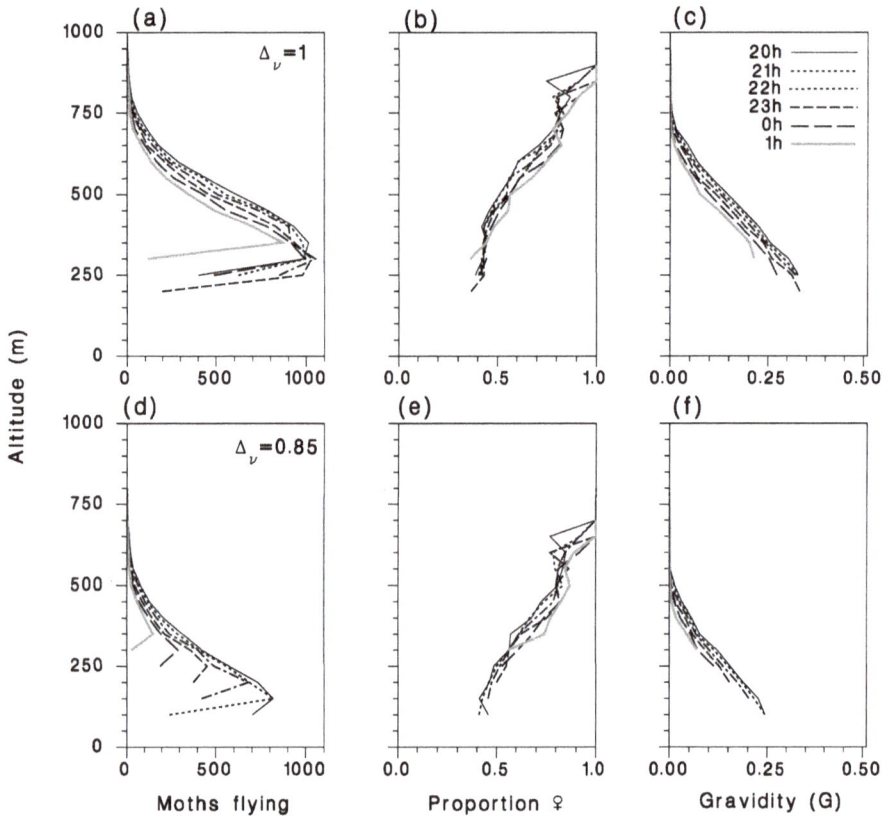

Figure 9. Hourly vertical density profiles of simulated migrating moths between liftoff at 20:00 (sunset) and 01:00 the next morning. Top row: $\Delta_\nu = 1.0$; bottom row: $\Delta_\nu = 0.85$. (**a,d**) Number of moths flying. (**b,e**) Proportion female. (**c,f**) Gravidity, the proportion of initial fecundity carried by females.

These simulation results are summarized over the entire simulation period in Figure 10. Most moths flew early in the evening while temperatures were highest, and their numbers diminished over the night as the temperature decreased throughout the specified boundary layer profile (Figure 10a). A smaller value of Δ_ν led to a decrease in the number of moths that remained airborne and the overall duration of the flight period (Figure 10a). The average altitude of airborne moths changed over time; heavier individuals, especially the most gravid females, were forced to land earlier as temperature decreased, leaving the males and less-gravid females in the flight concentration profile (Figure 10b). In addition, a smaller value of Δ_V led to a lower mean flight altitude through much of the night, as the airborne moths sought warmer air for sustained flight, until only the least-gravid females remained airborne near the top of the profile at the end of the night (Figure 10b). Overall, there are more airborne females than males, which corresponds to observations of a female-biased sex ratio among migrant moths [9]. However, the proportion of females among airborne moths in both simulations increased during the flight period, as males generally landed earlier than the less-gravid females (Figure 10c). This biased sex ratio is the result of different weight to wing surface area relationships in males and females. As the night progressed, the overall gravidity of migrating females decreased as heavier females were forced to land earlier. With a smaller value of Δ_V, egg loads carried by migrating females were even lower as the more-gravid females found sustained flight more difficult and landed earlier than their less-gravid counterparts (Figure 10d).

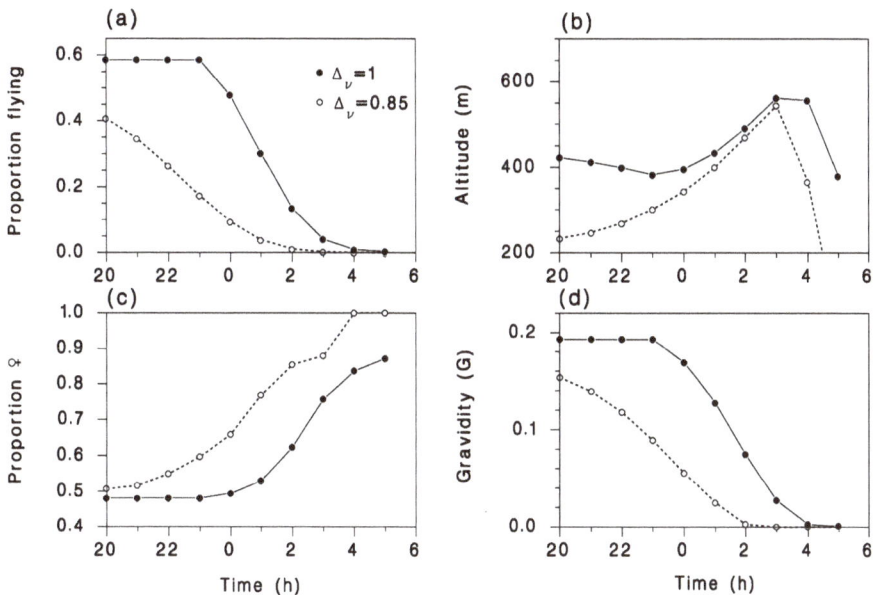

Figure 10. Summary of simulated migratory flights. (**a**) Proportion of moths in flight. (**b**) Mean flight altitude. (**c**) Proportion female among migrating moths. (**d**) Gravidity, the proportion of initial fecundity carried by migrating females. ●: $\Delta_V = 1.0$; ○: $\Delta_V = 0.85$.

4. Discussion

Our explicit use of empirically-derived biophysical relationships between SBW moth size and weight, gravidity, wingbeat frequency, and temperature in this model improves greatly upon previous individual-based models of SBW migratory flight [21]. Fundamental determinants of liftoff, cruising altitude, and descent are emergent properties of interactions between modeled moth flight behaviors and the evolving surface conditions and boundary layer temperature profile. The empirical analyses underlying our flight model are highly consistent with known SBW life history [9], and the consequent model provides additional insight into other aspects of its aerobiology that have remained poorly understood.

Our results suggest that fully gravid females have liftoff wingbeat frequencies $v_L > 50$ Hz (outside the reported range of $v = 25$–42 Hz in [9]), while spent females have much lower liftoff wingbeat frequencies ($v_L < 30$ Hz). This difference, which can be attributed to the weight of eggs carried by the female, explains why fully gravid females do not readily emigrate [13,14,39,42,43]. However, females from severely defoliated stands are smaller, lighter, and have lower fecundity than well-fed females. Those lighter females can fly soon after emergence without laying any eggs [15,16], an observation also consistent with our model formulation. Simulated vertical density profiles (Figure 9a) are further consistent with recent radar observations of SBW mass migration events [19]. Indeed, temperature responses by migrating insects remain among one of the most plausible factors of the atmosphere influencing the formation of flight layers in and above the atmospheric boundary layer across a broad range of insect taxa and flight seasons [6,18,44]. Temperature constraints on flight may further explain temperature dependence in the timing of liftoff relative to sunset and dusk [9], as we have explored in a companion paper [17].

These simulation results, despite their idealized setting, suggest several features of SBW dispersal that may be consequential to our understanding of SBW outbreak spatiotemporal dynamics. The fundamental premise of our individual-based model is that moth flight altitude is determined primarily by interaction between the air temperature profile in the atmospheric boundary layer and the individual moth's ability to maintain a sufficient wingbeat frequency to remain airborne. Depending on the temperature profile and its trend through the night, we found that most moths eventually land because of cool air at some time well before sunrise, as observed with radar [9]. We can thus conclude from our simulations that flight duration results from an interaction between air temperature and the moth's weight and size. Though we have not specified a boundary layer wind profile along with the simulation temperature profile, we can still draw some conclusions from the flight duration itself. For example, the most-gravid females carrying significant egg loads fly for comparatively short periods and hence short distances. Perhaps counterintuitively, some females fly higher (and longer) than males, but those females are less gravid with smaller egg loads (Figure 10c,d). These outcomes suggest some limitations on the ability of the adult SBW to disperse eggs over the full range of distances that can be reached by the source population in a given location. However, distance limitations can also change over the flight season, with late-emerging adults generally being lighter and thus able to attain a greater flight range under generally warmer (and thus even more favorable) temperature conditions. Source populations in severely defoliated regions will likewise emerge to be smaller and lighter, with more-gravid females having an increased ability to lift off and fly longer distances in those warmer conditions.

Absent from our simulations, and from the discussion thus far, are the potential effects of a variable wind profile in the atmospheric boundary layer on flight distances. Where temperature is a principal determinant of flight altitude, a demographic sorting occurs in the vertical profile of moth concentration. The most-gravid females generally remain at the top of the surface boundary layer inversion (Figure 9c,f) during their flight, while males and less-gravid females generally settle into sustained flight above that level, with males present throughout the profile. Boundary layer wind profiles in the same seasons that produce our idealized temperature profile have a generally similar shape. Winds are typically slower near the surface (due to surface friction, and not accounting here for possible directional changes), and increase with altitude to a possible maximum just above the top of the boundary layer inversion. There, a low-level jet can form in otherwise calm synoptic conditions on some nights [45,46] and with often diminishing wind speeds above that maximum.

The vertical demographic sorting effect of the ambient temperature profile is thus compounded by the boundary layer wind profile, resulting in further sorting over the distance travelled from the source region. More-gravid females fly at the top of the surface inversion layer, in or near the greatest wind speeds, and may have a greater flight range despite landing first because of overall boundary layer cooling. The largest concentration of males generally fly lowest, potentially below the wind maximum, and land within the shortest distance, but with considerable spread over the entire flight range as higher-flying males reach greater distances; less-gravid females, flying at and just above the greatest wind speeds, can reach the greatest distances from the source area before landing. In terms of egg deposition by migrants, a distinct pattern emerges from these (admittedly idealized) considerations. The highest concentration of eggs (as much as half of original fecundity) is deposited in the natal area by the gravid females who are too heavy to fly upon emergence. Adjacent to that natal area, there is a low concentration of egg deposition where migration landings are dominated by short-flying males, out to a distance where the most-gravid females land and form a secondary concentration peak in egg deposition, which diminishes slowly with distance as less-gravid females land and deposit their eggs at lower concentrations. The coherence of this pattern in any particular direction from the source region remains contingent on wind direction, its persistence through the evening and night, and its variability with height in the atmospheric boundary layer. The overall effects of these combined emergent properties on the spatiotemporal dynamics of SBW outbreaks are yet to be determined, but our individual-based model holds promise for the examination of the structure and consequences of dispersal events under both current and potential future conditions [47].

An example of the importance of demographic details (i.e., males and females, and relative gravidity of females) occurs in consideration of convergent atmospheric wind patterns [9]. Dispersal in general tends to dilute the concentrations of SBW from the initial source areas as moth landings occur across a much broader region. However, various features of weather, topography, and land–water interfaces often create wind convergence zones that can concentrate flying insects [48]. Examples of mesoscale concentration of migrating SBW moths associated with sea breezes have been observed via radar in a number of instances (e.g., the works of [49,50]), and smaller-scale convergence zones associated with topography have been shown to concentrate other species of flying insects (e.g., the work of [51]). If such concentrations are also associated with changes in boundary layer temperature, as often occurs in valley regions at night, the density of landing insects can be far greater in those locations and the enhanced concentration of deposited eggs can lead to greater defoliation and potential for outbreak initialization in subsequent years [51]. Given a general assumption of dispersal spread with migration distance, mesoscale convergence needs to occur close to primary source areas for such concentrations to occur. Localized convergence associated with topography occurs most strongly close to the ground [48], generally where greater numbers of more-gravid females can fly, and, as we have demonstrated, flight distance from the source can effectively select for critical demographic factors such as male/female ratios and female gravidity.

5. Conclusions

This model of SBW migratory flight, combined with a model predicting the circadian rhythm of liftoff times [17], constitutes significant progress in our understanding of the interactions between SBW and its environment and the emergent effects of those interactions on SBW dispersal and migration. Investigations of model behavior in real weather conditions, over actual terrain, and in comparisons with various sources of observations such as Doppler radar [19] and trap networks are needed to advance the development, calibration, and validation of our model framework and its assumptions. Of the several model parameters estimated here, Δ_v remains the most uncertain; traditional observations are not sufficient for its estimation, leading us to specify its value in the flight simulations described above. Its value could be defined better by comparing sets of observed and simulated vertical distributions of migrating moths. The parameters of Equation (9) were obtained somewhat indirectly, from field observations of emigrating moths [40]. A better understanding of the shape of the relationship between temperature and wingbeat frequency in Equation (9) for both male and female SBW could provide further significant improvement of our individual-based model. One simplifying assumption, in need of verification, is the absence of thermoregulation in SBW moths. It is possible that moths may warm their thoracic muscles so they can lift off and migrate in cold temperatures, and it is still possible (contrary to the work of [32]) that sustained flight further warms those muscles, changing the (still uncertain) efficiency and energy consumption of SBW physiological activity over the course of the flight. Finally, the morphometric data presented here (Figure 1) come from a single location (Lac des Huit Milles in Quebec, Canada) in 1989 and 1990, but we know of additional datasets relating wing size and weight in gravid and spent females from which we may

Forests **2019**, *10*, 802

obtain a better understanding of the interannual variation in that relationship that is not fully explained by defoliation. Incorporating these additional data would undoubtedly improve our model.

Despite these remaining uncertainties, we now have a well-founded process model to describe three of the principal aerobiological stages of moth migration for the SBW: launch/ascent, horizontal transport, and descent/landing [21,52]. Such a model can now be linked to a model of SBW phenology [46], functions defining circadian rhythm of migration flight behavior [17], maps of known populations and/or defoliation activity, and weather model outputs [53] to simulate the entire migration process in near-real-time. Comparison of the outputs of such a model with observations from ground-based trap networks and radar data could be used to calibrate the least understood model parameters, point to avenues for model improvement, and validate model predictions. Such an integrated model might ultimately be used to predict mass migration events and the distribution of SBW eggs to assist in the management of potential SBW outbreaks.

Author Contributions: Conceptualization, J.R., J.D., and R.S.-A.; Methodology, J.R. and R.S.-A.; Software, R.S.-A.; Validation, J.R., R.S.-A., and B.R.S.; Formal Analysis, J.R.; Investigation, J.R. and J.D.; Resources, J.D. and J.R.; Data Curation, J.R.; Writing—Original Draft Preparation, J.R., J.D., B.R.S., and M.G.; Writing—Review & Editing, J.R., B.R.S., J.D., M.G., and R.S.-A.; Visualization, J.R.; Supervision, J.R. and J.D.; Project Administration, J.R. and J.D.; Funding Acquisition, J.R., J.D., and B.R.S.

Funding: This research was funded by the natural resources and forest departments of Newfoundland, Nova Scotia, New Brunswick, Quebec, Ontario, Manitoba, Saskatchewan, Alberta, and British Columbia. Funds were also provided by the Atlantic Canada Opportunities Agency, and under the U.S./Canada Forest Health & Innovation Summit Initiative sponsored by the USDA Forest Service, Canadian Forest Service of Natural Resources Canada, and the U.S. Endowment for Forestry and Communities, with cooperation by SERG-International.

Acknowledgments: We thank Ariane Béchard, Alain Labrecque, and Pierre Therrien for their contributions to the field work, and Marc Rhainds (Canadian Forest Service, Atlantic Forestry Centre, Fredericton, NB) for discussions related to spruce budworm moth morphology and morphometry. M.G. wishes to acknowledge the support of Philip A. Townsend at the University of Wisconsin–Madison.

Conflicts of Interest: The authors declare no conflict of interest.

References

1. Johnson, C.G. *Migration and Dispersal of Insects by Flight*; Methuen & Co. Ltd.: London, UK, 1969; p. 763. [CrossRef]

2. Schowalter, T. *Insect Ecology: An Ecosystem Approach*, 4th ed.; Academic Press: London, UK, 2016; p. 774.

3. Den Boer, P.J. Spreading of risk and stabilization of animal numbers. *Acta Biotheor.* **1968**, *18*, 165–194. [CrossRef] [PubMed]

4. Dingle, H.; Drake, V.A. What is migration? *Bioscience* **2007**, *57*, 113–121. [CrossRef]

5. Chapman, J.W.; Drake, V.A.; Reynolds, D.R. Recent insights from radar studies of insect flight. *Ann. Rev. Entomol.* **2011**, *56*, 337–356. [CrossRef] [PubMed]

6. Drake, V.A. The vertical distribution of macro-insects migrating in the nocturnal boundary layer: A radar study. *Bound.-Layer Meteorol.* **1984**, *28*, 353–374. [CrossRef]

7. Riley, J.R.; Reynolds, D.R. A long-range migration of grasshoppers observed in the Sahelian zone of Mali by two radars. *J. Anim. Ecol.* **1983**, *52*, 167–183. [CrossRef]

8. Westbrook, J.K.; Nagoshi, R.N.; Meagher, R.L.; Fleischer, S.J.; Jairam, S. Modeling seasonal migration of fall armyworm moths. *Int. J. Biometeorol.* **2016**, *60*, 255–267. [CrossRef] [PubMed]

9. Greenbank, D.O.; Schaefer, G.W.; Rainey, R.C. Spruce budworm (Lepidoptera: Tortricidae) moth flight and dispersal: New understanding from canopy observations, radar, and aircraft. *Mem. Entomol. Soc. Can.* **1980**, *112*, 1–49. [CrossRef]

10. Gray, D.R.; MacKinnon, W.E. Outbreak patterns of the spruce budworm and their impacts in Canada. *For. Chron.* **2006**, *82*, 550–561. [CrossRef]

11. Régnière, J.; Delisle, J.; Pureswaran, D.S.; Trudel, R. Mate-finding Allee effect in spruce budworm population dynamics. *Entomol. Exp. Appl.* **2013**, *146*, 112–122. [CrossRef]

12. Régnière, J.; Cooke, B.J.; Béchard, A.; Dupont, A.; Therrien, P. Dynamics and management of rising spruce budworm populations. *Forests* **2019**, *10*, 748. [CrossRef]

13. Wellington, W.G.; Henson, W.R. Notes on the effects of physical factors on the spruce budworm, *Choristoneura fumiferana* (Clem.). *Can. Entomol.* **1947**, *79*, 168–170. [CrossRef]

14. Rhainds, M.; Kettela, E.G. Oviposition threshold for flight in an inter-reproductive migrant moth. *J. Insect Behav.* **2013**, *26*, 850–859. [CrossRef]

15. Blais, J.R. Effects of the destruction of the current year's foliage of balsam fir on the fecundity and habits of flight of the spruce budworm. *Can. Entomol.* **1953**, *85*, 446–448. [CrossRef]

16. Van Hezewijk, B.; Wertman, D.; Stewart, D.; Beliveau, C.; Cusson, M. Environmental and genetic influences on the dispersal propensity of spruce budworm (*Choristoneura fumiferana*). *Agric. For. Entomol.* **2018**, *20*, 433–441. [CrossRef]

17. Régnière, J.; Garcia, M.; St-Amant, R. Modeling migratory flight in the spruce budworm: Circadian rhythm. *Forests* **2019**. in revision.

18. Reynolds, D.R.; Chapman, J.W.; Edwards, A.S.; Smith, A.D.; Wood, C.R.; Barlow, J.F.; Woiwod, I.P. Radar studies of the vertical distribution of insects migrating over southern Britain: The influence of temperature inversions on nocturnal layer concentrations. *Bull. Entomol. Res.* **2005**, *95*, 259–274. [CrossRef] [PubMed]

19. Boulanger, Y.; Fabry, F.; Kilambi, A.; Pureswaran, D.S.; Sturtevant, B.R.; Saint-Amant, R. The use of weather surveillance radar and high-resolution three dimensional weather data to monitor a spruce budworm mass exodus flight. *Agric. For. Meteorol.* **2017**, *234*, 127–135. [CrossRef]

20. Anderson, D.P.; Sturtevant, B.R. Pattern analysis of eastern spruce budworm *Choristoneura fumiferana* dispersal. *Ecography* **2011**, *34*, 488–497. [CrossRef]

21. Sturtevant, B.R.; Achtemeier, G.L.; Charney, J.J.; Anderson, D.P.; Cooke, B.J.; Townsend, P.A. Long-distance dispersal of spruce budworm (*Choristoneura fumiferana* Clemens) in Minnesota (USA) and Ontario (Canada) via the atmospheric pathway. *Agric. For. Meteorol.* **2013**, *168*, 186–200. [CrossRef]

22. Dudley, R. *The Biomechanics of Insect Flight*; Princeton University Press: Princeton, NJ, USA, 2000.

23. Harvey, G.T. Mean weight and rearing performance of successive egg clusters of eastern spruce budworm (Lepidoptera: Tortricidae). *Can. Entomol.* **1977**, *109*, 487–496. [CrossRef]

24. Eveleigh, E.S.; Lucarotti, C.J.; McCarthy, P.C.; Morin, B.; Royama, T.; Thomas, A.W. Occurrence and effects of *Nosema fumiferanae* infections on adult spruce budworm caught above and within the forest canopy. *Agric. For. Entomol.* **2007**, *9*, 247–258. [CrossRef]

25. Nealis, V.G.; Régnière, J. Fecundity and recruitment of eggs during outbreaks of the spruce budworm. *Can. Entomol.* **2004**, *136*, 591–604. [CrossRef]

26. Anderson, T.W.; Darling, D.A. Asymptotic theory of certain "goodness-of-fit" criteria based on stochastic processes. *Ann. Math. Stat.* **1952**, *23*, 193–212. [CrossRef]

27. Sanders, C.J. A summary of current techniques used for sampling spruce budworm populations and estimating defoliation in eastern Canada. *Can. For. Serv. Info. Rep.* **1980**, *O-X-306*.

28. Gupta, A.K.; Nadarajah, S. *Handbook of Beta Distribution and Its Applications*; CRC Press: Boca Raton, FL, USA, 2004.

29. Byrne, D.N.; Buchmann, S.L.; Spangler, H.G. Relationship between wing loading, wingbeat frequency and body mass in homopterous insects. *J. Exp. Biol.* **1988**, *135*, 9–23.

30. Deakin, M.A.B. Formulae for insect wingbeat frequency. *J. Insect Sci.* **2010**, *10*, 96. [CrossRef] [PubMed]

31. Farnsworth, E.G. Effects of ambient temperature, humidity, and age on wing-beat frequency of *Periplaneta* species. *J. Insect Physiol.* **1972**, *18*, 827–839. [CrossRef]

32. Farmery, M.J. The effect of air temperature on wingbeat frequency of naturally flying armyworm moth (*Spodoptera exempta*). *Entomol. Exp. Appl.* **1982**, *32*, 193–194. [CrossRef]

33. Unwin, D.M.; Corbet, S.A. Wingbeat frequency, temperature and body size in bees and flies. *Physiol. Entomol.* **1984**, *9*, 115–121. [CrossRef]

34. Oertli, J.J. Relationship of wing beat frequency and temperature during takeoff flight in temperate-zone beetles. *J. Exp. Biol.* **1989**, *145*, 321–338.

35. Foster, J.A.; Robertson, R.M. Temperature dependency of wing-beat frequency in intact and deafferented locusts. *J. Exp. Biol.* **1992**, *162*, 295–312.

36. Snelling, E.P.; Seymour, R.S.; Matthews, G.D.; White, C.R. Maximum metabolic rate, relative lift, wingbeat frequency and stroke amplitude during tethered flight in the adult locust *Locusta migratoria*. *J. Exp. Biol.* **2012**, *215*, 3317–3323. [CrossRef] [PubMed]

37. Huang, J.; Zhang, G.; Wang, Y. Effects of age, ambient temperature and reproductive status on wing beat frequency of the rice leafroller *Cnaphalocrocis medinalis* (Guenée) (Lepidoptera: Crambidae). *Appl. Entomol. Zool.* **2013**, *48*, 499–505. [CrossRef]

38. Chuine, I.; Régnière, J. Process-based models of phenology for plants and animals. *Annu. Rev. Ecol. Evol. Syst.* **2017**, *48*, 159–182. [CrossRef]

39. Sanders, C.J.; Wallace, D.R.; Lucuik, G.S. Flight activity of female spruce budworm (Lepidoptera: Tortricidae) at constant temperatures in the laboratory. *Can. Entomol.* **1978**, *110*, 627–632. [CrossRef]

40. Schaefer, G.W.; Allsopp, K. *Analysis and Interpretation of Ground-Based and Airborne Radar Data Collection for the Spruce Budworm Dispersal Studies 1973–1976 Inclusive*; Final Report No. 4. Part B. The Effects of Canopy Temperature and Meteorological Factors on the Number of Moths Entering the Air Space; Ecological Physics Research Group, Cranfield Institute of Technology: Bedfordshire, UK, 1978.

41. Malingowski, J.; Atkinson, D.; Fochesatto, J.; Cherry, J.; Stevens, E. An observational study of radiation temperature inversions in Fairbanks, Alaska. *Polar Sci.* **2014**, *8*, 24–39. [CrossRef]

42. Wellington, W.G. The light reactions of the spruce budworm, *Choristoneura fumiferana* Clemens (Lepidoptera: Tortricidae). *Can. Entomol.* **1948**, *80*, 56–82. [CrossRef]

43. Henson, W.R. Mass flights of the spruce budworm. *Can. Entomol.* **1951**, *83*, 240. [CrossRef]

44. Wood, C.R.; Clark, S.J.; Barlow, J.F.; Chapman, J.W. Layers of nocturnal insect migrants at high-altitude: The influence of atmospheric conditions on their formation. *Agric. For. Entomol.* **2010**, *12*, 113–121. [CrossRef]

45. Angevine, W.M. Transitional, entraining, cloudy, and coastal boundary layers. *Acta Geophys.* **2008**, *56*, 2–20. [CrossRef]

46. Angevine, W.M.; Tjernström, M.; Žagar, M. Modeling of the coastal boundary layer and pollutant transport in New England. *J. Appl. Meteorol. Climatol.* **2006**, *45*, 137–154. [CrossRef]

47. Régnière, J.; St-Amant, R.; Duval, P. Predicting insect distributions under climate change from physiological responses: Spruce budworm as an example. *Biol. Invasions* **2012**, *14*, 1571–1586. [CrossRef]

48. Pedgley, A.D.E. Concentration of flying insects by the wind. *Philos. Trans. R. Soc. Lond. B* **1990**, *328*, 631–653. [CrossRef]

49. Schaefer, G.W. An airborne radar technique for the investigation and control of migrating pest species. *Philos. Trans. R. Soc. Lond. B* **1979**, *287*, 459–465. [CrossRef]

50. Dickinson, R.B.B.; Haggis, M.J.; Rainey, R.C. Spruce budworm moth flight and storms: Case study of a cold front system. *J. Clim. Appl. Meteorol.* **1983**, *22*, 278–286. [CrossRef]

51. Pedgley, D.E.; Reynolds, D.R.; Riley, J.R.; Tucker, M.R. Flying insects reveal small-scale wind systems. *Weather* **1982**, *37*, 295–306. [CrossRef]

52. Isard, S.A.; Gage, S.H.; Comtois, P.; Russo, J.M. Principles of the atmospheric pathway for invasive species applied to soybean rust. *BioScience* **2005**, *55*, 851–861. [CrossRef]

53. Benjamin, S.G.; Weygandt, S.S.; Brown, J.M.; Hu, M.; Alexander, C.R.; Smirnova, T.G.; Olson, J.B.; James, E.P.; Dowell, D.C.; Grell, G.A.; et al. A North American hourly assimilation and model forecast cycle: The Rapid Refresh. *Mon. Weather Rev.* **2016**, *144*, 1669–1694. [CrossRef]

![forests logo] *forests*

MDPI

Article

Modeling Migratory Flight in the Spruce Budworm: Circadian Rhythm

Jacques Régnière [1,*], Matthew Garcia [2] and Rémi Saint-Amant [1]

[1]   Natural Resources Canada, Canadian Forest Service, 1055 PEPS street, Quebec, QC G1V 4C7, Canada; Remi.Saint-Amant@Canada.ca

[2]   Department of Forest and Wildlife Ecology, University of Wisconsin-Madison, Madison, WI 53706, USA; Matt.E.Garcia@gmail.com

*   Correspondence: Jacques.Regniere@Canada.ca; Tel.: +1-418-648-5257

Received: 30 July 2019; Accepted: 30 September 2019; Published: 5 October 2019

Abstract: The crepuscular (evening) circadian rhythm of adult spruce budworm (*Choristoneura fumiferana* (Clem.)) flight activity under the influence of changing evening temperatures is described using a mathematical model. This description is intended for inclusion in a comprehensive model of spruce budworm flight activity leading to the simulation of mass migration events. The model for the temporal likelihood of moth emigration flight is calibrated using numerous observations of flight activity in the moth's natural environment. Results indicate an accurate description of moth evening flight activity using a temporal function covering the period around sunset and modified by evening temperature conditions. The moth's crepuscular flight activity is typically coincident with the evening transition of the atmospheric boundary layer from turbulent daytime to stable nocturnal conditions. The possible interactions between moth flight activity and the evening boundary layer transition, with favorable wind and temperature conditions leading to massive and potentially successful migration events, as well as the potential impact of climate change on this process, are discussed.

Keywords: moths; migration; forest protection; spruce budworm; *Choristoneura fumiferana* (Clem.); early intervention strategy; modelling; circadian rhythm

---

1. Introduction

Migration is a biological imperative observed in many animals and generally well understood in ecological and evolutionary terms [1]. This paper focuses on observations of circadian rhythms of migratory flight activity in the adult spruce budworm (*Choristoneura fumiferana* (Clem.)) (Lepidoptera: Tortricidae) with consideration of near-surface and boundary layer meteorological processes. In a companion paper, we discussed the basic physics of spruce budworm (SBW) flight and developed a mathematical framework with which we can model adult moth migration [2]. The ultimate objective of this work is the development of a simulation model to allow both historical analysis and real-time prediction of SBW population dispersal though moth migration and oviposition. We aim for a model that accounts for source population distribution and density, seasonal phenology, circadian rhythms of activity, and interactions with meteorological conditions at the surface (in the host forest stands) and within the atmospheric boundary layer.

Numerous activities in moths and butterflies have been related to daily cycles of behavior and are often classified into diurnal (daytime), nocturnal (nighttime), and crepuscular (twilight, at dawn and/or dusk) rhythms [3,4]. These rhythms may be truly circadian, driven by an internal (endogenous) biological clock, or diel rhythms that are driven by exogenous cues such as light and temperature cycles [5]. The primary difference between these two types is that circadian rhythms continue, though not necessarily on a 24-h cycle, regardless of external cues, while diel rhythms may shift with a change in exogenous conditions or cease entirely in the absence of such cues. Circadian rhythms may be entrained by exogenous cues, such as the diurnal cycle of light intensity and temperature, and might otherwise drift, but these cycles remain active even in the absence of external cues [5]. In insects, the entraining cues may include the availability of food resources [6], light intensity [7–10], temperature [11], humidity [5], sex-related differences such as pheromone release and its own dependence on temperature [12–14], and predation risk [15–17].

The SBW moth's daily pattern of flight activity is a typical case of crepuscular circadian rhythm [18–20]. The adult SBW is poikilothermic [21] with its activity dependent on an optimal range of temperatures in the host forest. Temperature affects the timing of diel periodicities in the behavior of many lepidoptera [16], though a tendency toward purely diurnal or nocturnal activity may be suppressed by high or low temperatures during those times of day, respectively. In the adult SBW, male and female emigration flight is centered around sunset, with peak activity for males about 1.5 h later than for females, depending often on weather-related influences from temperature, wind speed, humidity, cloudiness, rainfall, and barometric pressure [20]. It is important here to distinguish SBW migratory flight behavior from activities such as mating. Environmental conditions that may not be conducive to flight above the forest canopy may still promote levels of flight activity sufficient for mating within the forest canopy, and typically on the same crepuscular cycle [18] but also displaying a minor peak in the morning hours [20]. However, successful migration flights depend on an alignment of both the timing of SBW moth activity and favorable wind and temperature conditions in the atmospheric boundary layer above the forest canopy. Because mating occurs soon after adult emergence [10] and mated and fully gravid females are usually too heavy to fly until at least their first cycle of oviposition [10,22], we may infer two things when we witness migration events. First, conditions in both the forest canopy and the atmospheric boundary layer favor flight activity at that time. Second, conditions on prior evenings may not have been favorable for flight above the forest canopy but at least supported mating activity and initial oviposition that also depend on temperature, so that lighter, less-gravid females can undertake migration.

The time of day at which SBW moths depart on their migratory flights therefore depends on crepuscular circadian rhythms and determines, to a large extent, the atmospheric conditions under which this movement is attempted and the distance that the insects travel [10]. We should perhaps differentiate here between migration on clear nights, which the mathematical model described below addresses, and migration flights that may be triggered by cues such as the passage of a cold front [23,24] or thunderstorm surface outflow [8,25]. Both of these potential triggers are commonly observed over land areas in the late afternoon and evening hours in the summer, when the crepuscular activity of SBW moths is at its peak. Historical observation of a SBW migration flight triggered by a solar eclipse suggests that cues other than time of day can influence the timing of migratory flight in this species [11]. Observations using both entomological and meteorological radar [26–33] have revealed the behavior of many migrating insects; such observations may enable the detection of moth dispersal events triggered under particular weather conditions (e.g. storm winds and the accompanying precipitation). However, it is not clear whether the high turbulence associated with fronts and storm outflows is a trigger for moth migration or simply facilitates the liftoff of a larger proportion of moths in strong updrafts. In many cases, the moths in flight can be drawn into frontal circulations and convective cells and subjected to low temperatures or wetting in rain events. Such conditions cause the SBW moth to fold its wings and drop to the surface [8], ending the migration flight and potentially leading to high mortality. Massive and successful migration events have been observed on generally calm evenings following a hot afternoon, progressing late into the night, with favorable (i.e., mild-to-strong) and coherent winds above the forest canopy [32].

For our effort at flight modeling, we relate the more general crepuscular circadian rhythm of the SBW in its adult stage, including behaviors such as mating within the forest canopy, to its eventual evening liftoff, flight above the canopy, and potentially successful nocturnal migration. In this paper we develop a mathematical model to relate SBW crepuscular activity to time of day and temperature, and we calibrate this model with observations of SBW moth flights made from scaffolds in North

American boreal forest stands under natural conditions. The applications of these findings to our model of SBW migratory flight are then discussed.

2. Materials and Methods

2.1. Mathematical Model

We consider two factors as determinants of migratory flight of the SBW in this model: Time of day and temperature. Temperature directly affects the flight activity of SBW moths through wingbeat frequency [2]. The literature also suggests that migratory flight may be inhibited when temperature remains above a certain threshold well into the night, and that migratory flight does not occur much after midnight [10].

We express the circadian rhythm of moth liftoff activity as a quasi-Gaussian pattern centered near (though not necessarily at) the time of sunset t_s. This pattern can be shifted by ambient temperature conditions. We thus develop a time transformation τ around the central time t_c

$$\tau = \frac{t - t_c}{t_c - t_0} \text{ if } t \le t_c \text{ or } \frac{t - t_c}{t_m - t_c} \text{ if } t > t_c \tag{1}$$

with the start and end of the liftoff period denoted t_0 and t_m, respectively, and the full likelihood distribution of moth liftoff described by

$$f_t = \begin{cases} \left(1 - \tau^2\right)^2 & \text{when } -1 \le \tau \le 1 \\ 0 & \text{otherwise} \end{cases} \tag{2}$$

where $f_t = 1$ is the maximum at $t = t_c$ or $\tau = 0$ (Figure 1).

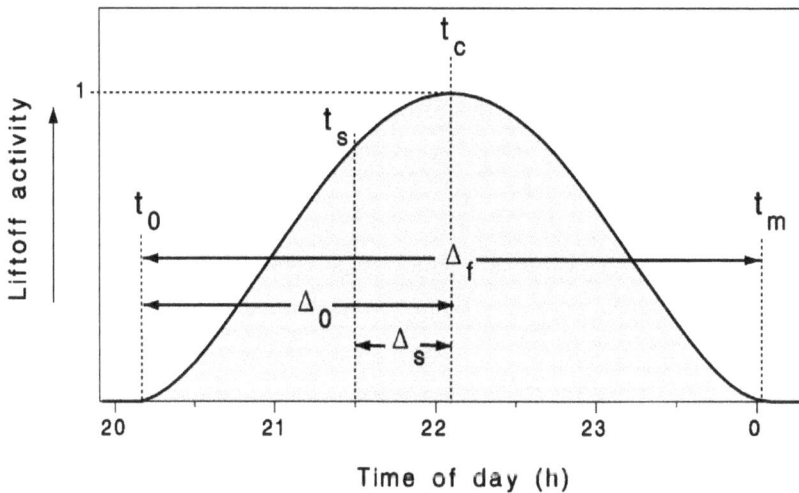

Figure 1. Circadian model of crepuscular migration liftoff activity (Equation (2)) illustrating terms used to define τ in Equation (1), with t_s: Time of sunset; Δ_s: Time between sunset t_s and peak flight t_c; Δ_0: Time between first flight t_0 and peak flight t_c; Δ_f: Duration of the migration liftoff period; and t_m: End of flight period.

The time difference between the onset of the moth liftoff period t_0 and the time of peak likelihood t_c is used to define Δ_0 (in hours):

$$\Delta_0 = t_0 - t_c \tag{3}$$

The end of the liftoff period t_m can be used with the onset time t_0 to define the full duration of liftoff activity Δ_f (in hours):

$$\Delta_f = t_m - t_0 \qquad (4)$$

The function in Figure 1 is not necessarily symmetric about the peak time t_c, by virtue of Equation (1). Expecting the peak of the moth liftoff period to occur after the time of local sunset, we can define the time difference Δ_s (in hours):

$$\Delta_s = t_c - t_s \qquad (5)$$

where the sunset time t_s depends on latitude and time of year.

Equation (2) can be viewed as a probability density function for moth liftoff depending on time-of-day and subject to the offsets and durations defined in Equations (3)–(5). To use this in a discrete time simulation context, we need to express Equation (2) as a cumulative distribution function ranging from 0 at t_0 to 1 at t_m so that we can calculate the probability of moth liftoff in a short interval. This definite integral is given by:

$$F_t = \int_{-1}^{1} f_t d\tau = \int_{-1}^{1} \left(1 - \tau^2\right)^2 d\tau = \frac{1}{2C}\left(C + \tau - \frac{2}{3}\tau^3 + \frac{1}{5}\tau^5\right) \qquad (6)$$

where $C = 1 - 2/3 + 1/5 = 0.533333$ is a constant of integration ensuring that F_t remains strictly in the range $(0,1)$ for τ (defined by Equation (1)) in the range $(-1, 1)$. The probability of a moth taking off in an interval $\Delta\tau$ is therefore:

$$p(\tau, \tau + \Delta\tau) = \frac{1}{2C}\left\{\Delta\tau - \frac{2}{3}\left[(\tau + \Delta\tau)^3 - \tau^3\right] + \frac{1}{5}\left[(\tau + \Delta\tau)^5 - \tau^5\right]\right\} \qquad (7)$$

To completely specify this model, we must estimate the values of the three duration and offset parameters: Δ_0 for Equation (3), Δ_f for Equation (4), and Δ_s for Equation (5).

The circadian likelihood of a moth attempting liftoff in a time interval $\Delta\tau$ is given by Equation (7), but the actual physical capability for flight is determined by temperature with respect to the moth's weight and forewing area. It was established in [2] that wingbeat frequency is a sigmoid function of ambient temperature T (°C),

$$v(T) = \frac{72.5}{1 + \exp^{-0.115(T-23)}} \qquad (8)$$

and that the wingbeat frequency required for moth liftoff is a function of its weight M and (single) forewing surface area A,

$$v_L = 167.5\frac{\sqrt{M}}{A} \qquad (9)$$

The actual diel pattern of migration on any given evening is therefore a potentially complex interaction between the circadian rhythm of activity (given by Equation (7)) and the effect of decreasing evening temperature on flight capability (Equations (8) and (9)). Because we are concerned with the deposition of eggs at the end of a migratory flight, leading to the next generation of SBW larvae in potentially new locations, we also calculate the proportion of initial fecundity carried away by emigrating females (G) based on their mass and forewing area (see [2]):

$$G = \frac{\ln(M) + 6.4648 - 2.14A}{0.9736 + 1.3049A} \qquad (10)$$

2.2. Observations and Model Calibration

The time of local sunset and the ambient temperature are external inputs to this system, but several terms defining Equations (3)–(5) (Δ_0, Δ_f, and Δ_s, respectively) must be determined from observations. For this purpose, we assembled 22 historical datasets of diel SBW moth flight observations. Two of these datasets were published previously by Greenbank et al. [10]. The other 20 datasets were obtained from similar observations made in the vicinity of Sault Ste. Marie and Black Sturgeon Lake, ON, by the late C.J. Sanders and his team in 1976 (two datasets), 1987 (eight datasets), and 1989 (10 datasets). In all cases, the number of moths ascending into emigration flight was observed over short periods of time (typically 5 min) from tall scaffolds above the canopy and recorded at intervals of 15 min from early evening (prior to sunset) to total darkness. On some nights, night-vision (infrared) devices were used after total darkness to observe late flying moths, primarily to ensure that those were not numerous. Air temperature was recorded at the scaffold location at the same frequency. Sunset times for each location and observation date were obtained from www.timeanddate.com/sun/canada/ (last accessed 1 October 2019). Because these canopy-level observations did not distinguish between males and females, and because Greenbank et al. [10] reported that the sex ratio of moths caught during emigration liftoff did not vary much with time of night, we made no attempt to model explicitly sex differences in the diel periodicity of SBW flight.

We used all of these datasets to estimate Δ_0 and Δ_s from the time at which the first (t_0) and the median (50%; t_c) migrating moths were observed on each observation night. The value of Δ_f (duration of the diel liftoff period) was determined from the time of first (t_0) and last (t_m) emigration. The values of Δ_s were related by regression analysis to T_{19h30} the ambient air temperature (°C) at 19h30 local DST, prior to the earliest observed emigration flights:

$$\Delta_s = p_1 + p_2 T_{19h30} \tag{11}$$

Observed values of Δ_0 were related by regression analysis to the observed values of Δ_s:

$$\Delta_0 = p_3 + p_4 \Delta_s \tag{12}$$

The observed duration of the flight period (Δ_f in hours from first to last moth seen emigrating) was related to the observed values of Δ_0. Given pronounced heteroscedasticity in these observations, we expressed this relationship as

$$\frac{\Delta_f}{\Delta_0} = p_5 \tag{13}$$

However, because cooling evening temperatures have as a consequence the apparent reduction of the flight period, we used an optimization algorithm to estimate the value of a sixth parameter k_f that, by multiplying with the value of p_5, maximized the R^2 between observed and simulated proportions of moths taking flight (with a precision of ±5%). The potential duration of the flight period is then given by

$$\hat{\Delta}_f = k_f p_5 \Delta_0 \tag{14}$$

2.3. Simulations

To obtain the simulated proportions of moths taking flight we simulated 10,000 individual SBW moths (50% female) using the method described in [2], with mean and variances of wing surface areas and weights corresponding to the moths caught in canopy traps above a forest stand near Lac des Huit-Milles near Amqui, QC, during the 1989 and 1990 SBW flight seasons. Equation (7) was used to determine the proportion of individuals that could potentially lift off at a given time, and then Equations (8) and (9) were used to determine which of those moths could actually lift off given the ambient temperature and their individual mass and wing area. The proportion and sex ratio of moths taking off and the proportion of total fecundity carried away by emigrating females (Equation (10)) were compiled at intervals of 15 min.

3. Results

We found a significant linear relationship between T_{19h30} and the median timing of moth emigration Δ_s (Figure 2a) as described by Equation (11). There was also a significant negative relationship between the timing of the onset of emigration Δ_0 and the value of Δ_s (Figure 2b), as described by Equation (12). The average value of parameter p_5 in Equation (13) provided a good description of the relationship between observed flight duration Δ_f and the time Δ_0 between first flight and peak flight (Figure 2c). Using the calibrated values for model parameters p_1 through p_5 (Table 1), the goodness-of-fit (R^2) between observed and simulated cumulative liftoff patterns improved from 0.72 to an asymptotic maximum of 0.775 using $k_f = 1.35$ (Figure 2d) in Equation (14). The resulting simulated patterns of liftoff timing corresponded well with field observations (Figure 3).

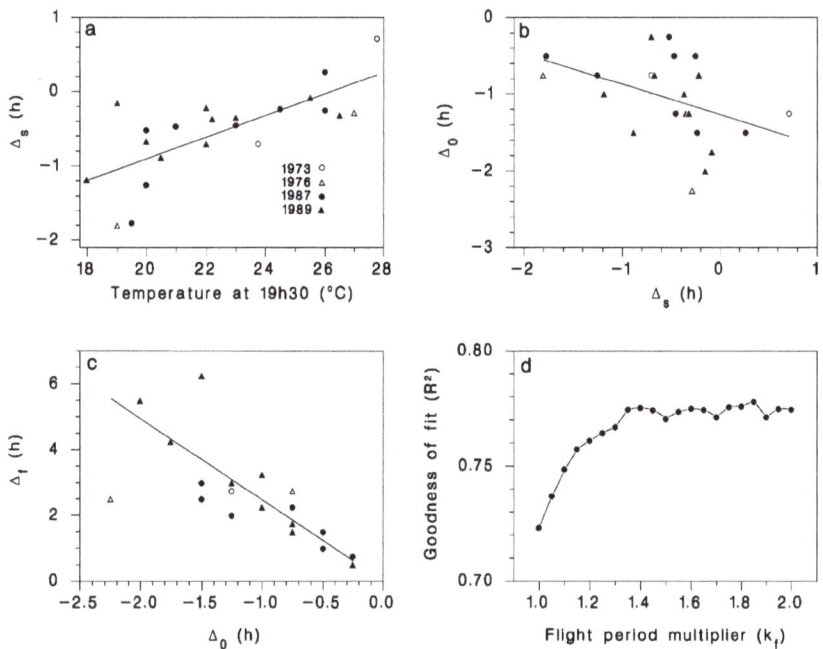

Figure 2. (**a**) Relationship between T_{19h30} the temperature at 19h30 local DST and Δ_s the shift between the time at which 50% of migrating moths were observed and the time of sunset t_s. Line is Equation (11). (**b**) Relationship between Δ_s and the shift between the first migrating moth and sunset Δ_0. Line is Equation (12). (**c**) Relationship between Δ_0 and the observed duration of the moth migration period Δ_f. Line is Equation (13). (**d**) Optimization (maximum R^2) of the value of parameter k_f in Equation (14). The value $k_f = 1.35$ yielded $R^2 = 0.775$ between observed and simulated cumulative proportion of moths emigrating over 22 evenings (see Figure 3).

Table 1. Model parameter calibration and regression results.

Parameter	Calibration Value	Equation	Regression Statistics
p_1 p_2	-3.8 ± 0.7 h 0.145 ± 0.031 h/°C	(11)	F = 22.5; df = 1,20; R^2 = 0.529; $p < 0.001$
p_3 p_4	-1.267 ± 0.146 h -0.397 ± 0.187	(12)	F = 4.5; df = 1,20; R^2 = 0.183; $p < 0.047$
p_5	-2.465 ± 0.152	(13)	
k_f	1.35 ± 0.025	(14)	$R^2 = 0.775$

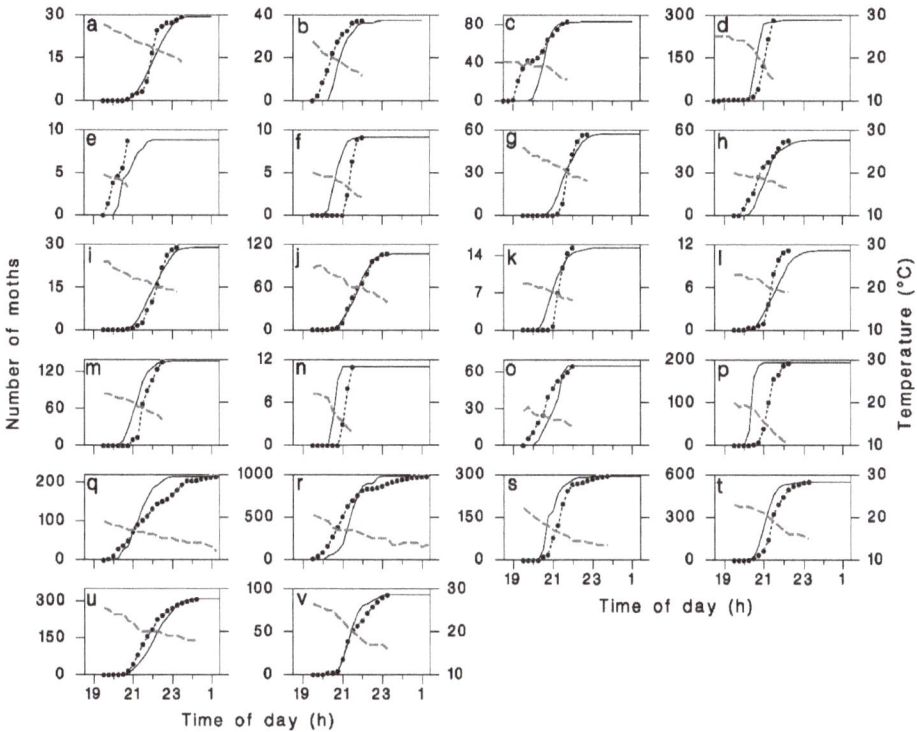

Figure 3. Observed (closed circles with dotted lines) and simulated (solid lines) cumulative moths flights with observed evening temperatures (dashed grey lines). (**a**) and (**b**) 9 and 16 July 1973 using data from [19]. (**c**) and (**d**) 2 and 5 July 1976, from observations by C.J. Sanders near Sault Ste. Marie, Ontario. (**e–v**): 18 individual nights in July 1987 and 1989, from observations by C.J. Sanders near Black Sturgeon Lake, Ontario.

The circadian pattern of migration flight is strongly affected by temperature in two ways. As ambient temperature increases, the entire curve illustrated in Figure 1 shifts later relative to sunset and increases in overall duration (Figure 4a). Based on this model, the likelihood of crepuscular flight should peak nearly 0.5 h before sunset for a relatively cool evening (e.g., $T_{19h30} = 20\,°C$), nearly 0.5 h after sunset for a warmer evening (e.g., $T_{19h30} = 25\,°C$), and even later, nearly 1.5 h after sunset, for a relatively hot evening (e.g., $T_{19h30} = 30\,°C$). However, the relationships between the wingbeat frequency and temperature, body weight, and wing area also determine how many moths can lift off at any given time. The resulting migration pattern, and its amplitude, are the composite of these effects (Figure 4a). Interestingly, as a result of the liftoff wingbeat relationship (Equations (8) and (9)), the sex ratio (proportion females) of emigrating moths varies with temperature and time during the evening (Figure 4b). Females make up the larger portion of migrants on cooler evenings, and later in the evening, contradicting the observation reported by Greenbank et al. [10] that the sex ratio of moths caught during migration liftoff did not vary much with time of night. The fecundity carried away by emigrating females also varies as a function of temperature and time through the evening (Figure 4c). Heavier females (i.e., those with a greater proportion of their remaining fecundity) can only lift off at warmer temperatures and thus typically earlier in the evening, while lighter females emigrating in cooler air at the end of the evening tend to have lower gravidity.

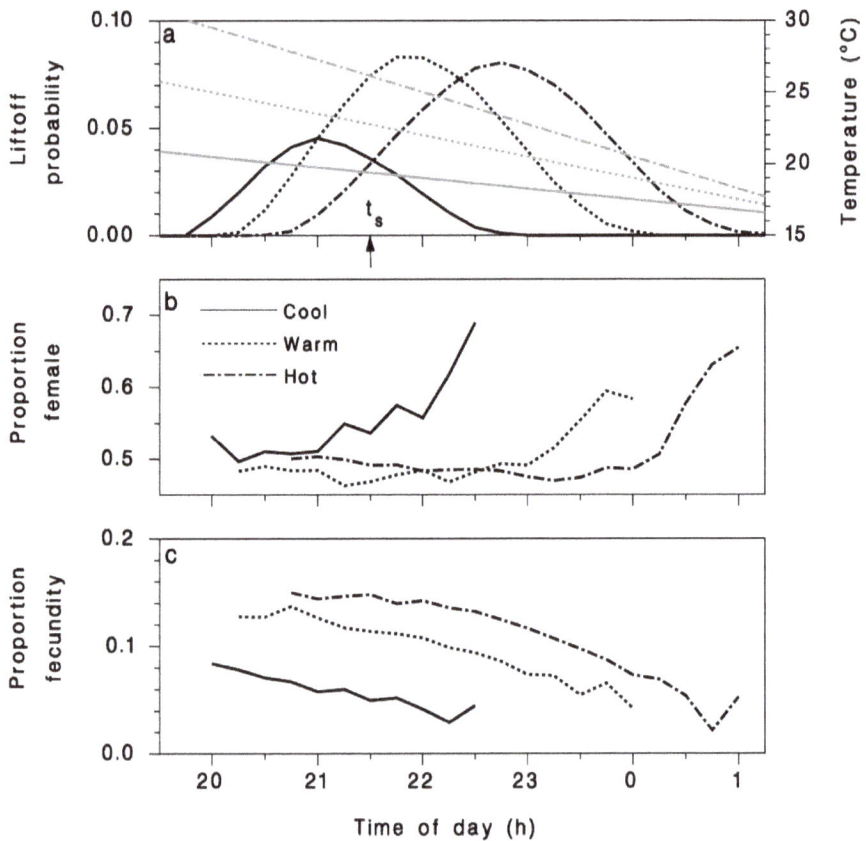

Figure 4. (**a**) Proportion of moths (black curves) emigrating under three temperature regimes (gray lines) on cool (solid), warm (dotted), and hot (dash-dotted) nights. (**b**) Proportion female among the moths emigrating at each time step. (**c**) Average proportion of initial fecundity carried away by emigrating females (Equation (10)).

4. Discussion

The interaction of a crepuscular circadian rhythm of activity around sunset and the temporal variation of ambient temperature determines the diel pattern of emigration flight in adult SBW moths. These moths are believed capable of flight at temperatures between 14 and 30 °C [2,10,34]. Below that range, the moths are lethargic and cannot fan their wings fast enough to lift their body weight in flight. Above that range, the moths probably enter a state of torpor that protects them against heat damage [34]. As with circadian rhythms in many other species, temperature also affects the onset and duration of diel activity [13,14]. On cool evenings, the diel pattern of emigration starts and peaks early relative to sunset, and on warm evenings the pattern is shifted to later hours. This dependence of activity on temperature is likely adaptive, allowing migration to occur under a wider variety of atmospheric conditions. It may also ensure that emigration is synchronous for the moths in a given area, serving to keep the migrants together. Such mass flights are important to the population dynamics of this insect species: immigrants, when numerous enough, can produce progeny in numbers that overwhelm the various natural enemies keeping populations at endemic levels during the subsequent generation [35]. In addition, the migrating moths that begin their flights later on warm nights may still have sufficient time to travel long distances to new host areas, especially with favorable wind support, without running the risk of overly long flight periods and exhausted energy reserves. It is generally believed that migratory flights of SBW do not continue much past sunrise [10], although this assumption needs empirical support.

The total number of moths that emigrate from a given location, and the number of eggs that females can carry away by migration, are both strongly affected by evening temperatures on nights favorable to flight activity. Coherent wind patterns in the atmospheric boundary layer can develop from various phenomena under both stormy and clear conditions, but to our knowledge the meteorological support for insect migration events on otherwise clear nights has not been examined in prior studies. The evening transition from the turbulent daytime atmospheric boundary layer to the often-stable nighttime boundary layer over land, starting generally 1–2 h before sunset, is a subject of particular interest to both scientific and engineering communities [36,37]. The structure of the stable, stratified boundary layer has been examined in extensive detail [38]. However, describing the evening transition near sunset to that stable state remains a complex task, primarily because both the spatially heterogeneous land surface and the free atmosphere 1–2 km above the surface can exert complex dynamic influences on the lowest several hundred meters of the atmosphere. Detailed work [39–43] has described a three-stage evening transition of the boundary layer from convective (daytime) to stratified (nighttime) conditions, generally through the development of an inversion layer (where temperature increases with height) that grows upward from the surface after sunset. As the growth of the stable boundary layer eventually ceases in the hours after sunset, a coherent layer of high wind speeds known as the low-level jet often develops at the top of the temperature inversion in association with both the large-scale synoptic pattern and surface physiography (e.g., coastlines, mountain barriers, land cover heterogeneity) [37,44].

As the surface cools quickly in post-sunset hours, the migrating SBW moths ascend above the forest canopy and eventually concentrate [2,9,10] at the top of the inversion layer with higher temperatures (favorable to higher moth wingbeat frequencies) and stronger winds. Shear-based turbulence beneath the low-level jet may intermittently disrupt the upper levels of the stable layer well into the night [45–47]. Researchers have identified that turbulence associated with areas and layers of strong winds can serve as an indicator that helps orient flying insects in a common downwind direction [48]. In the absence of additional disruption by an overnight storm or frontal passage, the stable nocturnal boundary layer generally persists until its turbulent erosion shortly after sunrise the next morning. Insect crepuscular activity has evolved to take advantage of these dynamic conditions in the sunset transition of the atmospheric boundary layer for nocturnal migration flights. Similar patterns of behavior have been observed and studied in various species, and under similar weather conditions, on six continents around the world: Africa [49], Australia [26,27,50], East Asia [51,52], Europe [28,29,53–56], South America [57], and North America [8,10,31–33].

5. Conclusions

Our analysis of evening crepuscular flight activity in SBW moths is an essential component of our larger model of migration activity, deepening our understanding of spatiotemporal patterns and processes in the course of SBW population fluctuations and outbreak events. It is in the evening development and overnight persistence of the stable nocturnal boundary layer, amid otherwise calm synoptic conditions, that we find strong support for coherent, long-distance SBW migration events resulting from the temporal alignment of crepuscular circadian activity and favorable boundary layer temperature and wind conditions above the forest canopy. Where a nocturnal low-level jet develops above the boundary layer, greater shear turbulence is produced just below that jet in the warm inversion, with those combined conditions favorable to coherent long-distance migratory flight for large numbers of insects. We are thus interested in continuing our examination of the evening crepuscular conditions favorable to liftoff from the forest canopy that can result in mass migration events. We also intend to focus on the dynamic conditions in the evening transition and nocturnal boundary layer above the forest canopy that lead to the coherent layering [26,28,29,50,53–55,58], orientation [9,10,58–60], and long-distance transport [27,61] of migrating insects.

Climate change may affect the circadian rhythm of SBW emigration. However, the coarse representations of the atmospheric boundary layer provided by current climate models [62] do not allow us to predict those effects with any accuracy using the emigration model described here. The most obvious evidence of changing climate has been found in an increase of daily minimum temperatures [63], with less pronounced increases of daily maximum temperatures. In our model, SBW moths attempt to optimize their liftoff time with respect to evening temperature to remain in the warmer air that supports the wingbeat of heavier moths before the cooler nocturnal boundary layer is established and flight activity is curtailed. All else remaining constant, increasing nighttime minimum temperature allows for slower evening cooling and a corollary lengthening of the migration flight period. Additional data of the type used in the present analysis may show such a change over the past several decades and in the future. Warmer nighttime temperatures may also destabilize the nocturnal boundary layer to some extent [64]. An overall warmer and well-mixed nocturnal boundary layer would allow SBW moths to fly at lower altitudes, which would otherwise remain too cold for flight activity. A less stable nocturnal boundary layer implies greater turbulence, which can bolster some moths flying in updrafts while driving others toward the surface in downdrafts. Finally, diminished nocturnal boundary layer stability would likely result in a more diffuse upper boundary of the near-surface inversion layer, which may not provide sufficient thermal gradients for the formation of the low-level jet. Without the aid of high wind speeds in and near that nocturnal low-level jet, far shorter SBW flight distances could be expected in mass migration events. We look forward to the opportunity to examine such changes in boundary layer dynamics using detailed numerical models (e.g., [65]) and the potential effects of changing evening transition and nocturnal boundary-layer conditions on our insect flight modeling efforts.

Author Contributions: J.R. developed the model and parameter estimates; M.G. helped with parameter estimation, literature review and writing; R.S.-A. helped with programming and writing. Conceptualization, J.R. and R.S.-A.; Methodology, J.R.; Software, R.S.-A.; Validation, J.R., M.G. and R.S.-A.; Formal Analysis, J.R.; Investigation, J.R.; Resources, J.R.; Data Curation, J.R.; Writing—Original Draft Preparation, J.R. and M.G.; Writing—Review & Editing, J.R., M.G. and R.S.-A.; Visualization, J.R.; Funding Acquisition, J.R.

Funding: This research was funded by the natural resources and forest departments of Newfoundland, Nova Scotia, New Brunswick, Quebec, Ontario, Manitoba, Saskatchewan, Alberta and British Columbia. Funds were also provided by the Atlantic Canada Opportunities Agency, and under the US/Canada Forest Health & Innovation Summit Initiative sponsored by the USDA Forest Service, Canadian Forest Service of Natural Resources Canada, and the USA Endowment for Forestry and Communities, with cooperation by SERG-International.

Acknowledgments: The authors acknowledge Brian R. Sturtevant for his helpful comments on the manuscript. M.G. wishes to acknowledge the support of Philip A. Townsend at the University of Wisconsin–Madison.

Conflicts of Interest: The authors declare no conflict of interest.

References

1. Dingle, H. *Migration: The Biology of Life on the Move*, 2nd ed.; Oxford University Press: New York, NY, USA, 2014; 326p. [CrossRef]

2. Régnière, J.; Delisle, J.; Sturtevant, B.R.; Garcia, M.; St-Amant, R. Modeling migratory flight in the spruce budworm: Temperature constraints. *Forests* **2019**, *10*, 802. [CrossRef]

3. Hu, G.; Lim, K.S.; Reynolds, D.R.; Reynolds, A.M.; Chapman, J.W. Wind-related orientation patterns in diurnal, crepuscular and nocturnal high-altitude insect migrants. *Front. Behav. Neurosci.* **2016**, *10*, 32. [CrossRef] [PubMed]

4. Kawahara, A.Y.; Plotkin, D.; Hamilton, C.A.; Gough, H.; St Laurent, R.; Owens, H.L.; Homziak, N.T.; Barber, J.R. Diel behavior in moths and butterflies: A synthesis of data illuminates the evolution of temporal activity. *Org. Divers. Evol.* **2018**, *18*, 13–27. [CrossRef]

5. Rund, S.; O'Donnell, A.; Gentile, J.; Reece, S. Daily rhythms in mosquitoes and their consequences for malaria transmission. *Insects* **2016**, *7*, 14. [CrossRef]

6. Broadhead, G.T.; Basu, T.; von Arx, M.; Raguso, R.A. Diel rhythms and sex differences in the locomotor activity of hawkmoths. *J. Exp. Biol.* **2017**, *220*, 1472–1480. [CrossRef] [PubMed]

7. Wellington, W.G. The light reactions of the spruce budworm, *Choristoneura fumiferana* Clemens (Lepidoptera: Tortricidae). *Can. Entomol.* **1948**, *80*, 56–82. [CrossRef]

8. Henson, W.R. Mass flights of the spruce budworm. *Can. Entomol.* **1951**, *83*, 240. [CrossRef]

9. Schaefer, G.W. Radar observations of insect flight. *Symp. R. Entomol. Soc. Lond.* **1976**, *7*, 157–197.

10. Greenbank, D.O.; Schaefer, G.W.; Rainey, R.C. Spruce budworm (Lepidoptera: Tortricidae) moth flight and dispersal: New understanding from canopy observations, radar and aircraft. *Mem. Entomol. Soc. Can.* **1980**, *112*, 1–49. [CrossRef]

11. Kipp, L.R.; Lonergan, G.C.; Bell, W.J. Male periodicity and the timing of mating in the spruce budworm (Lepidoptera: Tortricidae): Influence of population density and temperature. *Environ. Entomol.* **1995**, *24*, 1150–1159. [CrossRef]

12. Cardé, R.T.; Roelofs, W.L. Temperature modification of male sex pheromone response and factors affecting female calling in *Holomelina immaculata* (Lepidoptera: Arctiidae). *Can. Entomol.* **1973**, *105*, 1505–1512. [CrossRef]

13. Cardé, R.T.; Comeau, A.; Baker, T.C.; Roelofs, W.L. Moth mating periodicity: Temperature regulates the circadian gate. *Experientia* **1975**, *31*, 46–48. [CrossRef] [PubMed]

14. Comeau, A.; Cardé, R.T.; Roelofs, W.L. Relationship of ambient temperatures to diel periodicities of sex attraction in six species of Lepidoptera. *Can. Entomol.* **1976**, *108*, 415–418. [CrossRef]

15. Alerstam, T.; Chapman, J.W.; Bäckman, J.; Smith, A.D.; Karlsson, H.; Nilsson, C.; Reynolds, D.R.; Klaassen, H.G.; Hill, J.K. Convergent patterns of long-distance nocturnal migration in noctuid moths and passerine birds. *Proc. R. Soc. B Biol. Sci.* **2011**, *278*, 3074–3080. [CrossRef]

16. Krauel, J.J.; Westbrook, J.K.; McCracken, G.F. Weather-driven dynamics in a dual-migrant system: Moths and bats. *J. Anim. Ecol.* **2015**, *84*, 604–614. [CrossRef]

17. Krauel, J.J.; Brown, V.A.; Westbrook, J.K.; McCracken, G.F. Predator–prey interaction reveals local effects of high-altitude insect migration. *Oecologia* **2018**, *186*, 49–58. [CrossRef]

18. Sanders, C.J. Daily activity patterns and sex pheromone specificity as sexual isolating mechanisms in two species of *Choristoneura* (Lepidoptera: Tortricidae). *Can. Entomol.* **1971**, *103*, 498–502. [CrossRef]

19. Sanders, C.J.; Lucuik, G.S. Effects of photoperiod and size on flight activity and oviposition in the eastern spruce budworm (Lepidoptera: Tortricidae). *Can. Entomol.* **1975**, *107*, 1289–1299. [CrossRef]

20. Simmons, G.A.; Chen, C.W. Application of harmonic analysis and polynomial regression to study flight activity of *Choristoneura fumiferana* (Clem.) (Lepidoptera: Tortricidae) in the field. Abstracts, Forty-Seventh Annual Meeting, Eastern Branch Entomological Society of America. *J. N. Y. Entomol. Soc.* **1975**, *83*, 266.

21. Régnière, J.; Powell, J.; Bentz, B.; Nealis, V. Effects of temperature on development, survival and reproduction of insects: Experimental design, data analysis and modeling. *J. Insect Physiol.* **2012**, *58*, 634–647. [CrossRef]

22. Wellington, W.G.; Henson, W.R. Notes on the effects of physical factors on the spruce budworm, *Choristoneura fumiferana* (Clem.). *Can. Entomol.* **1947**, *79*, 168–170. [CrossRef]

23. Dickison, R.B.B.; Haggis, M.J.; Rainey, R.C. Spruce budworm moth flight and storms: Case study of a cold front system. *J. Clim. Appl. Meteorol.* **1983**, *22*, 278–286. [CrossRef]

24. Dickison, R.B.B.; Haggis, M.J.; Rainey, R.C.; Burns, L.M.D. Spruce budworm moth flight and storms: Further studies using aircraft and radar. *J. Clim. Appl. Meteorol.* **1986**, *25*, 1600–1608. [CrossRef]

25. Pedgley, D.E.; Reynolds, D.R.; Riley, J.R.; Tucker, M.R. Flying insects reveal small-scale wind systems. *Weather* **1982**, *37*, 295–306. [CrossRef]

26. Drake, V.A. The vertical distribution of macro-insects migrating in the nocturnal boundary layer: A radar study. *Bound. Layer Meteorol.* **1984**, *28*, 353–374. [CrossRef]

27. Drake, V.A. Radar observations of moths migrating in a nocturnal low-level jet. *Ecol. Entomol.* **1985**, *10*, 259–265. [CrossRef]

28. Reynolds, D.R.; Chapman, J.W.; Edwards, A.S.; Smith, A.D.; Wood, C.R.; Barlow, J.F.; Woiwod, I.P. Radar studies of the vertical distribution of insects migrating over southern Britain: The influence of temperature inversions on nocturnal layer concentrations. *Bull. Entomol. Res.* **2005**, *95*, 259–274. [CrossRef]

29. Reynolds, D.R.; Smith, A.D.; Chapman, J.W. A radar study of emigratory flight and layer formation by insects at dawn over southern Britain. *Bull. Entomol. Res.* **2008**, *98*, 35–52. [CrossRef] [PubMed]

30. Chapman, J.W.; Drake, V.A.; Reynolds, D.R. Recent insights from radar studies of insect flight. *Ann. Rev. Entomol.* **2011**, *56*, 337–356. [CrossRef]

31. Westbrook, J.K.; Eyster, R.S.; Wolf, W.W. WSR-88D doppler radar detection of corn earworm moth migration. *Int. J. Biometeorol.* **2014**, *58*, 931–940. [CrossRef]

32. Boulanger, Y.; Fabry, F.; Kilambi, A.; Pureswaran, D.S.; Sturtevant, B.R.; Saint-Amant, R. The use of weather surveillance radar and high-resolution three-dimensional weather data to monitor a spruce budworm mass exodus flight. *Agric. For. Meteorol.* **2017**, *234*, 127–135. [CrossRef]

33. Westbrook, J.K.; Eyster, R.S. Doppler weather radar detects emigratory flights of noctuids during a major pest outbreak. *Remote Sens. Appl. Soc. Environ.* **2017**, *8*, 64–70. [CrossRef]

34. Sanders, J.C.; Wallace, D.R.; Luicuik, G.S. Flight activity of female eastern spruce budworm (Lepidoptera: Tortricidae) at constant temperatures in the laboratory. *Can. Entomol.* **1978**, *110*, 627–632. [CrossRef]

35. Régnière, J.; Cooke, B.; Béchard, A.; Dupont, A.; Therrien, P. Dynamics and management of rising outbreak spruce budworm populations. *Forests* **2019**, *10*, 748. [CrossRef]

36. Baklanov, A.A.; Grisogono, B.; Bornstein, R.; Zilitinkevich, S.S.; Taylor, P.; Larsen, S.E.; Rotach, M.W.; Fernando, H.J.S. The nature, theory, and modeling of atmospheric planetary boundary layers. *Bull. Am. Meteorol. Soc.* **2011**, *92*, 123–128. [CrossRef]

37. Angevine, W.M. Transitional, entraining, cloudy, and coastal boundary layers. *Acta Geophysica* **2008**, *56*, 2–20. [CrossRef]

38. Mahrt, L. Stably stratified atmospheric boundary layers. *Ann. Rev. Fluid Mech.* **2014**, *46*, 23–45. [CrossRef]

39. Mahrt, L. The early evening boundary layer transition. *Quart. J. R. Meteorol. Soc.* **1981**, *107*, 329–343. [CrossRef]

40. Mahrt, L. Nocturnal boundary-layer regimes. *Bound. Layer Meteorol.* **1998**, *88*, 255–278. [CrossRef]

41. Mahrt, L. The near-surface evening transition. *Quart. J. R. Meteorol. Soc.* **2017**, *143*, 2940–2948. [CrossRef]

42. Acevedo, O.C.; Fitzjarrald, D.R. The early evening surface-layer transition: Temporal and spatial variability. *J. Atmos. Sci.* **2001**, *58*, 2650–2667. [CrossRef]

43. Sastre, M.; Yagüe, C.; Román-Cascón, C.; Maqueda, G. Atmospheric boundary-layer evening transitions: A comparison between two different experimental sites. *Bound. Layer Meteorol.* **2015**, *157*, 375–399. [CrossRef]

44. Angevine, W.M.; Tjernström, M.; Žagar, M. Modeling of the coastal boundary layer and pollutant transport in New England. *J. Appl. Meteorol. Climatol.* **2006**, *45*, 137–154. [CrossRef]

45. Nieuwstadt, F.T.M. The turbulent structure of the stable, nocturnal boundary layer. *J. Atmos. Sci.* **1984**, *41*, 2202–2216. [CrossRef]

46. Acevedo, O.C.; Mahrt, L.; Puhales, F.S.; Costa, F.D.; Medeiros, L.E.; Degrazia, G.A. Contrasting structures between the decoupled and coupled states of the stable boundary layer. *Quart. J. R. Meteorol. Soc.* **2016**, *142*, 693–702. [CrossRef]

47. Mahrt, L. Microfronts in the nocturnal boundary layer. *Quart. J. R. Meteorol. Soc.* **2019**, *145*, 546–562. [CrossRef]

48. Reynolds, A.M.; Reynolds, D.R.; Smith, A.D.; Chapman, J.W. Orientation cues for high-flying nocturnal insect migrants: Do turbulence-induced temperature and velocity fluctuations indicate the mean wind flow? *PLoS ONE* **2010**, *5*, e15758. [CrossRef]

49. Riley, J.R.; Reynolds, D.R.; Rainey, R.C. Radar-based studies of the migratory flight of grasshoppers in the middle Niger area of Mali. *Proc. R. Soc. Lond. Ser. B Biol. Sci.* **1979**, *204*, 67–82. [CrossRef]

50. Rennie, S.J. Common orientation and layering of migrating insects in southeastern Australia observed with a Doppler weather radar. *Meteorol. Appl.* **2014**, *21*, 218–229. [CrossRef]

51. Feng, H.; Wu, X.; Wu, B.; Wu, K. Seasonal migration of *Helicoverpa armigera* (Lepidoptera: Noctuidae) over the Bohai Sea. *J. Econ. Entomol.* **2009**, *102*, 95–104. [CrossRef]

52. Fu, X.; Zhao, X.; Xie, B.; Ali, A.; Wu, K. Seasonal pattern of *Spodoptera litura* (Lepidoptera: Noctuidae) migration across the Bohai Strait in northern China. *J. Econ. Entomol.* **2015**, *108*, 525–538. [CrossRef] [PubMed]

53. Wood, C.R.; Chapman, J.W.; Reynolds, D.R.; Barlow, J.F.; Smith, A.D.; Woiwod, I.P. The influence of the atmospheric boundary layer on nocturnal layers of noctuids and other moths migrating over southern Britain. *Int. J. Biometeorol.* **2006**, *50*, 193–204. [CrossRef] [PubMed]

54. Wood, C.R.; Reynolds, D.R.; Wells, P.M.; Barlow, J.F.; Woiwod, I.P.; Chapman, J.W. Flight periodicity and the vertical distribution of high-altitude moth migration over southern Britain. *Bull. Entomol. Res.* **2009**, *99*, 525–535. [CrossRef] [PubMed]

55. Wood, C.R.; Clark, S.J.; Barlow, J.F.; Chapman, J.W. Layers of nocturnal insect migrants at high-altitude: The influence of atmospheric conditions on their formation. *Agric. For. Entomol.* **2010**, *12*, 113–121. [CrossRef]

56. Dreyer, D.; El Jundi, B.; Kishkinev, D.; Suchentrunk, C.; Campostrini, L.; Frost, B.J.; Zechmeister, T.; Warrant, E.J. Evidence for a southward autumn migration of nocturnal noctuid moths in central Europe. *J. Exp. Biol.* **2018**, *221*, 179218. [CrossRef]

57. Wang, H.-H.; Grant, W.E.; Elliott, N.C.; Brewer, M.J.; Koralewski, T.E.; Westbrook, J.K.; Alves, T.M.; Sword, G.A. Integrated modelling of the life cycle and aeroecology of wind-borne pests in temporally-variable spatially-heterogeneous environment. *Ecol. Model.* **2019**, *399*, 23–38. [CrossRef]

58. Reynolds, A.M.; Reynolds, D.R.; Smith, A.D.; Chapman, J.W. A single wind-mediated mechanism explains high-altitude "non-goal oriented" headings and layering of nocturnally migrating insects. *Proc. R. Soc. Lond. B Biol. Sci.* **2010**, *277*, 765–772. [CrossRef]

59. Aralimarad, P.; Reynolds, A.M.; Lim, K.S.; Reynolds, D.R.; Chapman, J.W. Flight altitude selection increases orientation performance in high-flying nocturnal insect migrants. *Anim. Behav.* **2011**, *82*, 1221–1225. [CrossRef]

60. Riley, J.R. Collective orientation in night-flying insects. *Nature* **1975**, *253*, 113. [CrossRef]

61. Reynolds, A.M.; Reynolds, D.R.; Riley, J.R. Does a "turbophoretic" effect account for layer concentrations of insects migrating in the stable night-time atmosphere? *J. R. Soc. Interface* **2009**, *6*, 87–95. [CrossRef]

62. Holtslag, A.A.M.; Svensson, G.; Baas, P.; Basu, S.; Beare, B.; Beljaars, A.C.M.; Bosveld, F.C.; Cuxart, J.; Lindvall, J.; Steeneveld, G.J.; et al. Stable atmospheric boundary layers and diurnal cycles: Challenges for weather and climate models. *Bull. Am. Meteorol. Soc.* **2013**, *94*, 1691–1706. [CrossRef]

63. Davy, R. The climatology of the atmospheric boundary layer in contemporary global climate models. *J. Clim.* **2018**, *31*, 9151–9173. [CrossRef]

64. McNider, R.T.; Steeneveld, G.J.; Holtslag, A.A.M.; Pielke, R.A.; Mackaro, S.; Pour-Biazar, A.; Walters, J.; Nair, U.; Christy, J. Response and sensitivity of the nocturnal boundary layer over land to added longwave radiative forcing. *J. Geophys. Res. Atmos.* **2012**, *117*, D14106. [CrossRef]

65. Horvath, K.; Koracin, D.; Vellore, R.; Jiang, J.; Belu, R. Sub-kilometer dynamical downscaling of near-surface winds in complex terrain using WRF and MM5 mesoscale models. *J. Geophys. Res. Atmos.* **2012**, *117*, D11111. [CrossRef]

forests

MDPI

Article

The Impact of Moth Migration on Apparent Fecundity Overwhelms Mating Disruption as a Method to Manage Spruce Budworm Populations

Jacques Régnière [1,*], Johanne Delisle [1], Alain Dupont [2] and Richard Trudel [3]

1   Natural Resources Canada, Canadian Forest Service, PO Box 10380 Stn Ste-Foy, Quebec, QC G1V 4C7, Canada
2   Société de Protection des Forêts contre les Insectes et Maladies, 1780 Rue Semple, Quebec, QC G1N 4B8, Canada
3   Consultant en Entomologie RT Enr., 342 Saint-Joseph, Lévis, Québec, QC G6V 1G2, Canada
*   Correspondence: Jacques.Regniere@Canada.ca; Tel.: +1-418-648-5257

Received: 29 July 2019; Accepted: 4 September 2019; Published: 6 September 2019

Abstract: Aerial applications of a registered formulation of synthetic spruce budworm female sex pheromone were made in 2008, 2013 and 2014 to disrupt mating in populations of this forest insect pest in Quebec, Canada. Each year, the applications resulted in a 90% reduction in captures of male spruce budworm moths in pheromone-baited traps. A commensurate reduction in mating success among virgin females held in individual cages at mid-crown of host trees was also obtained. However, there was no reduction in the populations of eggs or overwintering larvae in the following generation (late summer and fall). The failure of this approach as a viable tactic for spruce budworm population reduction could have resulted from considerable immigration of mated females, as evidenced by high rates of immigration and emigration that caused steep negative relationships between apparent fecundity and the density of locally emerged adults.

Keywords: spruce budworm; moth; tortricidae; *Choristoneura fumiferana* (Clemens); forest protection; early intervention strategy; pheromone mating disruption; migration; dispersal

1. Introduction

The spruce budworm, *Choristoneura fumiferana* (Clemens) is an episodic tortricid defoliator of balsam fir, *Abies balsamea* (L.) Mill, and several members of the *Picea* genus, in particular white spruce *Picea glauca* (Moench) Voss, in the boreal forests of eastern North America [1,2]. Outbreaks are broadly regional, recurring every 30–40 years and lasting over 15 years [3,4]. In Canada, control methods are limited to two registered insecticides: The pathogenic bacterium *Bacillus thuringiensis* var. *kurstaki* (Btk) and the ecdysone agonist tebufenozide.

In 2007, after years of basic research, conceptual development, laboratory and field testing [5], a commercial product containing a synthetic formulation of the female sex pheromone for the spruce budworm (SBW) was registered in Canada for use against this insect: Disrupt Micro-Flakes® SBW (Hercon Environmental, Emigsville, PA, USA). This sex attractant formulation can be applied by aircraft to disrupt mating by interference with the sex pheromone emitted by females to attract males [6]. The mechanisms suggested to cause mating disruption include (1) false-plume (trail) following, (2) camouflage, (3) desensitization (adaptation and/or habituation), or (4) a combination of these [7–15].

In false-plume following, male moths are competitively attracted either to calling females or to pheromone dispensers; the latter decrease the limited search time of males and reduce mating encounters [9,14–18]. In camouflage, calling females occur within larger plumes of dispensers so that males cannot distinguish female plumes and locate the sources for mating. Desensitization includes adaptation and habituation in which high concentrations of pheromone cause neuronal fatigue so the insect becomes unresponsive to the pheromone for some time, again limiting effective search time and reducing chances of finding mates during the flight period [7,19–24].

Mating disruption (see [25] for a review) first proved useful in controlling cabbage looper moths, *Trichoplusia ni* (Hubner) [26]. It has since been used successfully on a number of insect pests and is a viable alternative to conventional insecticide programs for the control of several tortricids [27]. It offers many advantages, including reduced insecticide use, and thus conservation of natural enemies, decreased potential for the development of insecticide resistance, reduced insecticide residues on crops and in the environment, and reduced costs associated with worker protection and labor management [28].

In this paper, we report on the results of three field mating disruption trials with aerially-applied pheromone against adult spruce budworm, conducted in 2008, 2013 and 2014 in Quebec, Canada, under an array of different circumstances including forest composition, level of defoliation and the density of the target insect populations.

2. Materials and Methods

2.1. Sites

The 36 experimental sites used in these tests were located in eastern Quebec, Canada (Figure 1). Fourteen 50 ha plots were established in 2008 and were located north and east of Baie-Comeau, on the north shore of the St-Lawrence estuary. In 2013, twelve 30 ha plots were set-up on the south shore of the St-Lawrence river between Rimouski, Matane and Causapscal. In 2014, ten 100 ha plots were established in stands to the south and east of Rimouski. These sites were selected on the basis of forest composition and spruce budworm population density. Stands were mainly composed of balsam fir, white and black spruce, with a variable hardwood content (< 50% basal area) dominated by birches, aspen and maple. The North Shore area was already in a severe budworm outbreak in 2008, and was suffering its third year of defoliation. In the fall of 2007, population densities in the area of the North Shore where the mating disruption tests were conducted averaged 27.4 ± 1.1 overwintering larvae (L_2) per branch (SEM, $n = 42$). The spruce budworm populations in the Lower St-Lawrence (LSL) were at lower density in the fall of 2012, averaging 12.8 ± 2.3 L_2/branch (SEM, $n = 36$), and were either still at low density or in their first year of low to moderate defoliation in 2013. The populations in the LSL sites used in 2014 were at low density in the fall of 2013, averaging 5.8 ± 1.1 L_2/branch (SEM, $n = 150$), and had not caused significant defoliation in the years prior to the tests.

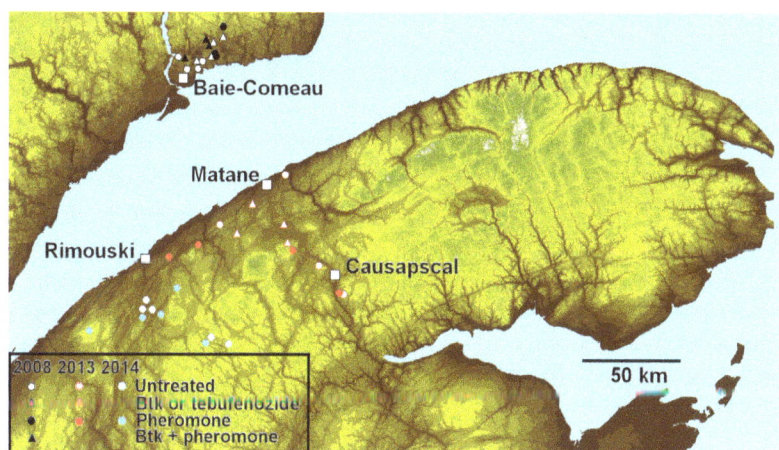

Figure 1. Elevation map of the eastern portion of southern Quebec, Canada (the Gaspé peninsula), showing the location of all plots in this study. Black: 2008. Red: 2013; Blue: 2014. Open squares: Selected city locations.

2.2. Treatments

In 2008, three treatments were applied: Btk double applications (5 days apart) in four plots, pheromone in three plots, Btk + pheromone in three plots. Four untreated plots were used as controls. The Btk (FORAY 76B, Valent BioSciences, Libertyville, IL, USA) was applied at the peak of the 4th instar at the rate of 1.5 L/ha (30 BIU/ha) at each application. At the beginning of the male moth flight season (11 and 12 July), Disrupt Micro-Flakes® SBW were applied by helicopter to three of these Btk-treated plots as well as three of the untreated plots. An ASTAR BA helicopter (Airbus Helicopter SAS, Marseille, France), equipped with the AG-NAV2® GPS navigation system (AG-NAV, Barrie ON, Canada), was used. A spreader was attached under the helicopter to apply the dry pheromone flake formulation at the label rate of 0.5 kg flakes/ha (50 g AI/ha). Both the aircraft and the helicopter flew at about 15 m above the tree canopy.

In 2013, two treatments were compared: single applications of tebufenozide (four plots), and pheromone (four plots). Four untreated plots were used as controls. Tebufenozide was applied at the rate of 70 g AI/ha (2 L/ha of Mimic 2LV®, Valent BioSciences, Libertyville, IL, USA), when larvae had reached the 5th instar (see [29] for details). Pheromone (same product, same application rate as in 2008) was applied when pheromone traps (see Section 2.3 below) reached < 2% cumulative moth catch, using a Cessna 188 (Textron Aviation, Wichita KS, USA), equipped with a flake spreader, flying about 15 m above the canopy.

In 2014, the pheromone (same product, same application rate as in 2013) was applied on five plots at the beginning of the moth flight period using the same procedure as in 2013. Five untreated plots were used as controls.

2.3. Pheromone Traps

SBW males were caught in each plot using five Multipher® (Biocom, Quebec City, QC, Canada) traps baited with the standard synthetic SBW pheromone lure (Biolure®, Contech Enterprises, Victoria, BC, Canada) in each plot. Traps were placed at the mid-crown level (8–12 m above ground) of five balsam fir trees spread across each plot at least 50 m from each other. Traps were emptied every 2 to 4 days throughout the moth flight season and the number of moths caught was counted.

2.4. Mating Success

Female spruce budworm moths do not tend to fly prior to mating, and thus it is difficult to adequately estimate mating success from samples of feral females in the field. This is why virgin females used in this study to estimate mating success were obtained from a SBW colony (Great Lake Forestry Centre, Sault Ste. Marie, ON, Canada) reared on an artificial diet [30]. We have previously demonstrated that the mating success as well as the pheromone gland content of laboratory-reared females were comparable to those of wild females [31]. Immediately after pheromone applications, 24- to 48 h-old virgin spruce budworm females were installed in individual plastic cages (Figure 2). In 2008, a single-opening cage containing a single virgin female was used (Figure 2a) [32]. In 2013 and 2014, this cage was modified by adding a second open screen funnel (Figure 2b). This modified cage is hereafter referred to as the "double-opening cage". The openings allowed males to enter the cage but were too small for the larger females to escape. Mating cages were placed at mid-crown (8–12 m above ground) in balsam fir trees using Multipher® traps as cage support. In 2008, 15 virgin-female cages were exposed in each plot at each interval. In 2013, sample size was increased to 30 cages per plot per installation, and in 2014, samples of 50 cages were used. The caged females were recovered and replaced every 3 to 4 days until the end of the male flight season. The number of males found in the cages was recorded, and females recovered from the cages were dissected to determine their mating status by the presence of spermatophores in their bursa copulatrix. In 2008, females were dissected from 7 of the 14 sample dates. In 2013 and 2014, all females were dissected.

(a) (b)

Figure 2. Cages used to assess mating success of female moths of spruce budworm. Top: Open cages; bottom: Closed cages. (**a**) Single-opening cage, used in 2008, containing one female and several trapped males. (**b**) Double-opening cage, used in 2013 and 2014. Cup diameter: 4 cm.

In 2011, under field conditions, the mating success of females held in both types of cages (single- and double-opening) was compared to that of tethered females (that we consider as close to feral females as possible). Tethered females were attached by a thin monofilament nylon wire with a drop of glue on their pronotum to a mesh wire stage about 12 cm in diameter installed on Multipher pheromone traps hung at mid-crown (8–12 m above ground) under branch tips of host trees. Three sets of 95, 128 and 126 females in single and double opening cages and tethered, respectively were available for this comparison. All were exposed simultaneously for 48 h, in three separate replicates at intervals of 8 days during the peak moth flight season.

2.5. Foliage Sampling

Foliage samples of 15 to 100 branch tips were taken in each plot from the mid-crown (8–12 m above ground) of dominant and co-dominant balsam fir and white spruce trees at several points in the insect's life cycle. In all three years, a sample of 45 cm branch tips was taken at the end of larval development (beginning of pupation). In 2013 and 2014, a second sample was taken at the end of the pupal stage, once adult emergence had started (44% emergence on average), to determine the density of emerging adults. Live pupae recovered from the foliage were reared to adult emergence to determine their survival, and the number of survivors was added to the number of emerged pupal cases found on the foliage. In 2008, a foliage sample was not taken at this stage. To obtain an estimate of emerging adult density from late-larval density measured in 2008, a linear regression between adult (A) and late-larval (P) densities measured in 2013 and 2014, was used. This regression was: $A = 0.337P - 0.0098$ ($R^2 = 0.646$, $n = 22$). A third sample of 45 to 75 cm branch tips was taken at the end of the oviposition period, once adults were no longer being caught in pheromone traps and eggs had hatched. Egg masses found on the foliage were counted. In 2013 and 2014, eggs in each mass were also counted. The number of eggs found on foliage in 2008 was estimated by multiplying the number of egg masses found by the average number of eggs per mass observed in 2013 and 2014 (17.7 eggs/mass). Current year shoots on the branches from these three foliage samples were counted, and density was expressed per shoot.

In 2008 and 2014, an additional foliage sample was taken in the fall (75 cm branch tips, about 0.16 m²/branch among samples collected in 2008 and 2014) to measure the density of overwintering larvae (L_2). The L_2 were extracted using the NaOH washing method [33]. Shoots on these branches were not counted. To express L_2 density per shoot, the data from the egg sample collected in the LSL in 2013 were used to estimate, by linear regression, the number of shoots on balsam fir branches of 0.16 m² (106 shoots = 61 shoots + 281 shoots/m² × 0.16 m²).

2.6. Analysis

Because the experimental designs varied from year to year (use of insecticides), and because the question we address is the impact of pheromone applications on mating success and reproduction of the target populations, we chose not to distinguish insecticide applications as distinct treatments in our analyses. Thus, assuming that insecticide applications in the early larval stages have no repercussions on surviving adults other than their density, "controls" include plots that received insecticides (either Btk or tebufenozide) and "Treated" include plots that received pheromone alone or in combination with a prior Btk application (2008). Doing this allows us to perform global analysis and make comparisons between years. However, in our Figures, we distinguish with different symbols the various treatment combinations used.

The comparison of mating success (presence or absence of a spermatophore in the recovered females) among tethered and caged females was done by logistic regression analysis using the confinement device (cage type or tethering) as factor.

The effect of pheromone treatment (*T*) and year (*Y*) on total capture of male spruce budworm moths per pheromone trap (*M*) was analyzed with a general linear model (GLM) using the density of emerging adults (*A*) as a covariate. In the absence of net migration of moths, this density is the main determinant of trap capture. No pupae were found in two plots in 2013. For analysis, these zero values were replaced by 0.00005 adults/shoot, below the observed minimum of 0.000055 pupae/shoot. The model used was:

$$\text{Log}(M) = a + bY + (c + dY)T + (e + fY + gT + hYT)\text{Log}(A) \qquad (1)$$

The effect of pheromone treatment (*T*) and year (*Y*) on the proportion of caged virgin females that successfully mated (*P*) in each plot was analyzed by logistic regression using the density of emerging adults (*A*) as a covariate. In total, 3777 females were exposed, of which 1114 were mated (31.5%). The model used was:

$$\text{Logit}(P) = a + bY + (c + dY)T + (e + fY + gT + hYT)\text{Log}(A) \qquad (2)$$

The effect of pheromone treatment (*T*) and year (*Y*) on the relationship between egg density and emerging adult density was tested by a GLM of the form:

$$E = a + bY + (c + dY)T + (e + fY + gT + hYT)A \qquad (3)$$

where *E* is egg density (eggs per shoot = egg masses × mean eggs/mass) and *A* is emerging adult density. One missing value of *E* (no egg masses found in one plot in 2014) was replaced by 0.04 eggs/shoot, below the observed minimum of 0.049 eggs/shoot.

The ratio of egg density to emerging adult density, *E*/*A*, is an expression of apparent fecundity at the population level. In the absence of moth migration, apparent fecundity equals realized fecundity (the portion of their potential fecundity that moths succeed in laying before death). But when migration occurs (which is nearly always in spruce budworm), it also represents the contribution made by immigrant moths the egg population, as well as the loss of eggs that are carried away by emigrant moths (see [34] for a thorough discussion of apparent fecundity in the study of spruce budworm ecology). Apparent fecundity (E/A) was related to emerging adult density (A), year (Y) and pheromone treatment (T), as defined above, by a GLM of the form:

$$Log(E/A) = a + bY + (c + dY)T + (e + fY + gT + hYT)Log(A) \qquad (4)$$

The effect of pheromone treatment (T) and year (Y) on the relationship between the density of overwintering larvae in the fall following treatment, L_2 (available in 2008 and 2014), and emerging adult density (A) was obtained by a GLM of the form:

$$L_2 = a + bY + (c + dY)T + (e + fY + gT + hYT)A \qquad (5)$$

Models (1) to (5) were reduced by dropping least-significant terms one at a time until all remaining terms were significant, model consistency allowing ($\alpha < 0.05$). Residuals were tested for normality using the Anderson-Darling test [35].

3. Results

3.1. Mating Success of Caged and Tethered Females

Mating success varied significantly with holding device ($\chi^2 = 28.5$; df: 2346; $p < 0.001$). It was much higher among tethered females (39%, $n = 126$) than among females held in single-opening cages (12%, $n = 141$) (odds ratio 3.6, 95% confidence interval (CI): (2.5, 8.6)). In double-opening cages, females had the same likelihood of mating (33%, $n = 82$) as tethered females (odds ratio: 1.3, 95% CI: (0.7, 2.3)), and were more than 3-fold as likely to mate as females in single-opening cages (odds ratio 3.6, 95% CI: (1.8, 7.1)).

3.2. Pheromone Traps

The daily capture rate of male moths in pheromone traps was reduced by nearly 90% from the moment pheromones were applied, in all three years (Figure 3a–c). The success of caged virgin females at attracting males (Figure 3d–f) and successfully mating (Figure 3g–i) was also reduced. By contrast, insecticide applications (either Btk or tebufenozide) had no effect on capture in pheromone traps, or on mating success of virgin females, justifying that those treatments not be distinguished further in analyses.

Results of the GLM analysis (Equation (1)) indicate that total catch per pheromone trap varied significantly among years, and was reduced following pheromone applications, by the same amount (90%), each year (Table 1; $R^2 = 0.94$; Figure 4a–c). Importantly, there was no significant relationship between the total capture in pheromone traps and emerging adult density per shoot. Residuals of the reduced model were normally-distributed (Anderson-Darling $AD = 0.537$, $n = 36$, $p = 0.158$). Pheromone trap catch in the LSL was lower than expected in 2014 given the high emerging adult density per shoot (Figures 3c and 4c). It is likely that emerging adult density was overestimated because the sample providing this estimate was taken early relative to adult emergence in 2014 [36].

Table 1. Effects of emerging adult density (A), Year (Y) and pheromone treatment (T) on total capture of males per pheromone trap, Equation (1) reduced.

Source	DF	Adj SS	Adj MS	F-Value	p-Value
Year, Y	2	12.8851	6.44253	146.48	< 0.001
Pheromone, T	1	8.4613	8.46130	192.38	< 0.001
Error	32	1.4074	0.04398		
Lack-of-Fit	2	0.2451	0.12256	3.16	0.057
Pure Error	30	1.1623	0.03874		
Total	35	24.8012			

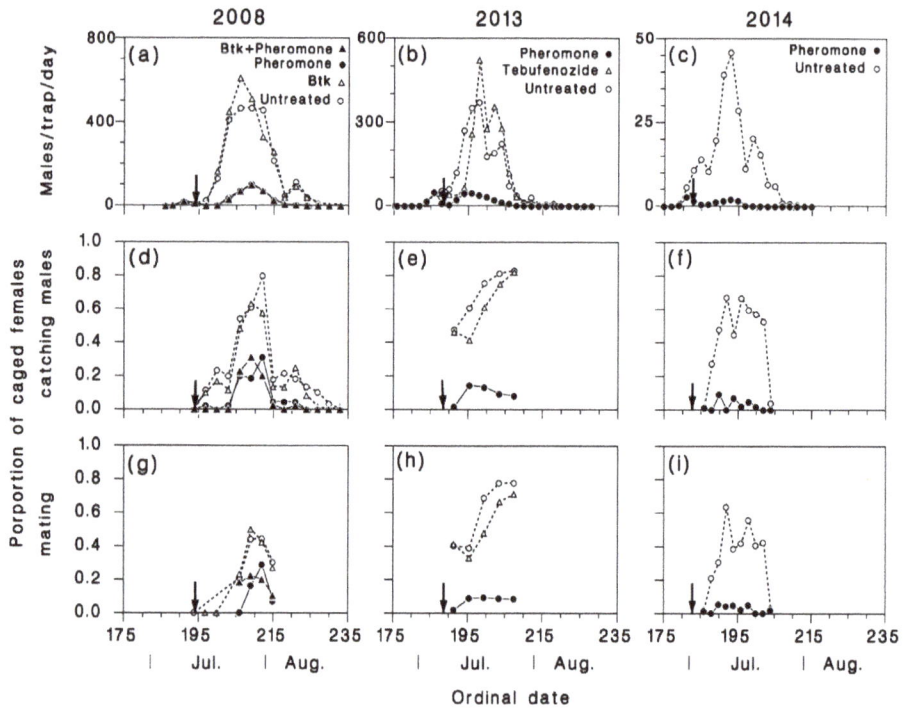

Figure 3. First row (**a,b,c**): Daily capture rates of male spruce budworm moths in pheromone traps. Second row (**d,e,f**): Success of caged virgin females at capturing males. Third row (**g,h,i**): Success of caged virgin females at getting mated. Left column: 2008. Center column: 2013. Right column: 2014.

3.3. Mating Success

The logistic regression model (Equation (2)) could not be reduced, because the 3-way interaction $Log(A) \times Y \times T$ was highly significant (Table 2). The overall fit was very good ($R^2 = 0.832$) (Figure 4d–f). The response to pheromone treatment (T) was clear in all years, but complex in the details because of its interactions with adult density, $Log(A) \times T$, and year, $Log(A) \times Y$ (Table 2). Overall, there was no clear relationship between mating success and emerging adult density, $Log(A)$, as it interacted with both treatment and year.

Table 2. Results of logistic regression of mating success among caged females as influenced by adult density (A), year (Y) and treatment with pheromone (T), Equation (2), full model.

Source	DF	Adj. Mean	Chi-Square	*p*-Value
Regression	11	90.2094	992.30	< 0.001
$Log(A)$	1	6.4617	6.46	0.011
Year, Y	2	28.2868	56.57	< 0.001
Pheromone, T	1	2.0262	2.03	0.155
$Log(A) \times Y$	2	7.1197	14.24	0.001
$Log(A) \times T$	1	10.4317	10.43	0.001
$Y \times T$	2	0.9070	1.81	0.404
$Log(A) \times Y \times T$	2	5.6419	11.28	0.004
Error	24	7.8267		
Total	35			

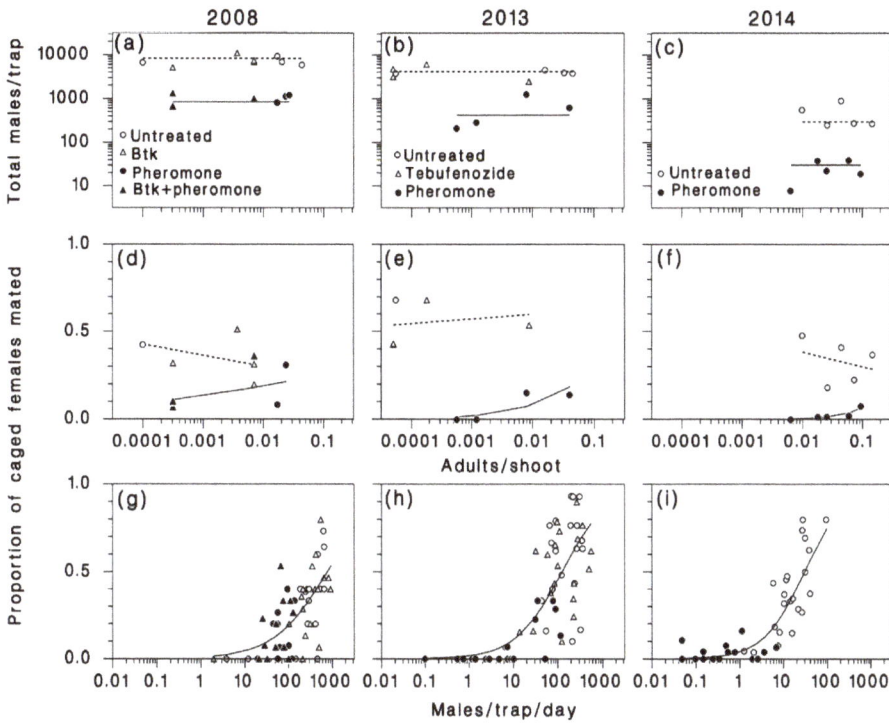

Figure 4. First row (**a,b,c**): Total capture of males per pheromone trap vs. adults per shoot, (lines: Equation 1, ——: With pheromone; ······: Without). Second row (**d,e,f**): Mating success vs. adults per shoot, (lines: Equation 2). Third row (**g,h,i**): Relationship between mating success of caged virgin females and the daily capture rate of males in pheromone traps during the exposure period (lines: Logistic regression). Left column: 2008. Center column: 2013. Right column: 2014.

The relationship between mating success among caged females and the daily capture rate of males in pheromone traps, however, corresponded closely with that reported by Régnière et al. [31] (Figure 4g–i).

3.4. Egg Density

Results of the GLM analysis (Equation (3)) indicate that egg density was unaffected by the pheromone treatment, *T*, and that its relationship with emerging adult density, *A*, varied significantly between years in both intercept and slope (Table 3; $R^2 = 0.68$; Figure 5a). Residuals of the reduced model were normally-distributed (Anderson-Darling $AD = 0.395$, $n = 36$, $p = 0.158$). Annual variations in the intercept and slope of the relationship between egg density and emerging adult density represent different levels of realized fecundity (slope) and immigration rates (intercept) (see [34] for a discussion). The immigration rate in 2008 was 0.234 eggs/shoot, the highest observed in this study (Baie-Comeau, area under full outbreak). It was intermediate (0.174 eggs/shoot) in 2013 (a high-migration year in a rising outbreak), and an order of magnitude lower (0.018 eggs/shoot) in 2014, farther to the west in the LSL region were the SBW infestation was still very patchy. Realized fecundity was highest on the North Shore in 2008 (13.4 eggs/moth), lowest during the high-migration year 2013 in the LSL (1.3 eggs/moth), and intermediate in 2014 (4.0 eggs/moth).

Table 3. General linear model (Equation (3)) of the effects of emerging adult density (*A*), Year (*Y*) and pheromone treatment (*T*) on the density of spruce budworm eggs. Reduced model.

Source	DF	Adj SS	Adj MS	F-Value	p-Value
Adults, *A*	1	0.42619	0.426188	54.08	< 0.001
Year, *Y*	2	0.12260	0.061300	7.78	0.002
A × *Y*	2	0.21352	0.106761	13.55	< 0.001
Error	30	0.23641	0.007880		
Lack-of-Fit	27	0.19422	0.007193	0.51	0.855
Pure Error	3	0.04219	0.014063		
Total	35	1.26226			

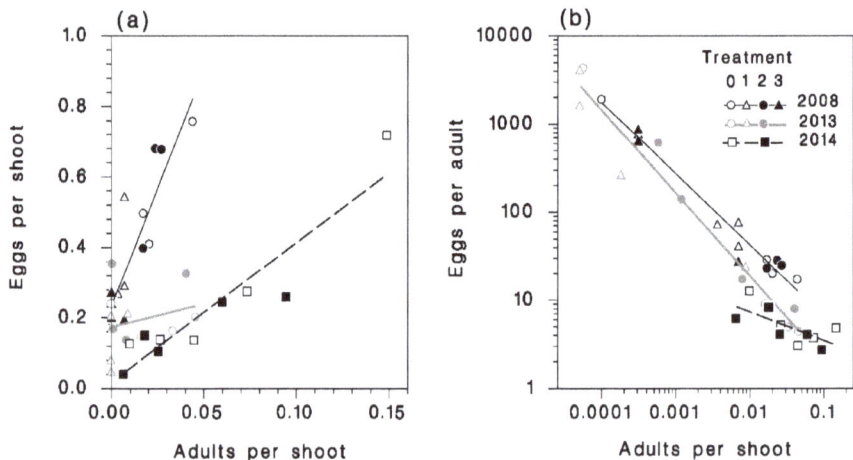

Figure 5. Relationship between adult density, and (**a**) egg density (lines: Equation (3)) and (**b**) apparent fecundity (eggs/adult) (lines: Equation (4)). Treatments: 0 is untreated, 1 is Btk or tebuzenozide, 2 is pheromone, 3 is Btk and pheromone.

3.5. Apparent Fecundity

Apparent fecundity was not affected by pheromone applications, but its relationship with adult density (Log(*A*)) varied significantly between years. The reduced model (Equation (4); Table 4) described the observations quite accurately (Figure 5b; R^2 = −0.965). Residuals were normally distributed (Anderson-Darling *AD* = 0.279, *n* = 36, *p* = 0.628).

Table 4. General linear model (Equation (4)) of the effects of emerging adult density (*A*), Year (*Y*) and pheromone treatment (*T*) on apparent fecundity, the ratio of eggs to adults (*E*/*A*). Reduced model.

Source	DF	Adj SS	Adj MS	F-Value	p-Value
Adult, Log(*A*)	1	5.7092	5.70918	173.78	< 0.001
Year, *Y*	2	0.4931	0.24654	7.50	0.002
Log(*A*) × *Y*	2	0.6170	0.30852	9.39	0.001
Error	30	0.9856	0.03285		
Lack-of-Fit	24	0.7913	0.03297	1.02	0.541
Pure Error	6	0.1942	0.03237		
Total	35	32.9479			

In 2013, egg density was independent of adult density and the slope of the relationship between Log(*E*/*A*) and Log(*A*) was −0.94 ± 0.06 (nearly −1), so that apparent fecundity was directly proportional to 1/*A* (grey line in Figure 5b), an indication of panmixis (see [34]). In 2008, the relationship between egg and adult density was significant (slope 13.4 ± 2.1 eggs/moth), with a high intercept indicating that considerable immigration was occurring in those populations (solid line in Figure 5a). Therefore, the slope of the relationship between apparent fecundity and adult density was also steep (-0.81 ± 0.04, no as close to -1 as in 2013). By contrast, in 2014, the intercept and slope of the relationship between egg and adult density were both lower than in 2008 (dashed line in Figure 5a). The slope of the relationship between apparent fecundity and adult density was consequently closer to zero (-0.32 ± 0.12). Those parameter values indicate that there was limited immigration into, but considerable emigration out of, those populations (dashed line in Figure 5b).

3.6. L_2 Density

The relationship between L_2 density and the density of emerging adults was very strong (Equation (5); $R^2 = 0.786$) (Figure 6a), and was unaffected by either pheromone applications or year (Table 5). Residuals were normally-distributed (Anderson-Darling $AD = 0.587$, $n = 24$, $p = 0.114$). Expressing the relationship in the form of the ratio L_2/A shows the same pattern as in apparent fecundity, where low-density populations have very high progeny to parent ratios when compared to higher-density populations (Figure 6b).

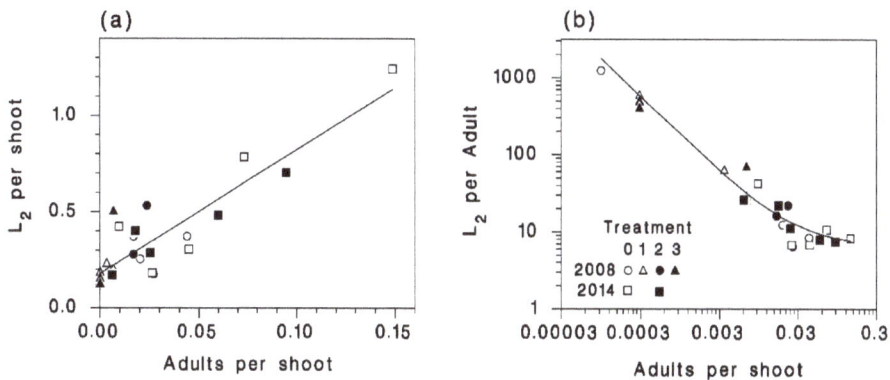

Figure 6. Relationships between adult density and (**a**) L_2 density, and (**b**) the ratio between L_2 density and adult density in 2008 and 2014. Lines: Equation (5). Treatments: 0 is untreated, 1 is Btk or tebuzenozide, 2 is pheromone, 3 is Btk and pheromone.

Table 5. General linear model (Equation (5)) of the effects of emerging adult density (A), Year (Y) and pheromone treatment (T) on L_2 density. Reduced model.

Source	DF	Adj SS	Adj MS	F-Value	p-Value
Adult, A	1	1.19871	1.19871	80.93	< 0.001
Error	22	0.32587	0.01481		
Lack-of-Fit	17	0.25502	0.01500	1.06	0.522
Pure Error	5	0.07085	0.01417		
Total	23	1.52458			

4. Discussion

Mating disruption has been successfully used against several moth pests with synthetic sex pheromone applied from the air or from the ground [18,27]. Aerially-applied pheromones incorporated in plastic laminate flakes have been successful in controlling low-density populations of the gypsy moth in the USA, even when those populations were surrounded by higher, untreated populations [37]. The facts that gypsy moth females do not fly and that migration rates in natural populations are very low have contributed to this success. Ground applications of wax-incorporated synthetic sex pheromones have also been effective in suppressing mating in the oriental fruit moth, *Grapholita molesta* [38], a tortricid.

Over the last four decades, considerable research has been devoted to understanding the concept and application of mating disruption with synthetic pheromones against the spruce budworm [5]. Most previous mating disruption trials were conducted in small plots, often with no replication (one plot per treatment) and provided no rigorous statistical reliability [5]. Thus, it is not surprising that much of this research was published in the grey literature. Trap shutdown is a common feature in most SBW mating disruption trials. However, no population reduction has been convincingly demonstrated, whether expressed as apparent fecundity (the ratio of eggs per locally-produced moth) or as reduced populations in the late summer or early fall. In contrast, our three large-scale mating disruption trials were fully replicated, with sample sizes ensuring accurate measurement of apparent fecundity (a ratio). This allows us to provide precise and unbiased estimates of the different parameters measured as a function of treatment, adult density and year. The 90% reduction in male attraction to pheromone sources (natural or synthetic) and similar reduction in the mating success of caged females were both highly promising results. However, we clearly demonstrated that pheromone applications had no impact on egg or L_2 counts or on apparent fecundity, and the use of mating disruption as a viable tool to control spruce budworm populations is doubtful.

The failure of pheromones to reduce egg or L_2 density or affect apparent fecundity could have resulted from a lack of effect on the mating success of feral females, which would suggest that mating success of caged females was not representative of the mating success of feral females. This is a hypothesis proposed in [5]. However, it is more likely that moth movement masked the limited effect of the pheromone applications. In the spruce budworm, several aspects of male and female behavior may affect the success of mating disruption with pheromones. Females calling from the top of highest trees in the canopy may still succeed in attracting males as most of the synthetic pheromone is released lower in the canopy or even on the ground [27]. We have indeed observed higher spruce budworm mating frequency in caged female placed at treetop compared to mid-crown in untreated populations (unpublished data). Daytime flight may also limit the success of mating disruption [26]. In spruce budworm, mating may occur during the daylight hours, which suggests that males can use visual cues to find females, especially when population densities are high [39]. However, other tortricid moths with daytime mating have been controlled successfully with mating disruption [27]. As a counterpart to these findings, we have seen no difference in mating success between females held in double-opening cages and tethered females. Nor has any difference in sexual attractiveness (amount of pheromone in the gland) been observed between feral and laboratory-reared females [31]. Therefore, considering the strong reduction observed in male trap captures and the mating success of caged-females, it seems unlikely that feral females in pheromone-treated plots would have succeeded in attracting mates, especially in low density populations, given the strong Allee effect associated with mate-finding in this species [31]. It is also worth noting that the positive relationship between mating success among caged females and the daily rate of capture of males in pheromone traps (Figure 4g–i) was still very apparent in pheromone-permeated air, suggesting that it is the abundance and ability of male moths to find the females that is important in determining mating success.

To the best of our knowledge, no study has provided clear evidence that the mating success of feral females is higher than that of caged or tethered females. However, even if this was the case for the spruce budworm, the huge apparent fecundities observed in low-density populations in the 2008 and 2013 trials (Figures 5b and 6b), could not possibly be attributed to egg laying by local females. Spruce budworm females are known to migrate after mating and after laying part of their egg complement [40,41] and so, given the strong negative relationship between the density of emerged adults and their apparent fecundity, moth movement is more likely the factor responsible for the failure of pheromone to reduce egg density. This failure was not the result of our treatment plots being too small. The scale of spruce budworm moth movement (in the 1000 km² range [40]) is of the same order as that of our test areas: in 2008 an area roughly 40 km in diameter (1250 km²), in 2013 an area of 100 km diameter (7850 km²) and in 2014 an area of 70 km diameter (3850 km²). Managing a forest insect with treatment areas of such a scale is physically impossible and probably ecologically undesirable.

Pesticide applications are not expected to have a direct impact on apparent fecundity, except perhaps because of their effects on the reproductive abilities of survivors to sub-lethal doses, as documented for both Btk [42] and tebufenozide [43]. In moths that migrate following mating, successful population control by mating disruption is difficult [27]. In such cases, area-wide population management is required, and only when sources of immigrants are reduced can mating disruption succeed.

The slopes of the relationship between apparent fecundity and adult density (log (*E/A*) to log (*A*)) have significant implications in population dynamics at the regional scale. While this topic is discussed elsewhere [34,36], certain features are worth noting here. In Figure 5b, we can see the main features of apparent fecundity in the dynamics of rising SBW population: Low populations get a significant fecundity boost from immigration, while high populations emit migrants (very low apparent fecundities). Interestingly, on an operational level, our results clearly show that sampling the L_2 rather than the eggs may be sufficient (and less labor intensive) to establish a relationship between apparent fecundity (L_2/adult) and adult density (Figure 6b). Another important feature is that moth movement is a broadly regional phenomenon [40], and as a result, pheromone treatments can have very little if any effect on apparent fecundity. This should be true regardless of plot size because treated areas cannot be as large as the scale at which moths mix by migration. In large part, because of the extent and scale of moth movement, apparent fecundity may reflect the net result of migration in different population density contexts at the regional scale. For instance, in a widespread outbreak, such as on the North Shore in 2008, moths leaving a stand are likely to be replaced by immigrants as indicated by the high immigration rate and realized fecundity recorded there (0.23 eggs/shoot and 13.4 eggs/moth). In a more patchy situation, such as occurred west of the growing LSL outbreak in 2014, departing moths are less likely to be replaced, leading to very low realized fecundity (4.0 eggs/moth). Immigration rates were also much lower (0.02 eggs/shoot) among the ten plots in 2014 than they were on the North shore in 2008 (0.23 eggs/shoot) or among the LSL sites in 2013 (0.17 ± 0.03 eggs/shoot).

5. Conclusions

The outbreak in our three test areas was already widespread when our trials took place, especially on the north shore of the St-Lawrence river near Baie-Comeau in 2008 and in the Lower St-Lawrence east of Rimouski in 2013 (Figure 1). In those two trials, the risk of moth immigration was high. In 2014, the risk of immigration into the ten sites of the trial, west and south of Rimouski, was lower, but high survival rates led to high population growth rates [36]. There was little chance that mating disruption would be effective under any of those circumstances.

There may still be a faint hope that mating disruption can be of practical use as an early intervention tool against spruce budworm, if it can be applied at the first sign of an outbreak over a vast enough area, in which significant immigration is unlikely. However, given that Btk and tebufenozide applications significantly reduce SBW population growth rates compared to untreated controls or pheromone-only treatments [36,44], we question the usefulness of pursuing an approach that has so consistently failed at delivering the necessary population reduction even in the most carefully conducted trials.

Author Contributions: Conceptualization, J.R. and J.D.; Methodology, J.R. and J.D.; Formal analysis, J.R.; Investigation, J.R, J.D., A.D. and R.T.; Resources, J.R , A D ; Data curation, J.R.; Writing—original draft preparation, J.R. and J.D.; Writing—review and editing, J.R. and J.D.; Visualization, J.R.; Supervision, J.R., A.D. and R.T.; Project administration, J.R. and A.D.; Funding acquisition, J.R. and A.D.

Funding: This research was funded by Forest Protection Ltd., the Société de protection des Forêts contre les Insectes et Maladies (SOPFIM), the provincial governments of Newfoundland, Nova Scotia, New Brunswick, Quebec, Ontario, Manitoba, Saskatchewan, Alberta, and the USDA Forest Service, through SERG-international.

Acknowledgments: We wish to acknowledge the assistance of Ariane Béchard in coordinating field work in 2013 and 2014.

Conflicts of Interest: The authors declare no conflict of interest.

References

1. Blais, J.R. Effects of the destruction of the current year's foliage of balsam fir on the fecundity and habits of flight of the spruce budworm. *Can. Entomol.* **1953**, *85*, 446–448. [CrossRef]
2. Sanders, C.J. Biology of North American spruce budworms. In *Tortricid Pests, Their Biology, Natural Enemies and Control*; van der Geest, L.P.S., Evenhuis, H.H., Eds.; Elsevier Science Publishers BV: Amsterdam, The Netherlands, 1991; pp. 579–620.
3. Royama, T. Population dynamics of the spruce budworm *Choristoneura fumiferana*. *Ecol. Monogr.* **1984**, *54*, 429–462. [CrossRef]
4. Jardon, Y.; Morin, H.; Dutilleul, P. Périodicité et synchronisme des épidémies de la tordeuse des bourgeons de l'épinette au Québec. *Can. J. For. Res.* **2003**, *33*, 1947–1961. [CrossRef]
5. Rhainds, M.; Kettela, E.G.; Silk, P.J. Thirty-five years of pheromone-based mating disruption studies with *Choristoneura fumiferana* (Lepidoptera: Tortricidae). *Can. Entomol.* **2012**, *144*, 379–395. [CrossRef]
6. Byers, J.A. Simulation of mating disruption and mass trapping with competitive attraction and camouflage. *Environ. Entomol.* **2007**, *36*, 1328–1338. [CrossRef]
7. Shorey, H.H. Manipulation of insect pests of agricultural crops. In *Chemical Control of Insect Behaviour: Theory and Application*; Shorey, H.H., McKelvey, J.J., Jr., Eds.; Wiley: New York, NY, USA, 1977; pp. 353–367.
8. Bartell, R.J. Mechanisms of communication disruption by pheromones in control of Lepidoptera: A review. *Physiol. Entomol.* **1982**, *7*, 353–364. [CrossRef]
9. Cardé, R.T. Principles of mating disruption. In *Behavior Modifying Chemicals for Pest Management: Applications of Pheromones and Other Attractants*; Ridgway, R.L., Silverstein, R.M., Eds.; Marcel Dekker: New York, NY, USA, 1990; pp. 47–71.
10. Valeur, P.G.; Löfstedt, C. Behaviour of male oriental fruit moth, *Grapholita molesta*, in overlapping sex pheromone plumes in a wind tunnel. *Entomol. Exp. Appl.* **1996**, *79*, 51–59. [CrossRef]
11. Cardé, R.T.; Staten, R.T.; Mafra-Neto, A. Behaviour of pink bollworm males near high-dose, point sources of pheromone in Þeld wind tunnels: Insights into mechanisms of mating disruption. *Entomol. Exp. Appl.* **1998**, *89*, 35–46. [CrossRef]
12. Evenden, M.L.; Judd, G.J.R.; Borden, J.H. Investigations of mechanisms of pheromone communication disruption of *Choristoneura rosaceana* (Harris) in a wind tunnel. *J. Insect Behav.* **2000**, *13*, 499–510. [CrossRef]
13. Gut, L.J.; Stelinski, L.L.; Thompson, D.R.; Miller, J.R. Behaviour-modifying chemicals: Prospects and constraints in IPM. In *Integrated Pest Management: Potential, Constraints, and Challenges*; Koul, O., Dhaliwal, G.S., Cuperus, G.W., Eds.; CABI Publishing: Cambridge, MA, USA, 2004; pp. 73–121.
14. Miller, J.R.; Gut, L.J.; de Lame, F.M.; Stelinski, L.L. Differentiation of competitive vs. non-competitive mechanisms mediating disruption of moth sexual communication by point sources of sex pheromone (part 1): Theory. *J. Chem. Ecol.* **2006**, *32*, 2089–2114. [CrossRef]
15. Miller, J.R.; Gut, L.J.; de Lame, F.M.; Stelinski, L.L. Differentiation of competitive vs. non-competitive mechanisms mediating disruption of moth sexual communication by point sources of sex pheromone (part 2): Case studies. *J. Chem. Ecol.* **2006**, *32*, 2115–2143. [CrossRef]
16. Daterman, G.E.; Sower, L.L.; Sartwell, C. Challenges in the use of pheromones for managing western forest Lepidoptera. In *Insect Pheromone Technology: Chemistry and Applications*; Leonhardt, B., Beroza, M., Eds.; American Chemical Society: Washington, DC, USA, 1982; pp. 243–254.
17. Mani, E.; Schwaller, R. Results of 12 years experience to control codling moth, *Cydia pomonella* L. by mating disruption. *IOBC/WPRS Bull.* **1992**, *15*, 76–80.
18. Stelinski, L.L.; Gut, L.J.; Pierzchala, A.V.; Miller, J.R. Field observations quantifying attraction of four tortricid moths to high-dosage pheromone dispensers in untreated and pheromone-treated orchards. *Entomol. Exp. Appl.* **2004**, *113*, 187–196. [CrossRef]
19. Bartell, R.J.; Roelofs, W.L. Inhibition of sexual response in males of the moth *Argyrotaenia velutinana* by brief exposures to synthetic pheromone or its geometric isomer. *J. Insect Physiol.* **1973**, *19*, 655–661. [CrossRef]
20. Kuenen, L.P.S.; Baker, T.C. Habituation versus sensory adaptation as the cause of reduced attraction following pulsed and constant sex pheromone pre-exposure in *Trichoplusia ni*. *J. Insect Physiol.* **1981**, *27*, 721–726. [CrossRef]

21. Figueredo, A.J.; Baker, T.C. Reduction of the response to sex pheromone in the oriental fruit moth, *Grapholita molesta* (Lepidoptera: Tortricidae) following successive heromonal exposures. *J. Insect Behav.* **1992**, *5*, 347–362. [CrossRef]

22. Rumbo, E.R.; Vickers, R.A. Prolonged adaptation as possible mating disruption mechanism in oriental fruit moth, *Cydia (Grapholita) molesta*. *J. Chem. Ecol.* **1997**, *23*, 445–457. [CrossRef]

23. Stelinski, L.L.; Miller, J.R.; Gut, L.J. Presence of long-lasting peripheral adaptation in the oblique-banded leafroller, *Choristoneura rosaceana* and absence of such adaptation in redbanded leafroller, *Argyrotaenia velutinana*. *J. Chem. Ecol.* **2003**, *29*, 405–423. [CrossRef]

24. Judd, G.J.R.; Gardiner, M.G.T.; Delury, N.C.; Karg, G. Reduced antennal sensitivity, behavioural response, and attraction of male codling moths, *Cydia pomonella*, to their pheromone (*E,E*)-8,10-dodecadien-1-ol following various pre-exposure regimes. *Entomol. Exp. Appl.* **2005**, *114*, 65–78. [CrossRef]

25. Miller, J.R.; Gut, L.J. Mating disruption for the 21st century: Matching technology with mechanism. *Environ. Entomol.* **2005**, *44*, 427–453. [CrossRef]

26. Shorey, H.H.; Gaston, L.K.; Saario, C.K. Sex pheromones of noctuid moths. XIV. Feasibility of behavioral control by disrupting pheromone communication in cabbage loopers. *J. Econ. Entomol.* **1967**, *60*, 1541–1545. [CrossRef]

27. Cardé, R.T.; Minks, A.K. Control of moth pests by mating disruption: Successes and constraints. *Annu. Rev. Entomol.* **1995**, *40*, 559–585. [CrossRef]

28. Thomson, D.; Brunner, J.; Gut, L.; Judd, G.; Knight, A. Ten years implementing codling moth mating disruption in the orchards of Washington and British Columbia: Starting right and managing for success. *IOBC/WPRS Bull.* **2001**, *24*, 23–30.

29. Van Frankeyhuyzen, K.; Régnière, J. Multiple effects of tebufenozide on the survival and performance of the spruce budworm (Lepidoptera: Tortricidae). *Can. Entomol.* **2017**, *149*, 227–240. [CrossRef]

30. McMorran, A. A synthetic diet for the spruce budworm, *Chroristoneura fumiferana* (Clem.) (Lepidoptera: Tortricidae). *Can. Entomol.* **1965**, *97*, 58–62. [CrossRef]

31. Régnière, J.; Delisle, J.; Pureswaran, D.; Trudel, R. Mate-finding Allee effect in spruce budworm population dynamics. *Entomol. Exp. Appl.* **2012**, *146*, 112–122. [CrossRef]

32. Delisle, J. *A Tool for Evaluating Mating Disruption in SBW: CFS Forges Ahead*; Canadian Forest Service: Quebec, QC, Canada, 2008; p. 43.

33. Sanders, C.J. *A Summary of Current Techniques Used for Sampling Spruce Budworm Populations and Estimating Defoliation in Eastern Canada*; Canadian Forest Service: Quebec, QC, Canada, 1980.

34. Régnière, J.; Nealis, V.G. Moth dispersal, egg recruitment and spruce budworms: Measurement and interpretation. *Forests* **2019**, *10*, 706. [CrossRef]

35. Anderson, T.W.; Darling, D.A. Asymptotic theory of certain "goodness-of-fit" criteria based on stochastic processes. *Ann. Math. Stat.* **1952**, *23*, 193–212. [CrossRef]

36. Régnière, J.; Cooke, B.; Béchard, A.; Dupont, A.; Therrien, P. Dynamics and management of rising outbreak spruce budworm populations. *Forests* **2019**, *10*, 748. [CrossRef]

37. Leonhardt, B.A.; Mastro, V.C.; Leonard, D.S.; McLan, W.; Reardon, R.C.; Thorpe, K.W. Control of low-density gypsy moth (Lepidoptera: Lymantriidae) populations by mating disruption with pheromone. *J. Chem. Ecol.* **1996**, *22*, 1255–1272. [CrossRef]

38. Stellinski, L.L.; Miller, J.R.; Ledebuhr, R.; Siegert, P.; Gut, L.J. Season-long mating disruption of *Grapholita molesta* (Lepidoptera: Tortricidae) by one machine application of pheromone in wax drops (SPLT-OFM). *J. Pest Sci.* **2007**, *80*, 109–117. [CrossRef]

39. Kipp, L.R.; Lonergan, G.C.; Bell, W.J. Male periodicity and the timing of mating in the spruce budworm (Lepidoptera; Tortricidae): Influences of population density and temperature. *Env. Entomol.* **1995**, *24*, 1150–1159. [CrossRef]

40. Greenbank, D.O.; Schaefer, G.W.; Rainey, R.C. Spruce budworm (Lepidoptera: Tortricidae) moth flight and dispersal: New understanding from canopy observations, radar and aircraft. *Mem. Entomol. Soc. Can.* **1980**, *110*, 1–49. [CrossRef]

41. Miller, C.A.; Greenbank, D.O.; Kettela, E.G. Estimated egg deposition by invading spruce budworm moths (Lepidoptera: Tortricidae). *Can. Entomol.* **1978**, *110*, 609–615. [CrossRef]

42. Moreau, G.; Bauce, E. Lethal and sublethal effects of single and double applications of *Bacillus thuringiensis* variety kurstaki on spruce budworm (Lepidoptera: Tortricidae) larvae. *J. Econ Entomol.* **2003**, *96*, 280–286. [CrossRef]

43. Dallaire, R.; Labrecque, A.; Marcotte, M.; Bauce, É.; Delisle, J. The sublethal effects of tebufenozide on the precopulatory and copulatory activites of *Choristoneura fumiferana* and *C. rosaceana*. *Ent. Exp. Appl.* **2004**, *112*, 169–181. [CrossRef]

44. MacLean, D.; Amirault, P.; Amos-Binks, L.; Cerleton, D.; Hennigar, C.; Johns, R.; Régnière, J. Positive results of an early intervention strategy to suppress a spruce budworm outbreak after five years of trials. *Forests* **2019**, *10*, 448. [CrossRef]

forests

MDPI

Article

Influence of a Foliar Endophyte and Budburst Phenology on Survival of Wild and Laboratory-Reared Eastern Spruce Budworm, *Choristoneura fumiferana* on White Spruce (*Picea glauca*)

Dan Quiring [1,*], Greg Adams [2], Leah Flaherty [1], Andrew McCartney [3], J. David Miller [4] and Sara Edwards [1]

[1] Population Ecology Group, Faculty of Forestry and Environmental Management, University of New Brunswick, Fredericton, NB E3B 6C2, Canada; flahertyl@macewan.ca (L.F.) sara.edwards@unb.ca (S.E.)
[2] J.D. Irving Limited, 181 Aiton Road, Sussex East, NB E4G 2V5, Canada; Adams.Greg@JDIRVING.com
[3] Maritime Innovation Limited, 181 Aiton Road, Sussex East, NB E4G 2V5, Canada; McCartney.Andrew@mfrl.ca
[4] Department of Chemistry, Carleton University, Ottawa, ON K1S 5B6, Canada; David.Miller@carleton.ca
*   Correspondence: quiring@unb.ca

Received: 3 May 2019; Accepted: 12 June 2019; Published: 13 June 2019

Abstract: A manipulative field study was carried out to determine whether the foliar endophyte fungus, *Phialocephala scopiformis* DAOM 229536, decreased the performance of eastern spruce budworm, *Choristoneura fumiferana* larvae developing on white spruce trees. Overwintered second-instar budworm larvae from a laboratory colony or from a wild population were placed on endophyte positive or negative trees one or two weeks before budburst. The presence of the endophyte in the needles reduced the survival of *C. fumiferana* from both a wild population and a laboratory colony. Survival for budworm juveniles up to pupation and to adult emergence was 13% and 17% lower, respectively, on endophyte positive trees. The endophyte did not influence the size or sex of survivors and budworm survival was not influenced by any two- or three-way interactions. Budworm survival was higher for wild than for laboratory-reared budworm and for budworm placed on trees a week before budburst. This may be the first field study to demonstrate the efficacy of an endophytic fungus against wild individuals of a major forest insect pest. The efficacy of the endophyte at low larval densities suggests that it could be a useful tactic to limit spruce budworm population growth in the context of an early intervention strategy.

Keywords: Pinaceae; endophytic fungi; plant tolerance; *Phialocephala scopiformis*; *Picea glauca*; spruce budworm; phenology; insect susceptibility

1. Introduction

Mutualistic interactions between fungi living within leaf tissues (endophytes) and their host plants are common [1]. Plant tissues provide endophytes with nutrients [2] and some endophytes provide plants with protection from herbivores and fungal diseases [1,3]. Although most previous work on endophyte–plant interactions has been carried out in grasses and other agricultural crops [1,4], endophytic fungi are common in foliage of many conifers and may play similar roles in these large, long-lived plants [5,6].

Previous studies carried out with potted seedlings under laboratory [7–9] and field conditions [10] demonstrated that the native rugulosin-producing endophyte, *Phialocephala scopiformis* DAOM

229536 Kowalski & Kehr (Helotiales:Ascomycota) reduced the growth of eastern spruce budworm, *Choristoneura fumiferana* Clemens (Lepidoptera:Tortricidae). Under nursery conditions, most of the effect was attributed to the presence of the anti-insect toxin rugulosin [10]. Building on those studies, we recently demonstrated a similar effect on budworm developing on white spruce trees that had been inoculated with the endophyte more than 10 years earlier [11]. The reduction in budworm survival was highest for larvae developing in the mid and upper crown of trees, the most important crown region for photosynthesis and tree growth. These results suggest that inoculation of white spruce trees with *P. scopiformis* could reduce tree susceptibility to spruce budworm during outbreaks.

In our previous study [11], laboratory-reared second-instar budworm were placed on trees in the field on a single date. Consequently, we do not know if the endophyte is as effective on wild as lab-reared budworm or during years when spring synchrony between larval emergence and budburst varies. Manipulative field studies carried out with lab-reared budworm reported that budworm survival is highest when second-instar budworm larvae emerge one to three weeks before budburst [12].

Here, we report results from a manipulated field study carried out to investigate the independent and interacting effects of the endophyte, *P. scopiformis*, larval source (wild or laboratory-reared), and budworm spring emergence–host plant budburst synchrony on the performance of spruce budworm. As in our previous study [11], the present study was carried out with relatively low budworm densities, and subsequently low levels of defoliation. The objective was to determine whether the endophyte would reduce budworm survival before a large outbreak occurred.

2. Materials and Methods

2.1. Study Site, Tree Selection, and Experimental Design

Field experiments were carried out near Havelock, New Brunswick in two adjacent "test plots" (45°587″ N, 65°26″ W) of approximately 10-year-old white spruce, *Picea glauca* (Moench) Voss, trees planted by JDI Limited from seedling stock in 2003. Test plots are described in our previous study [11]. Briefly, both untreated control and endophyte positive trees were interplanted at 2 m × 2 m spacing in each of two adjacent 0.12 ha plots. Study trees were grown in 2000 and 2001 at Sussex Tree Nursery and endophyte-inoculated trees were wound inoculated as described by Miller et al. [8] with cultures of *P. scopiformis*. Trees were tested for the presence of the endophyte prior to planting in the field in 2003 with a polyclonal antibody for mycelium, and by measuring the insect toxin rugulosin by HPLC [13].

In mid-April 2012, we selected 14 pairs of trees. Each tree pair consisted of one endophyte-inoculated and one control tree; trees within a pair were located <8 m from each other. Trees with noticeable browsing, defoliation, mechanical damage, or deformation due to spruce gall midge (*Mayetiola piceae* (Felt)) or spruce bud midge (*Rhabdophaga swainei* (Felt)) (Diptera:Cecidomyiidae) were excluded from the study. Presence of the endophyte in endophyte-inoculated trees and absence of the endophyte in uninoculated control trees was verified using the polyclonal antibody test [13]. We placed 15 wild larvae on one branch in the mid-crown on 21 April 2012 and another 15 larvae on an adjacent mid-crown branch on 28 April 2012, approximately 9 and 2 days before budburst started on the most phenologically advanced trees, and enclosed them within a sleeve cage. Fifteen laboratory-reared larvae were placed on an adjacent branch on each of the two dates, for a total of 4 sleeve cages per tree (i.e., 2 sources of larvae × 2 dates). Two of the 112 cages (i.e., 4 cages per tree × 28 trees) were damaged by winds and were not included in analyses. As the majority of buds burst 3–5 days after the first buds burst, most larvae in the phenology treatments were placed on trees approximately one or two weeks before budburst.

2.2. Insect Sources

The study was carried out with wild larvae collected in eastern Quebec and with laboratory-reared larvae. Disease-free second-instar budworm larvae were obtained from the rearing facility of the

Canadian Forest Service in Sault Ste. Marie, Ontario [14] and stored at 4 °C for 1–2 weeks before placement in the field.

To obtain overwintered wild second-instar larvae for use in experiments in spring 2012, we used pole pruners to collect branches from highly defoliated natural spruce/fir stands close to Baie Comeau, Quebec, in late July and early August 2011. Egg-bearing shoots were cut from branches and transported in coolers to the University of New Brunswick (UNB). Egg-bearing shoots were placed in metal trays and reared at 22 ± 1 °C and 65% ± 5% RH under a 14 h light: 10 h dark photoperiod. A piece of Parafilm™ with a smaller piece of cheesecloth attached to it, had been placed on the bottom of each tray and another larger piece was used to seal the top of each container. A black piece of cardboard, with an approximately 12 cm × 6 cm rectangular hole in the center, was placed over each tray. Following egg hatch, first instar budworm larvae spun hibernacula on the cheesecloth. The pieces of cheesecloth were removed two weeks later and placed in sleeve cages. We fixed the sleeve cages to the lower bole of spruce trees in the UNB woodlot in Fredericton where they overwintered. The cages and enclosed larvae were collected when needed for experiments.

2.3. Insect Rearing Procedures

The protocols were similar to those described in Quiring et al. [11]. Briefly, in spring 2012, pieces of cheesecloth on which the wild and lab-reared second instars had previously spun hibernacula were placed at 20 ± 1 °C, 75% RH under a 14 L/10 D photoperiod until the first larva emerged. Cheesecloth pieces with 15 hibernacula each were cut under a binocular microscope and transported to the field in a cooler. These were attached to each experimental branch with a pin. The branches were enclosed in a sleeve cage which then was attached to the branch. The cheesecloth pieces were removed from the cages two weeks later and the number of dead, second-instar larvae that had not left the cheesecloth recorded. Those remaining were not included in the survival calculations. We reattached the sleeve cages and monitored them weekly until the first pupa was observed. Juveniles were removed once most larvae had pupated, placed in aerated containers on moistened vermiculite, and reared under natural light in the laboratory at 20 ± 1 °C, 65% ± 5% RH. The few remaining larvae were provided foliage from the same branch on which they developed and pupated within several days of collection. All emerged adults were killed by freezing and sexed. One forewing of each female was measured under a binocular microscope with a micrometer. Female forewing length is positively correlated with fecundity [15]. At the end of summer, defoliation on current-year branches was visually estimated, as in [16].

2.4. Statistical Analysis

The independent and interacting effects of the endophyte, phenology, and larval source on larval survival (i.e., second instar to pupation), total survival (i.e., second instar to adult emergence), and adult sex ratio was evaluated using generalized linear mixed effects models with logit link functions and binomial probability distributions. Tree was included as a random factor. All generalized linear mixed effect models were carried out using the *glmer* function from the lme4 package (version 1.1.12) [17] of R (version 3.3.2) [18]. For these and subsequent analyses described below, we inspected residual plots of all models and found no obvious trends or heteroscedasticity. We used the *dispersion glmer* procedure from the blmeco package (version 2.1) [19] of R to verify that statistical models were not overdispersed.

We used likelihood ratio (LR) tests, obtained through the *anova* function in R, to evaluate the contribution of fixed effects. First, we evaluated the contribution of the interaction between endophyte and either larval source or phenology. When an interaction was not significant, the significance of main effects was determined by comparing models with one of the fixed effects to models with both fixed effects.

The effects of endophyte, phenology or larval source on the wing length of female survivors were examined using linear mixed effects models, with tree included as a random factor, using the *lmer* function in the lme4 package [17] of R. We subjected defoliation estimates, which were non-count proportion data, to logit transformation before analysis; we used the "empirical logit", log[(y + ε)/(1 − y + ε)], where ε is the smallest non-zero proportion observed because our data included values of 0 and 1 [20]. LR tests, described above, were used to test the significance of fixed factors. As expected, defoliation was very low (18.2 ± 1.2%, *N* = 110) and neither the independent nor interacting effect of endophyte was significant (*p* ≥ 0.3760).

3. Results

Survival of second-instar larvae until pupation or adult emergence (i.e., larval and total survival, respectively) was significantly influenced by the main effects of endophyte, budburst phenology and insect source but not by any two- or three-way interactions (Table 1). Total survival of larvae developing on endophyte-inoculated trees was lower than that of larvae on control trees (Figure 1b). A similar trend is evident for larval survival (Figure 1a) but the effect of endophyte was marginally insignificant (Table 1). Larval and total survival was reduced by approximately 12% and 17% when developing on endophyte-inoculated compared to endophyte-free trees (Figure 1).

Table 1. Summary of generalized linear mixed models evaluating the influence of a native endophytic fungus, larval source, and phenology on larval (i.e., second instar to pupa) and total (i.e., second instar to adult emergence) survival, adult sex ratio and female wing lengths of eastern spruce budworm reared on 14 white spruce trees with and 14 trees without the endophyte in 2012. Second-instar larvae from a laboratory colony or field population (insect source) were placed in the mid-crown of study trees approximately one or two weeks before budburst (phenology). Tree was included as a random variable in the mixed effects models (either GLMM with logit link or LMM).

Response Variable	Source of Variation	df	X^2	*p*
Larval Survival	Endophyte	1	3.3672	0.0665
	Insect source	1	21.3920	**<0.0001**
	Phenology	1	9.6566	**0.0019**
	Endophyte:Insect source	1	0.5317	0.4659
	Endophyte:Phenology	1	1.3282	0.2491
	Insect source:Phenology	1	0.0090	0.9244
	3-way interaction	1	1.0769	0.2994
Total Survival	Endophyte	1	9.0715	**0.0026**
	Insect source	1	25.1450	**<0.0001**
	Phenology	1	9.2577	**0.0023**
	Endophyte:Insect source	1	0.0045	0.9468
	Endophyte:Phenology	1	2.4841	0.1150
	Insect source:Phenology	1	0.6497	0.4202
	3-way interaction	1	3.3404	0.0676
Sex Ratio	Endophyte	1	0.0367	0.8480
	Insect source	1	0.3793	0.5380
	Phenology	1	1.104	0.2942
	Endophyte: Insect source	1	0.1281	0.7204
	Endophyte:Phenology	1	0.4997	0.4797
	Insect source:Phenology	1	0.9684	0.3251
	3-way interaction	1	2.8481	0.0915
Female Wing Length	Endophyte	1	0.7665	0.3813
	Insect source	1	7.9002	**0.0049**
	Phenology	1	0.0073	0.9321
	Endophyte:Insect source	1	2.6551	0.1032
	Endophyte:Phenology	1	1.0289	0.3104
	Insect source:Phenology	1	1.1638	0.2807
	3-way interaction	1	0.0809	0.7761

Note: Significant *p* values are presented in bold type.

Both larval and total survival of wild budworm was significantly higher than that of lab-reared budworm (Figure 2, Table 1). Larval and total survival were approximately 26% and 33% higher for wild than lab-reared budworm (Figure 2). Larval and total survival of budworm placed on trees approximately a week before budburst was approximately 15% and 16.5% higher, respectively, than that for larvae placed on trees two weeks before budburst (Figure 3, Table 1).

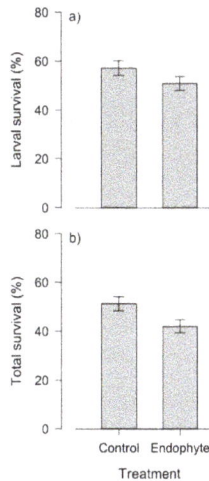

Figure 1. Mean (± SE) survival of second-instar eastern spruce budworm (**a**) to pupation (i.e., larval survival) and (**b**) to adult emergence (total survival) on white spruce trees with (Endophyte) or without (Control) a native endophytic fungus. $n = 14$ control and 14 endophyte trees.

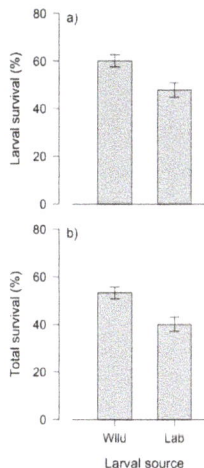

Figure 2. Influence of larval source on mean (± SE) survival of second-instar eastern spruce budworm (**a**) to pupation (i.e., larval survival) and (**b**) to adult emergence (total survival) on white spruce trees. Second instars were obtained from a wild population (Wild) or from a laboratory colony (Lab). $n = 14$ control and 14 endophyte trees.

The sex ratio of emerged adults was not influenced by the main or interacting effects of endophyte, phenology or insect source (Table 1). Similarly, the wing lengths of emerged females was not influenced by the main or interacting effects of endophyte and phenology. However, the wing lengths of wild budworm females were slightly but significantly longer than those from the lab-reared colony (1.24 ± 0.01 versus 1.21 ± 0.01 cm, Table 1).

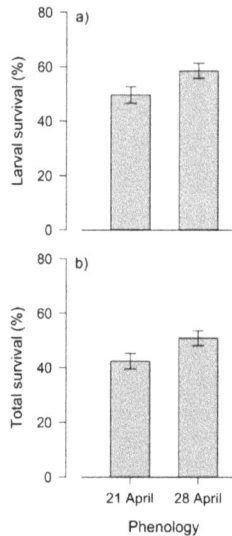

Figure 3. Influence of spring phenological synchrony between the date of emergence of eastern spruce budworm and of budburst of white spruce trees on mean (± SE) survival (**a**) to pupation (i.e., larval survival) and (**b**) to adult emergence (total survival) on white spruce trees. Second-instar larvae were placed on 28 trees approximately one (28 April 2012) or two (21 April 2012) weeks before budburst.

4. Discussion

Inoculation of study trees with a native endophytic fungus >10 years prior to the current study increased tree defense against a major forest pest. Most importantly, the endophyte was as effective against larger, wild budworm as it was against budworm from a laboratory colony. Reductions of approximately 17% in total survival of wild and lab-reared budworm, under two different phenological conditions, was similar to that reported in a study carried out with lab-reared budworm in the same study plots the two previous years [11]. The majority of budworm mortality attributable to the endophyte occurred during larval development and the presence of the endophyte did not influence adult size or sex ratio.

The present data indicate that budworm survival was not influenced by interactions between the endophyte and budburst phenology or insect source. In the earlier study, in which budworm were placed on the tree at one time point, interactions between the endophyte and crown level or insect density influenced budworm survival. The lack of an interaction in the present study is presumably not due to a lack of sufficient variation in these two variables because both insect source and budburst phenology independently influenced budworm survival.

The endophyte was as efficient in reducing the survival of wild larvae as it was in reducing the survival of laboratory-reared larvae, as indicated by the lack of an interaction between insect source and endophyte. This suggests that the endophyte may be effective against a range of budworm phenotypes. Wild and lab-reared budworm in the current study originated from different budworm populations and had experienced different environmental conditions prior to the field study.

Higher survival for larvae that were placed on study trees approximately one week before budburst than for those placed two weeks before budburst is probably due to either reduced success choosing and mining old needles or reduced nutritional quality of old needles until budburst. Second-instar budworm mine into old foliage in spring, where they feed and obtain some nutritive benefit [21], and remain there until budburst, when they move to feed on the bursting buds [12]. The study plots received approximately 2 cm of rain on 21 April 2012, the first date that budworm were placed on study trees, and 2 cm the next day, but did not receive any precipitation on 28 April 2012, the second date when budworm were placed on study trees [22]. Second-instar budworm are very small, and driving rain against the sleeve cages may have dislodged some from the branch surface or water entering the cages may have drowned others.

Following a manipulated field study carried out with laboratory-reared budworm on white spruce, Lawrence et al. [12] reported that budworm survival was highest when second instars were placed on buds 1–3 weeks before budburst, and that the survival of individuals was slightly higher when placed on trees two rather than one week before budburst. Thus, we speculate that the lower survival of larvae placed on trees two, as opposed to one, week before budburst was primarily due to reduced needle colonization success, due to inclement weather.

5. Conclusions

The present study extends previous field experiments carried out with lab-reared larvae and demonstrates that a native endophytic fungus reduces the survival of wild individuals of the major pest of coniferous trees in eastern North America. Interestingly, although budworm survival was influenced by spring larval emergence/host tree budburst synchrony and whether juveniles were wild or from a laboratory colony, the endophyte reduced budworm survival regardless of spring emergence/budburst synchrony and regardless of whether individuals were wild or laboratory-reared. Importantly, the endophyte was effective at relatively low larval densities and, thus, could offer a complementary tactic for hindering spruce budworm population growth in the context of an Early Intervention Strategy [23].

Author Contributions: D.Q., S.E., L.F. and J.D.M. contributed to the general study conceptualization; L.F., J.D.M., S.E. and D.Q. supervised and conducted the laboratory and/or field observations and measurements; D.Q. and S.E. carried out the statistical analyses and made figures; and D.Q. was responsible for funding acquisition and project administration. All authors interpreted the data and contributed substantially to manuscript preparation, development and revision.

Funding: This research was funded by a research grant from J.D. Irving, Limited to D.Q.

Acknowledgments: We thank L. Chase, A. Graves, L. May, B. Fitch and E. Owens for technical assistance; Rob Johns (CFS), Jacques Regnière (CFS) and Pierre Therrien (Quebec Ministry of Natural Resources) for help locating sites with spruce budworm eggs; and R. Johns, M. Stasny and three anonymous reviewers for comments on an earlier version of the manuscript. G Parker is thanked for the endophyte antibody analyses in 2012.

Conflicts of Interest: G.A. and A.M. respectively are employees of J.D. Irving, Limited (JDI) and Maritime Innovation Limited (a wholly owned subsidiary of JDI). The company provided funding for this research.

References

1. Rodriguez, R.J.; White, J.F., Jr.; Arnold, A.E.; Redman, R.S. Fungal endophytes: Diversity and functional roles. *New Phytol.* **2009**, *182*, 314–330. [CrossRef] [PubMed]
2. Clay, K. Fungal endophytes of grasses: A defensive mutualism between plants and fungi. *Ecology* **1988**, *69*, 10–16. [CrossRef]
3. Carroll, G.C. Fungal endophytes in stems and leaves: From latent pathogen to mutualistic symbiont. *Ecology* **1988**, *69*, 2–9. [CrossRef]
4. Busby, P.E.; Ridont, M.; Newcombe, G. Fungal endophytes: Modifiers of plant disease. *Plant Mol. Biol.* **2016**, *90*, 645–655. [CrossRef] [PubMed]
5. Miller, J.D. Foliar endophytes of spruce species found in the Acadian forest: Basis and potential for improving the tolerance of the forest to spruce budworm. In *Endophytes of Forest Trees: Biology and Applications*; Forestry Series; Pirttilä, A.M., Frank, A.C., Eds.; Springer: Berlin, Germany, 2011; pp. 237–249.
6. Tanney, J.B.; McMullin, D.R.; Miller, J.D. Toxigenic foliar endophytes from the Acadian Forest. In *Endophytes of Forest Trees*; Forestry Sciences 86; Pirttilä, A.M., Frank, A.C., Eds.; Springer International Publishing AG: Cham, Switzerland, 2018; pp. 343–381. [CrossRef]

7. Calhoun, L.A.; Findlay, J.A.; Miller, J.D.; Whitney, J.D. Metabolites toxic to spruce budworm from balsam fir needle endophytes. *Mycol. Res.* **1992**, *96*, 281–286. [CrossRef]

8. Miller, J.D.; MacKenzie, S.; Foto, M.; Adams, G.W.; Findlay, J.A. Needles of white spruce inoculated with rugulosin-producing endophytes contain rugulosin reducing spruce budworm growth rate. *Mycol. Res.* **2002**, *106*, 471–479. [CrossRef]

9. Sumarah, M.W.; Puniani, E.; Sørensen, D.; Blackwell, B.A.; Miller, J.D. Secondary metabolites from anti-insect extracts of endophytic fungi isolated from *Picea rubens*. *Phytochemistry* **2010**, *71*, 760–765. [CrossRef] [PubMed]

10. Miller, J.D.; Sumarah, M.W.; Adams, G.W. Effect of a rugulosin-producing endophyte in *Picea glauca* on *Choristoneura fumiferana*. *J. Chem. Ecol.* **2008**, *34*, 362–368. [CrossRef] [PubMed]

11. Quiring, D.T.; Flaherty, L.; Adams, G.; McCartney, A.; Miller, J.D.; Edwards, S. An endophytic fungus interacts with crown level and larval density to reduce the survival of eastern spruce budworm, *Choristoneura fumiferana* (Lepidoptera: Tortricidae), on white spruce (*Picea glauca*). *Can. J. For. Res.* **2019**, *49*, 221–227. [CrossRef]

12. Lawrence, R.K.; Mattson, W.J.; Haack, R.A. White spruce and the spruce budworm: Defining the phenological window of susceptibility. *Can. Entomol.* **1997**, *129*, 291–318. [CrossRef]

13. Sumarah, M.W.; Miller, J.D.; Adams, G.W. Measurement of a rugulosin-producing endophyte in white spruce seedlings. *Mycologia* **2005**, *97*, 770–776. [CrossRef] [PubMed]

14. Roe, A.D.; Demidovich, M.; Dedes, J. Origins and history of laboratory insect stocks in a multispecies insect production facility, with the proposal of standardized nomenclature and designation of formal standard names. *J. Insect Sci.* **2017**, *18*, 1–9. [CrossRef] [PubMed]

15. Thomas, A.W.; Borland, S.A.; Greenbank, D.O. Field fecundity of the spruce budworm (Lepidoptera: Tortricidae) as determined from regression relationships between egg complement, fore wing length, and body weight. *Can. J. Zool.* **1980**, *58*, 1608–1611. [CrossRef]

16. Moreau, G.; Quiring, D.; Eveleigh, E.; Bauce, E. Advantages of a mixed diet: Feeding on several foliar age classes increases the performance of a specialist insect herbivore. *Oecologia* **2003**, *135*, 391–399. [CrossRef] [PubMed]

17. Bates, D.; Maechler, M.; Bolker, B.; Walker, S. Fitting linear mixed-effects models using lme4. *J. Stat. Softw.* **2015**, *67*, 1–48. [CrossRef]

18. R Core Team. *R: A Language and Environment for Statistical Computing*; R Foundation for Statistical Computing: Vienna, Austria, 2016; Available online: https://www.R-project.org/ (accessed on 3 March 2019).

19. Korner-Nievergelt, F.; Roth, T.; von Felten, S.; Guelat, J.; Almasi, B.; Korner-Nievergelt, P. *Bayesian Data Analysis in Ecology Using Linear Models with R, BUGS and Stan*; Elsevier: Amsterdam, The Netherlands, 2015.

20. Warton, D.I.; Hui, F.K.C. The arcsine is asinine: The analysis of proportions in ecology. *Ecology* **2011**, *92*, 3–10. [CrossRef] [PubMed]

21. Trier, T.M.; Mattson, W.J. Needle mining by the spruce budworm provides sustenance in the midst of privation. *Oikos* **1997**, *79*, 241–246. [CrossRef]

22. Government of Canada. Daily Data Report for April 2012 (Parkdale, New Brunswick). 2018. Available online: http://climate.weather.gc.ca/climate_data/daily_data_e.html?timeframe=2&Year=2012& Month=4&Day=21&hlyRange=%7C&dlyRange=1983-07-01%7C2017-02-28&mlyRange=1983-01-01% 7C2006-02-01&StationID=6219&Prov=NB&urlExtension=_e.html&searchType=stnProv&optLimit= yearRange&StartYear=2012&EndYear=2012&selRowPerPage=25&Line=26&lstProvince=NB (accessed on 22 February 2019).

23. Pureswaran, D.S.; Johns, R.; Heard, S.B.; Quiring, D.T. Paradigms in eastern spruce budworm (Lepidoptera: Tortricidae) population ecology: A century of debate. *Environ. Entomol.* **2016**, *45*, 1333–1342. [CrossRef] [PubMed]

MDPI

St. Alban-Anlage 66

4052 Basel

Switzerland

Tel. +41 61 683 77 34

Fax +41 61 302 89 18

www.mdpi.com

Forests Editorial Office

E-mail: forests@mdpi.com

www.mdpi.com/journal/forests

www.ingramcontent.com/pod-product-compliance
Lightning Source LLC
Chambersburg PA
CBHW051843210326

41597CB00033B/5756